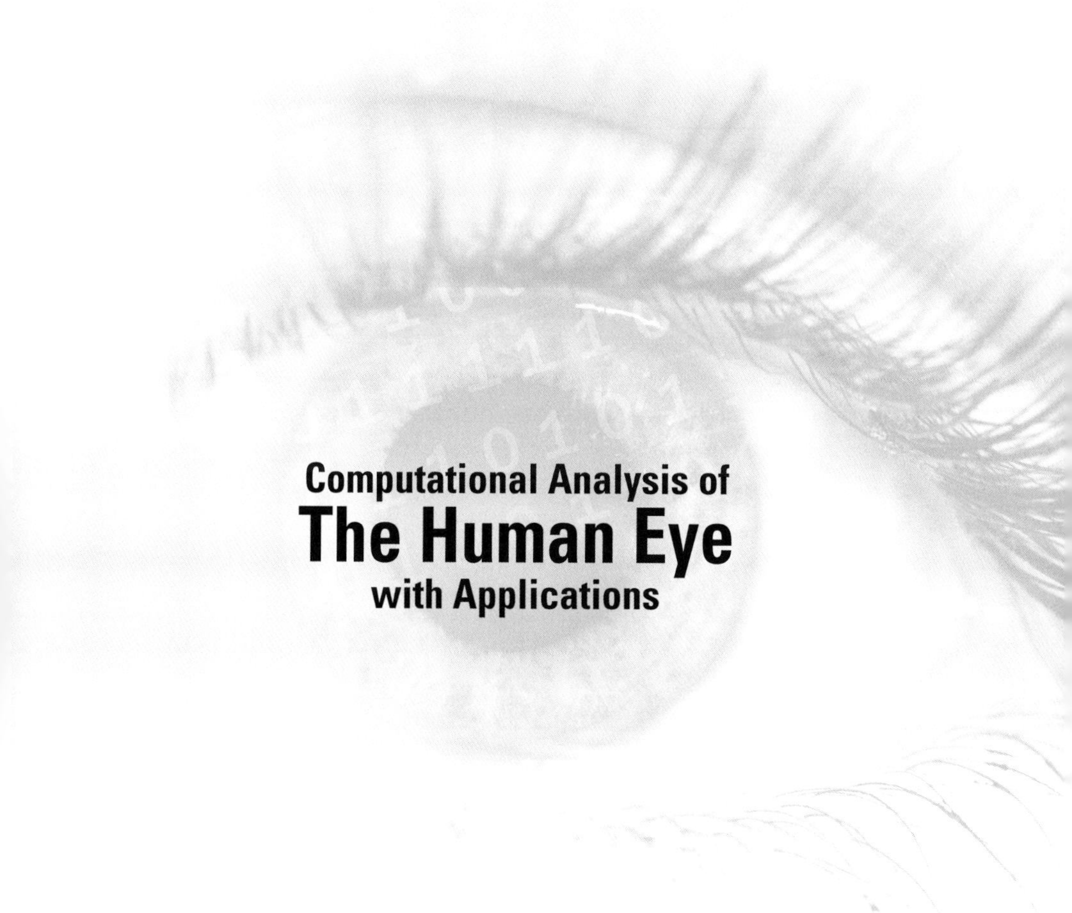

Computational Analysis of

The Human Eye

with Applications

Computational Analysis of
The Human Eye
with Applications

Sumeet Dua
Louisiana Tech University, USA

Rajendra Acharya U
Ngee Ann Polytechnic, Singapore

E Y K Ng
Nanyang Technological University, Singapore

Editors

World Scientific

NEW JERSEY • LONDON • SINGAPORE • BEIJING • SHANGHAI • HONG KONG • TAIPEI • CHENNAI

Published by

World Scientific Publishing Co. Pte. Ltd.
5 Toh Tuck Link, Singapore 596224
USA office: 27 Warren Street, Suite 401-402, Hackensack, NJ 07601
UK office: 57 Shelton Street, Covent Garden, London WC2H 9HE

British Library Cataloguing-in-Publication Data
A catalogue record for this book is available from the British Library.

COMPUTATIONAL ANALYSIS OF THE HUMAN EYE WITH APPLICATIONS

ISBN-13 978-981-4340-29-8
ISBN-10 981-4340-29-4

Typeset by Stallion Press
Email: enquiries@stallionpress.com

Printed in Singapore by Mainland Press Pte Ltd.

Contents

Chapter 1
The Biological and Computational Bases of Vision

Hilary W. Thompson*

1.1. Introduction to the Eye

The human visual system, from photoreceptors to higher-order processing of signals in the brain, is a remarkable example of the power of evolutionary processes acting over vast periods to develop a finely tuned system capable of complex recognition, recall, and execution of highly accurate tasks in space and time. In terms of genetics, physiology, and structure, the evolutionary pathway to higher vertebrate vision can be traced.[1] In fact, it has been hypothesized that the development of a complex visual system preceded and provided the evolutionary substrate for the evolution of brain expansion that made possible higher cognitive functions in primates.[2] Vertebrate vision is one of the most profound and illuminating examples illustrating the operation of evolutionary principles.[3]

From using chipped stones to the technology of spacecraft and global computer networks, humans have relied on the exceptional powers of vision to survive, prosper in, predict, and, to some degree, control and reshape an often-challenging environment. The primacy of vision in the human experience is evidenced in our language: to "see" also means to understand, as does "to have the eyes opened." Such definitions occur, not only in the English language, but also in the metaphors and figures of the speech of other languages as well. Humans are visual animals, and the loss of sight

*LSU Eye Center, LSUHSC School of Public Health, Department of Ophthalmology, New Orleans, LA.

is a profound tragedy, which presents challenges to both the individual and society. Compassionate concern about devising applications to save and restore vision has driven biomedical vision researchers as much as the desire to understand the pure science of how something as complicated and wonderful as our vision can work. The computational scientist has also taken lessons from what we know of biological vision systems to base the development of algorithms for detecting three-dimensional (3D) features of the external world as detected by a two-dimensional (2D) array of detectors, whether they are solid-state photocells or biological cellular transducers of light energy. The understanding of biological vision from the viewpoints of these diverse themes has been one of the great intellectual adventures of our species. Vision research has resulted in a number of Nobel prizes in the medicine and physiology category, as well as significant therapies for sustaining our vision and novel algorithms for machine vision.

This chapter will introduce the functional anatomy, sensory physiology, and neuronal information processing involved in human vision. It is our hope that this introduction will provide a background of fundamental insights and resources about biological vision to computer scientists who are interested in computer vision and computational image analysis. A number of important advances in thinking and methods in computational image analysis relate to our understanding of biological visual systems. Many more advances in computer vision methods are possible if we can understand and learn from the highly optimized image processing in biological visual systems.

1.2. The Anatomy of the Human Visual System

As we gaze into our own eyes in a mirror or into the eyes of another human being, we are aware of particular characteristic features. The clear curve of the cornea, set into the white of the conjunctiva, the wrinkled surface of the colored iris behind the clear cornea, the protective lids, and their eyelashes sweeping over the globe in periodic blinks. Handling a human eye removed from its bony orbit (something few of us, except for vision scientists, pathologists, and anatomists ever do), its toughness, and the feel of the globe as pliable but rigid due to fluid contained in its tough outer covering is very striking. The color of the iris and the size of the clear cornea dominate its appearance when the eye is in place; removed, the cornea and the iris behind it, with the opening (pupil) behind it, seem, in the isolated human eye, much too small for the size of the globe. Straggling off the rear, slightly off to one side (the nasal or side closest to the nose) is the tough sheath surrounding the optic nerve, the eye's output line to the rest of the brain and carrier of the complex visual processing the cellular circuits of the retina achieves. The optic nerve also conducts the nutritive lines, arteries, and veins, which supply the metabolic demands of the eye. Not visible from the outside of the globe (unless using optics to look through the cornea and lens) is the retina, where photoreception and the initial stages of visual processing occur.

Another notable aspect of the outside of the isolated human eye is the ocular muscles, which have attachment points on the tough outer white of the eye (sclera). Their number and orientation are responsible for the coordinated motion of the globes in the bony eye-sockets of the skull. Ocular movements are coordinated in the central nervous system (CNS) and interact with visual stimuli using intricate feedback from visual pathways.

These most evident and outward features of the human eye all play important roles in vision, including the processing of information and control of light required for image formation and binocular vision. The possibility of binocular vision in an animal is evident in the placement of the eyes in the front of a somewhat frontally flattened face, so that part of the fields of light detection of the two eyes can overlap. The connections and crossing-over of the retinal output cells to the next area of processing in the brain (see below) are another important aspect of binocular vision. Let us consider in more detail the layers of the mammalian eye. These layers include the retina where the photoreceptor cells reside, as well as other layers responsible for mechanical and nutritive supports. Within the retinal layer are neurons, which are part of the initial information-processing system to extract visual information even before neuron signals leave the globe of the eye via the optic nerve.

The tough outer "white" of the eye is a connective tissue layer known as the sclera. The intraocular pressure makes the sclera rigid enough to keep the optical length of the eye constant and support the movement of the eye under muscular control. The sclera's cellular makeup and protein molecule configuration change to form the cornea, the clarity of which equals that of inorganic glasses or crystals. This clarity of the cornea is due to a different size range, spacing, and orientation of the protein molecules (collagen) that make up the cornea as compared to the sclera. Corneal clarity is maintained by metabolic work of cell layers on the backside of the cornea (the corneal endothelial cell layer) and the outer surface (the corneal epithelium). These cell layers use cellular energy to maintain the ionic (and, hence, water) content of corneal tissue; this work prevents corneal swelling, maintains the spacing of collagen molecules and, thereby, keeps the tissue transparent. The corneal epithelium is densely innervated with nerve endings. This density of nerve endings is evident to us humans when "something is in my eye!" i.e. when debris gets onto the corneal surface. The refractive properties of the cornea account for a great percentage of the optical properties of the eye, far more than the lens. The importance of the cornea as a refractive tissue, bending light to help form an image on the photoreceptor layer of the retina, makes modern refractive surgery possible. Such surgery uses knife cuts and lasers to reshape the cornea to alter and correct its optical properties.

The middle of the layers of the eye, besides the retina and the outer sclera, are the several structures called the uveal tract. The choroid is a part of the uveal tract and is the capillary supply of the retina. This capillary system arises and returns from the blood vessels evident on the retinal surface and supplies the highly metabolically active neural tissue of the retina. In human eyes, the cells of the choroid layer contain a light-absorbing pigment, melanin. Placement of a pigment layer under the active regions of the photon receptor areas of the eye enhances visual acuity by absorbing stray photons that would reflect back on the photoreceptors, confounding spatial resolution. We often note that, when caught in the headlights of our cars at night, the eyes of cats, dogs, deer, and many other animals (in particular those with nocturnal habits) shine brightly. The eyes of these animals with eye shine have less melanin in the choroid than the human eye, but also have a reflective layer that enhances low-light vision by bouncing stray photons back to the photoreceptors, rather than sharpening visual acuity by absorbing them. This reflection produces low-light sensitivity at the cost of spatial resolution, and provides the eye shine that we see when light hits the eyes of these animals in the dark.

The ciliary body at the front of the eye is muscle tissue around the lens of the eye. This muscle is capable of changing tension on the lens to change lens shape and thereby adjust the focus of the image on the retina in near and distance vision (accommodation). Part of the ciliary body, the ciliary process, produces the aqueous humor, the fluid that fills the anterior chamber of the eye (area in front of iris and lens and behind the cornea). This constant fluid production maintains the fluid pressure and the rigidity of the eye, but fluid leakage balances production. This fluid leakage occurs at specialized meshwork structures at the limbus (the juncture of the cornea, the iris, and the white of the eye). Blockage of this drainage system increases the pressure in the eye (intraocular pressure) and can cause tunnel vision and eventual complete blindness by restricting blood flow and damaging the optic nerve in the diseased glaucoma. The iris is the wrinkly colored portion of the eye that we see through the cornea. The iris has muscle layers that control pupil dilation or contraction under feedback from the CNS. Pupil size adapts to light levels and, like the diaphragm of a camera, changes diameter according to light levels impinging on the eye. Between the lens and the retina is most of the fluid volume of the eye, the vitreous humor. The vitreous is not a simple fluid filling a chamber but a gel with complex protein structure that aids in the protection of the retina. It contains phagocytic cells that can remove cellular debris that might affect retinal image formation. The lessening of the ability of these phagocytic cells to remove all debris results in floaters in the vision, which are the shadows of debris projected on the retina. Both the visual accommodation and the power of phagocytic cells of the vitreous to scavenge debris decline as we humans age. As a result, we may wear bifocals and more often notice floaters in our vision should we be as fortunate as to age.

The inner sensory receptor and initial information-processing area known as the retina contains the photoreceptors (rods and cones). In addition, several other types of retinal nerve cells called interneurons form part of an initial information-processing circuit between the photoreceptor input and the retinal ganglion cells, which provide the retina's output to the brain. Interneurons are neurons that interpose between primary sensory receptor cells and nervous system outputs such as motor neurons; in the CNS, the term interneuron denotes neurons that act locally in CNS nuclei, as opposed to those that send out axons to form tracts connecting to other CNS areas. The retinal ganglion cells have axons that form the optic nerve and carry output to the next stage of the visual system. We will focus on the role of the retina in the visual system in the section below on retinal information processing. The complexity of the neuronal circuits in the retina is unique for a peripheral sense organ. This complex retinal circuitry arises during the embryonic development of the eye, during which the retina forms from an out pocketing of the growing brain.

The retinal photoreceptors are essentially at the bottom of the several layers of nerve cells of the retina. This location makes the human (primate) retina essentially "inverted," in the sense that the photons detected by the visual system travel through the output neurons (retinal ganglion cells) that form the optic nerve first, then through several layers of interneurons to finally reach the photoreceptors, the tips of which are just above the pigment epithelial layer of the choroid. The eyes of higher cephalopods (octopus and squids) represent a parallel evolution with the human eye, with many of the cellular features similar, except not "inverted." The photoreceptors of the cephalopod eye are on the outermost layer, and, therefore, the phototransduction apparatus receives the light first without it having to travel through multiple cellular layers prior to hitting the receptors as in our eyes. It is apparent that vision has arisen independently in various forms multiple times in evolution.[1] In the light of this parallel evolution in visual systems, the "inversion" in the human retina may, in addition to enhancing the sharpening of spatial information by photon absorption, also allow enhanced nutrient supply and support for turnover and restoration of photopigment molecules of photoreceptors by the cells of the retinal pigmented epithelium.

The retinal photoreceptors are of two general classes, rods and cones. These photoreceptors are the specialized sensory receptor cells where the transformation of the energy of photons into neuron signals takes place. This process is sensory transduction. In different sensory systems, the process of sensory transduction shows intricate molecular mechanisms specific for each sensory modality: vision, taste, and smell (the chemical senses), touch and hearing (the mechanoreceptive senses). We will discuss some details of visual transduction in Sec. 1.5 of this chapter. Nerve cells meet and communicate at specialized structures called synapses (Sec. 1.4 of this chapter). A number of synapses are involved in the circuitry of the retina; first, many of the rods and cones synapse onto bipolar cells directly, a neuron in the retina, which connects between photoreceptors and retinal ganglion cells, forming the most direct information pathway in the retina. Two layers of interneurons modulate the circuitry of this direct pathway in the retina. One layer is composed of the horizontal cells that synapse on photoreceptors at the pre-synaptic elements of their synapses on bipolar cells; this level contains additional synaptic complexity of the retinal circuitry and there are the amacrine cells, which interpose into the bipolar/retinal ganglion cell synapses. Finally, there are the interplexiform cells, which carry modulatory information from the amacrine cells back to the photoreceptor synaptic regions. All of these synapses and cell types are important in retinal function, and we will discuss this function in the section on retinal information processing (Sec. 1.6).

The retinal ganglion cells, as the output elements of the retina and its complex information processing, form the optic nerve. The 2D topology of the retina and the spatial orientation of the image projected on it by the optical system of the eye (cornea and lens, modulated by pupil size) are maintained in the optic nerve and its first synapse with the CNS, and in a number of subsequent higher CNS visual information-processing areas (Note: a "synapse" is a point of communication between two neurons, discussed in detail in Sec. 1.4). In addition, considerable specialization of the outputs of the retinal photoreceptors are also maintained, since there is luminance level, color, fine detail, motion, and shape information, all of which must be routed to specific processing mechanisms further "upstream" in the brain. There is spatial concentration of cones in a central area of the retina where we have the sharpest visual acuity as well as the best color vision. This area is called the fovea (Latin for "pit"), and it is recognizable when we look into the eye and see the surface of the retina. The retina is part of the CNS, and visualization of it is the only part of the CNS that is viewable directly in a noninvasive manner. This noninvasive viewing is via a hand-held instrument called an ophthalmoscope, which medical students and other clinical professionals learn to use when studying ophthalmology and physical examination. It allows a trained observer to look directly into the eye and see the retina, which often can reveal significant pathologic features that may be indicative of systemic diseases such as diabetes or high blood pressure, as well as ocular diseases. The fovea appears as a yellowish spot on the retina where there is a notable absence of larger blood vessels. The paucity of blood vessels goes along with the role of the fovea in our most acute and color-sensitive central vision. There are densely packed cones at the fovea, with rods absent from the most central part of the fovea.

Another obvious landmark on the retina is the optic nerve head. The blood vessels that spread out over the retina originate at this point, since they enter the eye through the center of the optic nerve, which forms up at the optic nerve head. Since the optic nerve joins with the retina at this point, there are no photoreceptors in the region, and it forms a blind spot in our vision. This blind spot can be explored by closing one eye and fixating the gaze on an "X" on a white piece of paper. A pencil moved away from the point the eye is fixated on disappears at the blind spot. We do not see the blind spot in our ordinary binocular vision due to "filling in" by the higher processing accomplished in the visual system.

There is a remarkable "crossover" of the retinal areas as the retinal ganglion cells project to the next level of the visual system. This crossing-over is important in information processing for color and fine detail, as well as for binocular vision. The crossover occurs at a structure called the optic chiasm in the optic nerve before it synapses with the CNS. About 60% of the retinal ganglion cells from the left eye cross over to the right side of the brain with the remaining 40% continuing onto the left side of the brain. A reciprocal crossover occurs with the retinal ganglion cells from the right eye crossing-over to the left and right side of the brain. The axons of the retinal ganglion cells, after crossing the optic chiasm, become part of the optic tract. Once they are part of the optic tract, they are no longer referred to as the optic nerve. Here, the temporal segments of the left eye retina and the nasal segment of the right eye retina enter the left optic tract, and the nasal segment of the left eye retina and the temporal segment of the right eye retina enter the right optic tract. The functional importance of this crossing-over is to establish a comparative signal for binocular vision and stereoscopic perceptions of depth.

In addition, it is interesting to realize that the optical properties of the cornea and lens result in an inversion of the environmental image, so topographic information from the environment is inverted and projected to the CNS, upside-down with respect to the visualized world. The maintenance of this upside-down representation of topography in the retinal ganglion cell projections throughout the visual system in the CNS is called retinotopic organization. Maps of the retina that maintain the spatial information of the image falling on the retina occur at multiple levels and in several places within a level in the CNS. The existence of these maps was confirmed by electrophysiological experimentation, as well as studies of the visual defects suffered by humans with focal lesions or damage of specific brain regions.

The first synapse of the retinal ganglion cells in the CNS, after crossing-over from the optic nerve to form the optic tract, is in the thalamus, a relay structure for sensory and motor information to the higher cortical centers. The structure where the retinal ganglion cells make their first synapse in the brain is the lateral geniculate nucleus. Geniculate relates in word origins to the word "genuflect" and indicates the knee-like bend of the structure. Note also that the word "nucleus" in this context refers to an area of the CNS where one area of the brain synapses with another. It may also contain the cell bodies of the neurons involved in the brain region. The term is not to be confused with the neuron's (or any cell's) "nucleus," which is a sub-cellular organelle that contains the cell's DNA and parts of the gene expression apparatus. Tracts or bundles of myelinated axons connect brain nuclei. These tracts act as communication pathways between processing centers in the brain.

There are three other sites of projection of retinal ganglion cells to areas in the brain in addition to the main visual processing areas. One is the projection to an area known as the pretectum, which projects to the Edinger-Westphal nucleus. This nucleus connects to the oculomotor nerve that controls the constrictor muscle in the iris. Information on light intensity falling on the retina allows this nucleus to control pupil constriction with increases of light into the eye. Another brain area that receives non-visual retinal output is the suprachiasmatic nucleus of the hypothalamus. This area receives output from those retinal ganglion cell axons that control day–night cycles or circadian rhythms. Via this connection, specialized retinal ganglion cells, which have their own photoreceptive pigments and do not need connection to rods or cones, drive rhythmic behavior in response to light–dark cycles.[4,5] Finally, there is the structure known as the superior colliculus that uses retinal input to control the movements of the head and eye. This chapter will discuss in detail only the function of the visual processing areas, not the aspects of the motor-eye control brain regions to which retinal ganglion cells project.

The main visual pathway in the CNS goes from the synapses of the lateral geniculate nucleus to the visual cortex of the occipital lobe. Areas of the brain are named for the bones of the skull that cover them; in this case, the occipital bone at the back of the head encases the visual cortex of the occipital lobe. Visual cortex is also known as the striate cortex for the banded striate appearance of its histology. The lateral geniculate nucleus neurons send their axons to the visual cortex via structures called the optic radiations. These connections go via two separate courses of the optic radiation, which travel within the internal capsule. The connections, which represent the inferior retina, take a pathway called Myer's loop, which go on an extravagant turn under the temporal lobes and end up in the superior visual cortex. The other bundles of axons from the lateral geniculate nucleus representing the superior retina take a more direct route under the parietal lobe of the brain to the inferior part of the visual cortex. Once again, the retinotopic organization of the retinal ganglion cells maintains its inverted spatial topography in these higher order projections of neurons in the visual system.

Beyond the striate or visual cortex, extrastriate visual processing areas lie in the occipital, parietal, and temporal lobes. Each of these areas again contains a retintopically organized map of the visual field. In all of these areas, responses may be specialized to a particular aspect of the visual scene, such as color, fine spatial resolution, motion, and the direction and velocity of motion. While research on these extrastriate cortical visual processing areas is ongoing, more than 30 such areas have been discovered to date.[6,7] All these many parts of the visual processing areas of the brain, with their functions of processing separate (simultaneous) aspects of the visual scene act as a parallel processing system. Adding to the potential complexity of processing is that each of the visual areas mentioned receive not only input from the previous layer of processing, but also synaptic connections from other brain areas. These represent feedback not only from other visual areas, but also from brain regions involved in nonvisual sensory processing, sensory integration, memory, and cognition.

Higher cortical centers not solely involved in visual processing may accumulate some of the results of these parallel visual system processes. In turn, decisions as to the nature of the visual content are made, so that the output of parallel processes must be associated somewhere in the brain. This occurrence in visual processing in the CNS is called the "binding" problem; it concerns how and where (if there is a "where") these separate parallel processes are bound together or combined into the complete impression (illusion?) of continuous visual information of a real outer world which the visual system provides our consciousness. It is interesting to note that, while it is beyond the scope of our present discussion, the same binding problem extends to the notion of a continuous self-consciousness that must involve parallel processes in many brain regions. Where does the final executive function that we recognize as our consciousness and the sense of personal continuity reside? Is it localized? Alternatively, is it a distributed emergent property of a complex of parallel processes? This deficiency in our understanding of these higher aspects of visual processing demonstrates that there is considerable knowledge to gain in biological vision research.

1.3. Neurons

A brief review of such an enormous area of current research as visual science must make demands on the reader. All the background required for full understanding, including chemistry and cellular biology cannot be provided in a single chapter. Basic texts on neuroscience provide in-depth treatment of these topics.[8] However, the basic properties of the neuron and signaling in neurons are essential to understand biological vision. Therefore, my goal is to provide a sound basic understanding of neuronal function, as it relates to neuronal signaling processes in vertebrate vision.

The neuron is the basic cell type of the nervous system of all animals. The idea that the neuron is the functional building block of nervous systems, and that signaling within and among neurons is dynamically polarized, indicating that a neuron and its connections to other neurons are a unidirectional communication pathway, is called the "neuron doctrine." While the neuron doctrine is giving way to a molecular-level understanding of neural function,[9] it is still a useful simplifying idea in gaining a basic concept of neuron physiology and structure.

There are other cells in the nervous system, glial cells, which outnumber neurons. Glial cells are concerned with provision of physical support, insulation, and metabolic support to neurons, including the absorption of neurotransmitters and buffering ions important in neuronal function. In the course of describing neural signal transmission, we will briefly discuss the insulating function of specialized glial cells in speeding neural conduction, but not the other functions of glia. It is the case, however, that due to the support and other functions of glia, nervous tissues such as the brain are not composed only of neurons.

A generalized neuron has four regions: a cell body (which contains the neuron nucleus and controls protein synthesis and other aspects of neuronal metabolism), dendrites (the input to the cell), the axon (a usually long signal conducting part), and presynaptic terminals (the connection to the next neuron or output element such as a muscle, in the neuronal circuit). While many specialized neurons vary considerably from this basic pattern of the generalized neuron, it does reflect a useful concept for understanding neuronal function. In neural signaling electrical current is conducted not by electrons *per se*, as in electrical circuits, but rather by atoms or molecules that carry net positive or negative charges in solution, usually in ionic form. Of course, these charges on ions in solution are due to the excess or deficient numbers of atomic or molecular electrons, but in neural function, we are not dealing with mobile electrons being current carriers as they are in metallic conductors. The common salts that form univalent cations (ions with a single unit positive charge), sodium (Na^+) and potassium (K^+), the divalent cation (an ion with a double unit positive charge), calcium (Ca^{++}), the univalent anion (negatively charged ion), chloride (Cl^-), and negative charges on amino acids and proteins, compose the charge carriers that provide the ionic basis of neural signaling.

The fundamentals of the ionic theory of nervous conduction were originally experimentally tested in invertebrates,[10] but research revealed that the basics of neuronal signaling work in a similar manner across the animal kingdom. When a fine recording electrode penetrates their cell membrane, neurons show a charge difference across that membrane, called the resting or membrane potential of around −65 millivolts (mV). This resting potential ranges over −40 to −80 mV in a variety of neurons. The standard convention is to declare the inside of the cell negative compared with the outside of the cell. The unequal distribution of charge across the neuron membrane is the driving force for all types of neuronal signals. Neuronal signals driven by this unequal transmembrane charge are of several distinct types. The diversity of neuron cellular architecture, the different types of neuronal circuits and the variety of interconnections among nerve cells, makes for an exceptionally large degree of variety and plasticity in neuron circuits.

Neuronal signaling involves shifts in the negative resting membrane potential to a transient positivity. This shift to the positive is due to an influx of positively charged ions, where the direction of current flow is the direction of the movement of positive charges. Increased negative potential is due to an influx of negatively charged ions (or an efflux of positive charges), increasing membrane polarization over the resting potential. The movement of ions across the neuron membrane occurs through membrane channels. These channels are protein molecules that span the lipid bilayers of the cell membrane, providing a conducting channel between the inside and outside of the cell. The properties of these membrane channels allow the selective control of ionic conductance and thereby the regulation of the ionic currents responsible for neuronal signals. Transmembrane channels are important in axonal signaling in neurons, signaling at synapses, and in the process of transduction. Transduction is the transfer of energy in a quantitative manner between environmental stimuli such as photons and ionic currents in sensory neurons. The transduction process is discussed in a Sec. 1.5 of this chapter.

Let us consider the signaling processes in sequence that go on in the parts of the idealized neuron and present the ionic basis of these signals. Neuronal signals are of two basic types: local currents, which decay over short distances of the neuronal membrane, and actively propagated potentials, which are called action potentials. These two kinds of signals are dependent on the distribution of membrane channels with different properties in the different parts of our basic neuron. The different signal types once again increase the possibilities for neuronal plasticity. Neurons can transition between enormous numbers of possible states that represent different signaling conditions and thus have great variation in input–output transfer functions. Furthermore, there are different types of signaling appropriate for local integration of multiple inputs and for unchanged propagation without signal drop across long lengths of neuron axon segments.

Potentials arising in the dendrites are initiated by a synaptic connection from another nerve cell or by some transduction process, which represents the transformation of environmental energy (photons in the case of vision) to a receptor potential. This input potential arises by the disruption of the resting membrane potential of the neuron in the dendritic or input region. The membrane potential is disturbed by an influx of positive ions into the cell causing a current flow and a disruption of the membrane potential from its negative resting value. This influx of positive charge carried by ions is via membrane pores (or channels) that are specific to ionic species. This selectivity is due to channel pore size and charges that exist on the inside of the protein channels, and the resistance of ionic species to the removal of their shell of water molecules that accompanies them in solution. This shell of hydration represents an energy barrier because the water molecules must be stripped away if the ion is to enter the channel. Forces larger than this energy barrier must drive ions into the channel stripping off this shell of hydration. These forces are the electrical and concentration gradients that are present across a neuron's cell membrane. Such considerations also guarantee the directionality of ionic current flow in membrane channels, and their selectivity for particular types of ions (i.e. ions of a given size, charge polarity, and magnitude).

Once the resting membrane potential is perturbed, made more positive (depolarizing) or more negative (hyperpolarizing), then if there is no active process that causes the membrane potential disturbance to be propagated to other parts of the membrane, the local potential decays as a charged capacitor would, with an exponentially decreasing time and distance function of charge reduction.[11] Hyperpolarizing and depolarizing potentials may occur together and allow time and spatial integration of inputs at the input segment of neurons. This integration occurs with a predictable algebraic summation. The integration of depolarizing and hyperpolarizing inputs occurs at an action potential initialization zone in neurons. Neurons of the brain may receive thousands of synaptic inputs, both excitatory and inhibitory, indicating the complex integration of inputs that may occur within one neuron, and in networks of interconnected neurons. Decrementing, not actively propagated potentials, are called local or graded potentials.

Signals are conducted down the lengths of axons without decrement by action potentials. The action potential initialization zone is downstream from the dendritic input segment of the idealized neuron. It is in this region, which is rich in voltage-controlled Na^+ membrane channels, that the integrated inputs to the neuron at the dendritic arborization have their potential impact on neuronal signaling. The result of an adequate (or threshold) level of depolarization at the initialization zone results in the generation of a new type of signal in the neuron, the action potential. Action potentials are transmitted over the long axon segments of neurons without any decline in amplitude or change in duration. This transmission without change in amplitude and duration occurs because the action potential is generated in a fall-forward active manner that occurs anew in each patch of neuron membrane. The fall forward process of excitation and action potential generation spreads from segment-to-segment along the axon in one direction. Generation of the potential depends on the membrane containing an adequate number and type of voltage gated ion channels.[12,13] Voltage-gated ion channels allow ionic currents to flow selectively in a particular direction across the neuron membrane as the membrane potential changes. The action potential has a characteristic time course of measurable current changes over time. An initial rising phase (rising means the depolarization of the negative membrane potential, i.e. movement to a more positive potential value) is mediated by a voltage-controlled Na^+ channel that allows an influx of Na^+ ions from the higher concentration outside the neuron membrane to the inside. This voltage-gated Na^+ channel is activated by a depolarizing current at the action potential initialization zone of a neuron. Once initialized, the rapid rise of the action potential depolarizes the voltage-gated Na^+ channels in the next patch of membrane. At the same time, the action potential in the initial segment of the neurons axon has undergone deactivation due to another aspect of the ion channel structure, one that physically closes conduction when a certain potential is reached. This closing of conduction explains why the action potential travel is in one direction; the patch of membrane in which it has just arisen is refractory due to the duration of inactivation, and backwards spread is inhibited.

The voltage-gated Na^+ channels open in large numbers when the membrane potential in the action potential initialization zone reaches a threshold level. During this rising phase of the action potential, the membrane potential is driven toward the Na^+ equilibrium potential. This potential level is that to which Na^+ will force membrane voltage when Na^+ conductance controls the membrane potential.[14] This potential, known as the equilibrium potential for Na^+, can be compared to a current from a battery, where the battery is the potential difference due to the unequal distribution of Na^+ across the neuron membrane. As the voltage-gated Na^+ ion channel becomes inactivated, another channel, the voltage-gated K^+ channel, opens and takes the control of the membrane potential, pushing it in turn back downwards toward the equilibrium potential of K^+. The K^+ voltage-controlled channel in turn is inactivated, and a slight undershoot at the end of the action potential occurs due to a brief voltage-controlled opening of Cl^- ion channels. The result is a return to the membrane resting potential. All of these events have electrical analogies, the initial unequal concentrations of ions across the resting membrane act like batteries with potential levels and polarities specific to the distribution of the particular ions in the resting membrane potential. The lipid bilayer and proteins of the neuron membrane act as a capacitor, storing the charges, and act as resistors reducing the ability of charge to cross the membrane. All of these ionic membrane events lead to an equivalent circuit that physically or computationally can be used to model neuron membranes and neuron signaling.

The speed of axonal conduction via action potentials is important in motor evocation of escape mechanisms when sensory inputs signal danger. In invertebrates, the greater speed of axonal conduction is accomplished by increasing axon diameter, resulting in "giant axons" that are involved in sensory to motor system connections. Increases of axonal conduction speed in vertebrates are accomplished in another manner, insulation. Insulation of axons is via specialized glial cell types that wind around axons forming a multilayer myelin sheath. This sheath is interrupted by "nodes" or areas in the myelin sheath that expose the neuron membrane. In these nodal regions, the concentration of voltage-gated Na^+ channels is very high, and the action potential is speeded up by "saltatory" or jumping conduction, where rather than separately activating each patch of membrane along the axon, only the nodes need to generate action potentials, which in turn jump to the next node.[15]

1.4. Synapses

Synapses are the points at which individual neurons communicate with other neurons to provide the inputs and outputs for neurons. The communication between neurons in neuronal circuits is through the actions of synapses, and synaptic actions have a great deal of flexibility to modulate information flow between neurons. At the synapse, the signal transfer functions of connections between neurons are altered in a way that shapes information transmission in the nervous system. Synapses and changes in synaptic efficacy are an important part of the mechanisms by which memories and other information are encoded and stored in the CNS. The plasticity that synapses impart to neuronal circuits allows learning, conditioned responses, and makes them important targets for drug development for neurological and mental diseases.

Synapses are characterized by distinctive anatomical and cellular structures that reflect the capacity of the neurons for manifold variations in communication and control, integration, and local processing of information. They are of two basic types, electrical and chemical; with electrical synapses providing direct coupling of neurons and chemical synapses acting over a synaptic cleft, a space that chemical neurotransmitters diffuse over to act on post-synaptic neurons. Synapses utilize the main ionic current-control mechanisms described for signaling within neurons. These include voltage-controlled ion channels that respond to and integrate incoming signals and establish outgoing signals, and initial differential ion concentrations inside and outside the membrane, resulting in a resting potential, and resting channels that allow ions to move down an electrical and concentration gradient to maintain and reestablish the resting potential. In addition, chemical synapses have novel mechanisms that allow the amplification and control of the time duration of neuronal signaling, as well as enhance the number of possible neuron response types and response flexibility.

Electrical synapses are connections between nerve cells that represent a physical connectivity of the cytoplasm within one neuron to the neuron with which it electrically synapses. The function of electrical synapses was first described in crayfish.[16] The connections of the neuronal membranes are made with adhesion structures called gap junctions. Channels made of proteins called connexins[17] form these gap junctions. Connexins form a partial channel on each side of the membrane and have subunits that match up to form the neuron-to-neuron channel. Connexin subunits can rotate with respect to each other, and can close off the connection, which may limit damage after trauma to the neurons.

Chemical synapses operate over a measurable gap between neurons. Depolarization of a synaptic ending in a chemical synapse initiates release of membrane contained vesicles containing a neurotransmitter substance. The vesicles release their contents by fusing with the presynaptic membrane, and the neurotransmitter substance diffuses across the small gap between the pre- and postsynaptic cells, binds to postsynaptic receptors, which in turn affect a neuronal signal in the postsynaptic neuron by acting to induce ionic currents and causing postsynaptic potentials. Neurotransmitter substances are mostly small molecules such as the amino acids L-glutamate, arginine, or glycine, but include other molecules such as gamma aminobutyric acid (GABA), or acetylcholine. Peptides can be neurotransmitters as well, including such substances as vasoactive intestinal peptide, or calcitonin gene-related peptide and many others.[18] Peptide neurotransmitters differ from small molecule neurotransmitters in their metabolism, actions and release mechanisms. Inhibitory synapses in the CNS often involve GABA or glycine. These inhibitory actions depend on receptor channels in which the binding of the neurotransmitter molecules directly causes a change in the channel conductance, conducting Cl^- ions into the cell, which depolarize the postsynaptic cell, making it less likely to fire an action potential. Excitatory synapses involve glutamate-gated ion channels that conduct Na^+ and K^+ ions. An influx of positively charged ions depolarizes the membrane and makes generation (firing) of an action potential more likely. There are synapses in the retinal neural circuits that represent exceptions to GABA and glycine as inhibition-related neurotransmitters. In fact, it is not the transmitter *per se* that is inhibitory or excitatory, but rather the kinds of channels, which the binding of the neurotransmitter activates on the postsynaptic membrane. In addition, other modulatory substances that are in the extracellular milieu of the synaptic regions, but which are not neurotransmitters released from presynaptic endings, are present in the nervous system. Receptors for these substances were discovered by the study of the action of neurotoxic substances and named for them, and the molecules that bind these modulatory sites on synapses discovered later. For example, synapses that use glutamate as a neurotransmitter are all excitatory but fall into several classes based on these other molecules. Thus, we can classify glutamate receptors as those that directly gate ion channels and those that invoke a second messenger system to activate ion channels indirectly. The direct acting glutamate receptors are classified as AMPA, kainite, and NMDA types (alpha amino-3-hydroxy-5-methylisoxazole-4-proprionic acid, kainite, and

N-methyl-D-aspartate). These molecules are the toxic agents that allowed discovery of these synaptic subtypes.[8]

There is increasing knowledge from ongoing research on the properties of inhibitory and excitatory synapses, the drugs that affect them and the diseases that may involve alterations in the function of synaptic physiology. The present review will not address any further detail on the pharmacology or pathology of synaptic function.

The pre- and postsynaptic elements of communicating nerve cells can be of vastly different sizes. This size difference allows for further complexity of integration of neuronal signals in situations where a number of small presynaptic elements contact a much larger postsynaptic neuron. Inhibitory and excitatory synaptic inputs undergo spatial and temporal integrations, giving rise to enormous numbers of possible states for the neuron-to-neuron communication effected by synapses.

1.5. Vision — Sensory Transduction

All the information we have concerning the state of the world outside our bodies, as well as the position of our bodies in space, and the forces acting on and within our body, are due to the activity of neurons in sensory systems. Thus, all of the wonder of the world we experience, light, sound, touch, taste, and smell, all arise through neuronal processes that begin with the transduction of environmental energy into neuronal signaling, and subsequent processing of the signals from primary sensory neurons in other neuronal circuits. Survival dictates that these representations of the outside world be largely accurate. Food, threats, shelter, and other members of our and other species must all be located accurately in the visual space for us to interact with them successfully.

In the case of vision, photon energy is transduced into neuronal activity in a quantitative manner that informs the nervous system about the luminance, spectral differences, orientation, and motion of light energy that impinges on the photoreceptors of the retina. A long pathway of discovery in neuroscience led to our current detailed molecular level understanding of the beginning of the process of seeing: the mechanism of transduction of the energy of photons into neural signals.

Rods are elongate cells that are comprised of an outer segment connected to an inner segment by a thin ciliary process, subsequently attached to the cell body and synaptic outputs by a narrow waist. Cones are similar in structure with differences in shape (hence, the names rod and cone), and in the details of the organization of the membrane and photoreceptor protein apparatus. In broad detail, cones and rods are similar, and we will describe the events of phototransduction in the rod where transduction was first explored. Rod outer segments are filled with photoreceptor discs, organelles that concentrate and manage the photoreceptive pigment and protein and the associated signaling pathways.

In the dark, rods are depolarized.[19] This depolarization is the opposite of the physiology of excitation in most sensory neurons, where a stimulus leads to the generation of an action potential and subsequent output at the synaptic or output end of the idealized neuron. Thus, sound vibrations excite hair cells in hearing; chemical binding excites chemical receptors in the smell and taste systems, all producing depolarization and action potentials. However, by this common standard, dark seems like excitation to rods and light an inhibitory input. Light hyperpolarizes rods and cones; in the dark, the receptor is depolarized. The reason for this apparent reversible of the sense of sensory signaling in vertebrate photoreceptors remains unexplained.

The sequence of events that set phototransduction in motion causes the reduction of cyclic guanidine monophosphate (cGMP) when a photon is absorbed by a form of vitamin A (retinol) associated with the protein opsin. This visual pigment called rhodopsin is in the photoreceptor discs. The retinol part of the photopigment absorbs the photon and the photon's energy causes the transmission of a conformational change in the retinal molecule to the associated protein molecule or opsin. In turn, the altered conformation allows the new conformation of the opsin molecule to activate a series of signaling proteins that reduce the concentration of cGMP in the outer segment, closing the cGMP-gated channels and reducing the inward Na^+, Ca^{++} current.[20] The K^+ channels are thus able to overcome the influx and drive the membrane potential toward hyperpolarization.

Opsin acts differently depending on the wavelength of the energy absorbed and creates effects that are specific to particular wavelength ranges of photons. This behavior accounts for the wavelength sensitivity of photoreceptors and is the beginning of the mechanism of color sensitivity. Note, however, that although the cones may be "tuned" for a specific wavelength range of light, each receptor receives just as many photons as any other receptor, and neuronal circuits that allow comparison of the different classes of cone output must distinguish the spectral sensitivity differences of cones. In the section on retinal processing, we will see how these retinal circuits work and how the information is processed in higher visual areas within the brain.

As to the details of the process of phototransduction, the changes in the opsin protein in turn activate a protein known as transducin (named for its role in the process of sensory transduction). The activated transducin protein in turn activates an enzyme that breaks down cGMP and prevents its activation of cGMP-gated Na^+, Ca^{++} channels. This complicated sequence has the advantage of enhancing the sensitivity of photon detection by an amplification process. Absorption of a single photon can cause closure of a considerable percentage of the channels in a rod membrane. This closure accounts for early observations that the photon threshold of human vision is at the level of the absorption of single photons in five to seven rods in a dark-adapted eye.[21] Other mechanisms act as molecular brakes and restore the rhodopsin to its initial state to start the photoreceptor cascade over. After a number of cycles, the photopigment-containing discs of the rods shed from the ends of the rod cell, then are phagocytized ("cell eaten") by the retinal pigment epithelial cells and degraded. Some components are recycled back to the base of the inner segments and used in the formation of new disc membranes where they add to the stack of discs in the outer segment. Processes are similar in cones, except that cones have membrane folding rather than separate discs; and, as mentioned above, in the cones the proteins in the retinol–opsin complex respond to different wavelengths of light.

1.6. Retinal Processing

The cell types of the retina combine into different circuits that are specific for the detection of different features of the visual scene. Rod and cone photoreceptors are different in shape (hence their names), and differ in their distribution over the human retina. Cones are concentrated in high density at the fovea or focal point of the image on the retina. Rods are absent from the center of the foveal region where there is the greatest visual acuity and color sensitivity, and where the cones are most densely packed. Rods are concentrated in the extra-foveal peripheral retinal regions, and are present at the furthest periphery of the retina where cones are almost absent. The circuits that the rod and cone photoreceptors make with the other retinal neurons differ as well, not only from rods to different types of cones but over regions of the retina for a particular receptor type as well. Rods are poor at spatial resolution but specialized for sensitivity to low light levels. Cones, due to their high concentration at the fovea, and their low degree of convergence on the other cells in the retinal circuits, have a high spatial resolution that we depend on for our sharpest visual acuity. Subsequent retinal and CNS circuits build on and maintain these differences in rod and cone processing.

Over much of the retina, rods and cones converge on the same retinal ganglion cells. Individual retinal ganglion cells respond to both rods and cone inputs. However, the neuronal circuits that rods and cones are connected to differ significantly, and these differences account for differences in visual acuity, motion detection, and color processing both in retinal processing and in subsequent higher levels of CNS visual processing as well. Rods are maximized as low-level detectors that sacrifice spatial accuracy; cones are organized to have maximum spatial and spectral acuity while sacrificing optimum possible sensitivity. Rods synapse with bipolar cells as do cones, but many rods (15–30) synapse on one bipolar cell. Cones, particularly those in the foveal region, synapse on one bipolar cell, and these cone bipolar cells synapse directly with retinal ganglion cells. Bipolar cells connected to rods also differ in being connected not directly to retinal ganglion cells but rather to a particular type of amacrine cell, which makes both electrical synapses (gap junctions) and chemical synapses with cone bipolar cells as well. In turn, these amacrine cells synapse with retinal ganglion cells. Each rod bipolar cell contacts a number of rods and a number of rod bipolar cells contact each rod amacrine cell before it synapses with a retinal ganglion cell. Single cones, especially in the fovea, contact one bipolar interneuron that in turn synapses on one retinal ganglion cell. This high-degree of convergence makes the rod circuits of the retina designed for optimal light detection, but at the same time poor in spatial resolution due to this same convergence, since the stimulus that activates a rod circuit could have come from anywhere in a large area covered by a number of photoreceptors. The one-on-one cone to bipolar cell circuits of the cone system maximizes spatial acuity. Cone bipolar cells, while they synapse with a single cone, do not synapse in turn on only one retinal ganglion cell. The convergence on retinal ganglion cells of several different types of cones (with different spectral sensitivities) is what makes possible the detection of color; it is only in comparing the output of different types of cones that makes spectral discrimination (distinguishing colors) possible. The mechanism involves the center-surround mechanisms described in the next paragraph.

The output of the retina is not simple photoreception; much more sophisticated processing is done in the retina. Retinal ganglion cells, in experiments using electrophysiological recordings to determine the responses of these cells to focal light stimuli on a screen, viewed by the retina of an experimental animal demonstrated that there is a center-surround effect of retinal ganglion cell responses to their receptive fields.[22] Single cell recordings with fine electrodes can reveal the receptive field of a retinal ganglion cell, which is the area of the visual space; in this case, a screen in front of an experimental animal, within which a light spot or other visual stimulus can activate that retinal ganglion cell. The area of the receptive field is accounted for by the retinal ganglion cell synapsing with a number of rods (or cones), giving it a spatially distributed input pattern. Center-surround effect means that a spot of light on the center of the receptive field evokes more signal (more action potentials occurring together in time in a train or burst), than that same light spot in an area of the cell's receptive field that is away from the center in the surround of the cell's receptive field. Such is the case for an "on-center" retinal ganglion cell. "Off-center" retinal ganglion cells show an opposing pattern; they are suppressed (producing fewer action potentials than their spontaneous background rate) when a spot of light is projected on the center of their receptive field and stimulated (producing more action potentials) when the spot is projected on the periphery of their receptive field. Off-center and on-center retinal ganglion cells are present in about equal numbers in the retina. They involve both rods and cones, and their receptive fields overlap so that several on-center and off-center retinal ganglion cells cover each point in visual space. The function of this coverage at the level of the retina is to detect not the absolute level of light falling on the retina, but rather differences in illumination over a wide range of levels.[23] The on-center and off-center retinal ganglion cells have the distinct characteristics they possess due to differences in synaptic mechanisms with their input rods and cones, bipolar cells, and amacrine cells.

Color vision is also of the center-surround contrast-detection type at the level of retinal ganglion cells. In color vision, each type of cone is in effect color blind, as it alone cannot distinguish between wavelengths of light. Because the spectral sensitivities of the three receptors overlap, any given wavelength will stimulate all three receptor types to different degrees. Cones fall into three spectral sensitivity ranges, and are named not by color sensitivity but in terms of their wavelength sensitivity optima. There are so-called S cones (for short wavelength); these absorb maximally at 440 nm. M cones or medium wavelength cones absorb maximally at 530 nm. In addition, L cones, long wavelength cones absorb maximally at 560 nm,[24] S cones represent about 5–10% of cones, twice as many L cones as M cones; with the ratio of cone types being L:M:S = 10:5:1.[25] The center-surround circuitry for cones on the retinal ganglion cells allows the initial discrimination of color in the retina. Note that it is all too easy to say "color" as in color vision, when talking about cones, however, to be technically correct, color sensations are a perception and a naming process that involves higher cortical functions. We should be referring to spectral sensitivity rather than color at this initial level of visual processing.

Trichromatic cones provide one representation of the spectral information in the circuits of the retina. The cone systems: S, M, and L become represented as opposing pairs at the level of retinal ganglion cells: R/G, Bl/Y, and Bk/W. "R" here is for red, or what might be called red-sensitive cones (L) (except for our caveat above about color sensation depending on higher cortical processing), G (green) is for the M cones, Bl (blue) for the S cones. "Y" is for yellow, which is detection of the summation of the M and L cone activity. The Bk/W (black and white) is a center surround comparison of the output of rods and represents the overall "volume" of light, being excited by all wave lengths of light and inhibited by reduced light or vice versa. As is indicated by the above descriptions, these color opponent retinal ganglion cell types, like the rod center-surround retinal ganglion cells exist as two types, center-on and center-off. Consider what these two types are for the R/G opposing pairs: one with an $R+G-$ center, and $G+R-$ surround and another with $G+R-$ center, $R+G-$ surround. Red spectral range light on the center of the first type is excitatory, and green light inhibitory; green light is excitatory on the surround and red inhibitory on the surround. The other type of R/G color opponent cell is the vice-versa pairing of the first. The Bl/Y color opponent cell pairs are arranged in similar manner. In addition to enabling recognition of the spectral character of the light impinging on the retina, the R/G and Bl/Y systems provide useful contrast information to distinguish changes in color with illumination level (such as when there is shading of part of a colored surface). They also aid in the distinction these shadings of colors from surfaces that are actually two different colors.[26]

Other retinal ganglion cell receptive fields do not have the center-surround organization. They have instead an overlap of opposing inputs. The activity of such a retinal ganglion cell might be a function of the relative levels of short, medium, and long wavelength cone responses such as: $S-(L+M)$.

P cells are the name for the above-described color opponent retinal ganglion cells. P cells represent a class of cells sensitive to color contrast, but which have little (comparatively speaking) sensitivity to luminance contrast, higher sensitivity to spatial frequency (the number of dark bars in a variegated pattern falling on the retina), and lower sensitivity to temporal frequency (how fast a visual stimulus is turned on and off). The other major group in this classification of retinal ganglion cells is the M cells. These cells have the opposite characteristics of the P cells; they are not connected only to cones in such a way as to have sensitivity to color contrast, but instead have sensitivity to luminance contrast, lower spatial frequency sensitivity, and higher temporal sensitivity.[27] The size of the nuclei and axons are smaller in the P cells and larger in the M cells. All of these differences are maintained and even spatially segregated in the next area of the visual system. The M of M cell stands for magnocellular, with *magno-* as a Latin word root indicating "large"; P in turn stands for parvocellular and *parvo-* is a Latin word root meaning small. These differences are reflected in further segregation of these classes in higher centers of the nervous system.

1.7. Visual Processing in the Brain

The processing of visual information at higher levels of the brain continues when the optic tracts, containing retinal ganglion cell axons from both eyes reach the first brain area where they synapse. This process is (as described in the anatomy of the human visual system section above) the lateral geniculate nucleus. The majority (~90%) of the retinal ganglion cells from the optic tract synapse in the lateral geniculate nucleus. The retinotopic organization of the retinal ganglion cells is maintained in the lateral geniculate nucleus; however, an important segregation of retinal ganglion cells occurs here prior to visual information passing onto higher visual centers. Retinal ganglion cells are recognizable as distinct by the size of the cell nucleus and axon diameters. These size distinctions reflect the type of connections that the retinal ganglion cell received in the retinal circuits. The M-type retinal ganglion cells segregate to the lateral geniculate nucleus area known as the magnocellular layer; while the P-type retinal ganglion cells are specific to a lateral geniculate region known as the parvocellular layer. The separate action of these layers of the lateral geniculate nucleus was discovered in lesion studies[28] in which visual testing of color sensitivity and spatial and temporal frequency testings were conducted before and after the destruction of the layers in animal (primate) models. These and other experiments demonstrate that P cells are important for color vision and vision with high spatial resolution and low temporal resolution. M cells have the contrasting properties having no color sensitivity, low spatial sensitivity, and high temporal resolution.

In terms of receptive fields and retinal information processing, lateral geniculate nucleus cells have a similar physiology to that described for the retinal ganglion cells, having center-surround antagonistic organization and specialization for color and acuity or sensitivity to light levels and temporal sensitivity. The lateral geniculate nucleus is a six-layered structure with the magnocellular inputs synapsing in the ventral two layers (ventral in this case indicates "toward the bottom of the skull") and the parvocellular pathways projecting to the dorsal-most (top of skull) four layers of the lateral geniculate nucleus. This separation of different retinal ganglion cell types was not evident in the organization of the retina; although the separation into magno- and parvocellular pathways each separately maintain the topographic organization of the retina. It has been proposed[6,29] that the parvo- and magnocellular pathways in the lateral geniculate nucleus and visual cortex represent separate parallel processing channels for motion and depth perception (magno-) and color and shape in formation (parvo-).

The responses to light stimuli of the lateral geniculate nucleus neurons resemble those of the retinal ganglion cells with a center-surround organization, but the lateral geniculate nucleus neurons have larger receptive fields and a stronger opposing effect of the surround. The layers of the lateral geniculate nucleus receive input from one or the other eye alone, alternating between the two eyes. Cells that receive input from both eyes first appear in the primary visual cortex to which the lateral geniculate neurons project. The spatial organization or topography of the retinal cells is maintained in the lateral geniculate nucleus. If retinal ganglion cells are close on the retina, they project to adjacent regions of the lateral geniculate nucleus layers.

The function of the cells of the striate cortex to which the lateral geniculate nucleus neurons project via the optic radiations was examined by single neuron electrophysiological recording techniques.[30] It was discovered that there are distinct types of responses that characterize classes of visual cortex cells. The so-called "simple" cells had receptive fields with both excitatory and inhibitory regions, so that the response of simple cells to bars of light or moving edges or bars of light could be predicted from the cell's responses in all its receptive field areas. Subclasses of these simple cells were seen as edge detectors, others (with elongate receptive fields) were seen as line or bar detectors. The behavior of these cortical cells was related to the behavior of several grouped center-surround units from the lateral geniculate nucleus.

A more frequent type of cell in the striate cortex is the complex cell ($\sim$75%). These were initially hard to identify, since, unlike retinal ganglion cells, they do not respond to stationary small spots of light, but are highly responsive to moving lines or edges in their receptive fields that move in a preferred direction. The receptive fields of complex cells are larger than those of simple cells. The activity of complex cells was explainable by response integration from many simple cells. Logically, this connection explains why they can act as edge detectors sensitive to a particular direction of movement.

A third class of striate cortex cells is hypercomplex cells, which had more complex receptive fields than complex cells. While they are sensitive to the lines and edges of light in their visual fields, they are inhibited when a line extends beyond their receptive fields. Thus, the hypercomplex cells seemed to be candidates for line or edge "end" detectors. More recently, other research has suggested that end detection is a continuous rather than an all or none phenomena and that it is due to lateral inhibition from surrounding striate cortical cells.[31] The end stopping property could work by convergence of excitatory and inhibitory complex cell inputs.[32]

In terms of color processing the striate cortex has "double opponent cells."[33] Double opponent cells of the visual cortex consist of center-surround-organized cells of four classes: those with an R + G− center, an R−G+ surround, a Bl+Y− center, a Bl−Y+ surround, an R−G+ center, an R+G− surround, a Bl+Y− center, a Y+Bl− surround, and achromatic center-surround cells, with a Bk − W+ center, Bk + W− surround (and vice versa). These cells provide the basis for simultaneous color contrast effects and receive input from the color-sensitive parvocellular regions of the lateral geniculate nucleus.

The organization of cells of these various orientations and type in the striate cortex were studied by electrophysiological and metabolic techniques (neuron uptake of radioactive nonmetabolizable sugar) that labeled active neurons in large brain areas.[34] The rescaling to greater numbers of cells of the primary retinotopic map from fewer cells of the lateral geniculate nucleus to the cortical visual areas is cortical magnification. Cortical magnification is the name for the observed effect of the central retinal areas spread over a larger brain area (hence enlarged) and distorted relative to peripheral retinal areas. This phenomenon has its primary source in the greater packing density of receptors in the foveal region of the retina compared to the periphery. The cortical magnification can be seen in published images of striate cortex.[34] The so-called "ocular dominance" columns or slabs are observed in striate cortex.[35] These are interleaved maps from the two eyes (hence "ocular dominance") organized in columnar fashion down the structure of the striate cortex. However, in recent work,[36] the functional importance of the structures such as ocular dominance columns has been questioned, since there appears to be little consistency in their presence or absence in different species with similar or distinct ecological preferences (as nocturnal vs. diurnal) or even within the members of a species. The functional importance of structures such as ocular dominance columns for visual systems where they are present remains unclear; it is possible that data on similarities between similar species are lacking because they have not been adequately studied, or that the relevant experimental questions have not been asked as of yet.

Considerable evidence on the development of receptive fields and cortical cell types during the growth of animals has been accumulated, beginning with the work of Hubel and Wiesel in the 1960s.[37,38] I will not describe this work in detail here except to note that if the young animal is visually deprived, visual system cells will have impoverished connections, and the animal will have reduced visual capacities.

Recent research into the physiology of the visual system above the level of the striate cortex has used neuroimaging techniques such as magnetic resonance imaging (MRI), functional MRI (fMRI),[39] and tensor diffusion MRI (diffusion MRI).[40] These methods allow researchers to make a more global assessment of activation of cortical regions and their connectivity, and have been applied to various visual systems including humans.

A variety of methods for assessing the connections and physiology of the higher visual system cortical areas has been extensively applied in a variety of primate species.[41] Methods in such studies include an extensive array of neuroscience techniques for understanding neuronal circuits including single cell electrophysiology of receptive fields and detection characteristics of cortical cells, MRI and fMRI, and histological tract-tracing methods. Understanding the connectivity and the neurophysiology of the cells in higher cortical visual areas has not reached the level of our present understanding of visual area 1. The physiology of visual area 1 provides the understanding that adds the biological vision aspect to the next section on biological vision and computer vision algorithms.

1.8. Biological Vision and Computer Vision Algorithms

Our understanding of the logic of the circuits of the visual system has inspired the development of a number of computational algorithms for various facets of computer vision. Much of this research is not tied to direct knowledge of the processing inherent in the neuronal circuits of the visual system, but rather only based on the knowledge that biological visual systems accomplish a particular task (such as pursuit projection and shooting targets from the air), and postulating that the task may be accomplished algorithmically. Novel image-analysis algorithms have also been developed from the more direct knowledge of visual system physiology.

There are a number of areas of computational science, some related to our detailed physiological knowledge of the visual system. The related research and results could be a book topic in themselves. Here, I will address two examples, image-analysis methods, such as edge or line detection algorithms derived from biological visual circuits, and machine-learning methods, in particular unsupervised hierarchical Bayesian methods for target matching.

Knowledge of the function of the visual system, in particular knowledge of the primary visual system processing in the retina, lateral geniculate nucleus, and striate cortex, inspired computer scientists to program work-alike algorithms that accomplished basic tasks in image analysis. An example of this line of research was the Marr-Hildreth algorithm.

The motivation for the development of line and edge detection visual system mechanisms developed as computational algorithms came from the work of David Marr.[42] The kernel of Marr's concepts of visual systems was that the neuronal events in biological vision systems are essentially accomplishing a computation task, and thereby complex systems such as biological vision systems are information-processing systems. Marr proposed a general idea for the construction of theories of information-processing systems, in particular visual systems. He noted three stages of description in understanding complex information-processing systems such as the visual system. These are the computational, algorithmic, and implementation levels. Understanding all of these levels is important for understanding information-processing systems. The computational level is the informational constraints that apply to the mapping of input to output information in an information-processing system. Mathematical methods can be used to model and check the computational level for validity. The algorithmic level is conceptually similar to the role of a computer program in computer science. The algorithmic level is the manner in which the computational level mapping of input to output is accomplished. As with computer code, there are many algorithmic ways a given computational mapping might be realized. The implementation level is the physical hardware (or "bioware" in the case of biological vision systems) by which a given algorithm, which realizes the computational level mapping, can be physically realized. These levels represent different levels of description and understanding of the same phenomena; understanding at only one of the levels is insufficient for full understanding of any complex information-processing system.

According to Marr's formulation, all three levels are necessary for complete understanding of, and effective theory construction for an information-processing system such as the visual system. In addition, if one level is correctly understood, our understanding of the realization at that level may constrain the possibilities at the other levels. Thus, if we can see how the neuronal circuitry we have discovered in living organisms implies an algorithm, and we realize that algorithm in computer code or verify that its mathematical machinery can produce the observed result given the known input, we have made a considerable step toward validating observations of visual system processing. The hardware implied an algorithm, and the algorithm of the proposed character provides a reasonable semblance of the output of the biological visual system at some level, given the input image. Therefore, we have obtained a suggestive confirmation of our ideas about how neuronal circuits accomplish the tasks we observe them to accomplish. We thus have the potential for double gain here, as we have a way to evaluate ideas about how the nervous system accomplishes its information-processing tasks, and we may discover novel algorithms that can improve machine tasks such as computer vision, independent of our research concerns in biological vision.

The performance of the biological visual system is described (over and above subjective experience and introspection) by an area of psychology called psychophysics. Psychophysics is human sensory physiology by interview, or asking humans to respond verbally or by some task when there are changes in their sensory input that they recognize consciously. Psychophysics thus provides the performance specifications for the system, telling us what the human visual system can perform. Similar methods are used in animal research, except that rather than a verbal response the experiment is arranged so that the animal can respond with a task to which it has been trained. The knowledge we have thus gained of the bioware, or the neuronal circuits that process the system inputs represents the implementation level that the visual system uses, analogous perhaps to the hardware circuitry of a computer. Marr, in attempting to develop a computational theory to accomplish the tasks a given level of the visual system could perform was attempting to realize the algorithmic level of understanding the visual system. The performance of the algorithms he devised could then be compared to the observed performance (determined by electrophysiology and psychophysics) of the biological visual processor. In turn, in understanding the algorithmic level and improving our algorithms to fit the implementational data better, we can iteratively improve our ideas of how the biological processing occurs.

Computational scientists, in their approaches to the problem of vision, long ago realized that an effective first step in automated visual recognition of objects was the discovery of lines and edges.[43] Marr expressed this recognition by postulating that the initial stages of biological visual processing systems develop a "primal sketch." This primal sketch was the input to subsequent stages that processed and compared it to other information about the visual scene to extract dimensional information in several steps. The primal sketch represented one aspect of the initial information that Marr postulated the neuronal circuitry of the early visual system (striate cortex or visual area 1, often denoted as V1) extracted from the 2D images on the retina. Marr thus set out to devise an algorithm to implement center surround and edge detection similar to that used by striate cortex neurons, in order to discover what sort of image information could be available at this stage of processing, and how useful such information could be in later visual processing. The results of this approach[44] combined a number of then commonly used methods in computer image analysis for edge detection. Improvements in their algorithm by other workers produced more effective edge detection routines that form the basis of modern methods.[45,46]

The combination of edge detection methods that were proposed by Marr and Hildreth were combined in a way that not only did a passable job of edge detection, but also matched Marr's concepts of what the visual system cells of the striate cortex were doing. This notion of neurons in visual area 1 as edge detectors also was in line with the original suggestions Hubel and Wiesel had made for the nature of the visual area 1 cortical cell physiology they had observed.[30] The receptive fields of striate cortex cells were revealed by later research to be more complex than the early Hubel and Wiesel research suggested.[46]

However, after Hubel and Wiesel's 1959 work, there were alternative concepts put forward that proposed a different scheme for the significance of the striate cortical cell responses. These concepts involved ideas that became quite important in our understanding of image processing in the visual cortex. They involved the idea that the neurons of the cortical visual area 1 are performing transforms on the input from the retina that act as filters for image feature scale and orientation; in other words, the cells perform a basis function fit on the input. This notion of visual system filters has progressed as being conceived of as Fourier transforms[48,49] to windowed Fourier transforms such as Gabor transforms,[50,51] to sums of Gaussian functions,[52] and finally to wavelet transforms.[53]

If these models are correct, researchers have asked, why were wavelet type basic functions realized in the biological vision systems that have evolved? A compelling answer, suggestive of much further research is that the statistical distributions of values in natural scenes (trees, landscapes, rocks, grass, and so on) dictate that orthogonal (decorrelated) basic functions provide the best fit to the observed statistical properties of natural scenes.[54] Natural scenes present considerable redundancy and orthogonal wavelet transforms provide an economic means of reducing this redundancy. While we are accustomed to this use of wavelets for image compression, compression is not the point of visual system neural coding. Rather, it has been proposed[53] that wavelet transforms realized in neuronal receptive field shapes allow for the reduction of correlations between the responses of different neurons, and thereby allows any one neuron to encode maximal information about the stimulus. Wavelet-based codes would be selected for because their scales could match the localized band-restricted structure the values of the retinal images of natural scenes take on. With decorrelation, the activity of any given cell is relatively independent of the activity of other cells, and inspection of a bank of such neuron activity at the next higher level of neuronal processing would allow the detection and extraction of complex structures characteristic of the statistics of natural scenes in the input signal.

While we are familiar with the attributes and the advantages of wavelets in modern computer image analysis, it is clear that wavelets are an exciting aspect of computer image analysis that has great power to extend our knowledge of the algorithms and implementation of seeing systems further. In computational image analysis, in particular in applications such as automated biomedical image analysis and computer-aided manufacturing, numerous current successful applications of statistical learning algorithms are applied to make decisions that discriminate conditions that have been segmented from image spaces using wavelet transforms. The work described above suggests that wavelet transforms are an optimization built into biological vision systems. Visual science at present only poorly understands higher levels of cortical function that process output from lower levels of the visual processing system. Output of lower level visual processing is sent to higher cortical levels that in turn feedback to lower visual processing areas. Memory matching and output decisions based on those matching results occur as the observable results of visual perception and human behavior tell us. That is, our visual brain is very good at fast visual processing to respond to visual input, predict, and react to the predicted consequences of what we see. Is it possible that statistical methods used in image analysis and statistical decision processes are realized in biological neural systems for making effective visual discriminations based on visual system feature extraction?

The answer, it has been suggested, is yes. In addition to ideas on computational image analysis being drawn from biological vision systems, decision theory and the design of machine-learning algorithms have also been furthered by research on the visual cortex.[55,56] This area of research represents the potential for future convergence of our models of biological vision with optimized algorithms for computer vision. At present, it is recognized that primate vision systems accomplish rapid identification of objects in the noisy visual space of natural scenes at speeds and accuracy superior to any current computer vision algorithms.[57] It has been pointed out that visual processing experiments, which use artificial visual stimuli, may mislead us about visual function, and the complexity of natural images may confound simple edge detectors or other low-level feature extraction methods.[58]

Because of the hierarchical nature of the nervous system, hierarchical Bayesian analysis seems a natural fit to the problem of how higher visual processing areas might categorize the processing results of lower (cortical visual area 1) processing by recalling and matching abstracted properties of visual information held in memory in higher cortical centers. We recognize objects or familiar faces in many different levels of illumination, orientations, distances, and backgrounds, in motion or at rest. A comparison to hierarchical prior distributions of parameters representing different levels of feature abstraction from visual stimuli stored in higher cortical areas would make accurate classification of lower visual processing information possible, if the computational machinery of hierarchical Bayesian inference were realized in neuronal circuits.[58] The store of prior distributions of abstracted parameters of visual targets could be learned from experience and stored in relatively compact form in memory.

The possibility of hierarchical Bayesian methods being realized in the function of neuronal circuits has been argued convincingly by several authors.[55,56,58] Furthermore, the concept of hierarchical processing across levels of the visual system and higher cortical centers suggests a number of testable hypotheses, some of which have been evaluated by fMRI and behavioral experiments.[57] Overall, the most compelling conclusion of this line of research is that the brain is apparently using algorithms as sophisticated (and perhaps identical to) the most advanced methods used in current image analysis and computational science. A further exciting aspect is that the brain is likely using methods we have yet to learn that are even more intricate and efficient than the best of our current computational methods.[58]

In conclusion, the visual system is the best-studied sensory system. It has incredible facility for rapid recognition of objects in the visual field and evocation of appropriate responses. As our understanding of higher processing in the visual system advances, we have great opportunity for learning and profiting from the methods used by the optimized biological image processors that have resulted from eons of natural selection.

References

1. Horridge, G.A. The evolution of visual processing and the construction of seeing systems. *Proc R Soc Lond Biol Sic* **230**:279–292, 1987.
2. Barton, R.A. Binocularity and brain evolution in primates. *Proc Natl Acad Sci USA* **101**(27):10113–10115, 2004.
3. Steering Committee on Science and Creationism, National Academy of Sciences. *Science and Creationism: A View from the National Academy of Sciences*, Second edition, ISBN: 0-309-53224-8, 48 pages, National Academy Press, Washington, DC, 1999. Free PDF can be downloaded from: http://www.nap.edu/catalog/6024.html.
4. Berson, D.M. Strange vision: ganglion cells as circadian photoreceptors. *Trends Neurosci* **26**:314–320, 2003.

5. Hattar, S., Liao, H.W., Takao, M., Berson, D.M., and Yau, K.W. Melanopsin containing retinal ganglion cells: architecture, projections, and intrinsic photosensitivity. *Science* **295**:1065–1070, 2002.
6. DeYoe, E.A. and Van Essen, D.C. Concurrent processing streams in monkey visual cortex. *Trends Neurosci* **11**:219–226, 1988.
7. Van Essen, D.C. and DeYoe, E.A. Concurrent processing in the primate visual cortex. In: Gazzaniga, M.S. (ed.), *The Cognitive Neurosciences*, pp. 383–400, MIT Press, Cambridge, MA, 1995.
8. Kandel, E.R., Schwartz, J.H., and Jessel, T.M. *Principles of Neural Science.* Fourth edition, 1414 pp. McGraw-Hill, New York, NY, 2000.
9. Guillery, R.W. Observations of synaptic structures: origins of the neuron doctrine and its current status. *Phil Trans R Soc B* **360**:1281–1307, 2005.
10. Hodgkin, A.L. and Huxley, A.F. A quantitative description of membrane current and its application to conduction and excitation in nerve. *J Physiol* (Lond) **117**:500–544, 1952.
11. Hodgkin, A.L. and Ruston, W.A.H. The electrical constants of a crustacean nerve fibre. *Proc R Soc Lond Biol Sci* **133**:444–479, 1946.
12. Armstrong, C.M. Voltage-dependent ion channels and their gating. *Physiol Rev* **72**:S5–S13, 1992.
13. Armstrong, C.M. and Hille, B. Voltage-gated ion channels and electrical excitability. *Neuron* **20**:371–380, 1998.
14. Hodgkin, A.L. and Katz, B. The effect of sodium ions on the electrical activity of the giant axon of the squid. *J Physiol* (Lond) **108**:37–77, 1949.
15. Huxley, A.F. and Stampfli, R. Evidence for saltatory conduction in peripheral myelinated nerve fibres. *J Physiol* **108**:315–339, 1949.
16. Furshpan, E.J. and Potter, D.D. Mechanism of nerve-impulse transmission at a crayfish synapse. *Nature* **180**:342–343, 1957.
17. Goodenough, D.A., Goliger, J.A., and Paul, D.L. Connexins, connexons and intracellular communication. *Ann Rev Biochem* **65**:475–502, 1996.
18. Gordon, W.C. and Bazan, N.G. Retina. In: John J. Harding (ed.), *Biochemistry of the Eye*, Chapman and Hall, London, 1997.
19. Schnapf, J.L. and Baylor, D.A. How photoreceptor cells respond to light. *Sci Amer* **256**:40–47, 1987.
20. Stryer, L. Cyclic GMP cascade of vision. *Ann Rev Neurosci* **9**:87–119, 1986.
21. Hecht, S., Shlaer, S., and Pirenne, M.H. Energy at the threshold of vision. *Science* **93**:585–587, 1941.
22. Kuffler, S.W. Discharge patterns and functional organization of mammalian retina. *J Neurophysiol* **16**:37–68, 1953.
23. Schiller, P.H., Sanall, J.H., and Maunsell, J.H.R. Functions of the on and off channels of the visual system. *Nature* **322**:824–825, 1986.
24. Schnapf, J.L., Kraft, T.W., and Baylor, D.A. Spectral sensitivity of human cone receptors. *Nature* **352**:439–441, 1987.
25. De Valois, R.L. and De Valois, K.K. A multi-stage color model. *Vision Res* **33**:1053–1065, 1993.
26. Dacey, D.M. Circuitry for color coding in the primate retina. *Proc Natl Acad Sci USA* **93**:582–588, 1996.

27. Shapley, R. and Perry, V.H. Cat and monkey retinal ganglion cells and their visual functional roles. *Trends Neurosci* **9**:229–235, 1986.
28. Merigan, W.H. and Maunsell, J.H.R. Macaque vision after magnocellular lateral geniculate lesions. *Vis Neurosci* **5**:347–352, 1993.
29. Livingstone, M.S. and Hubel, D.H. Segregation of form, color, movement, and depth: anatomy, physiology, and perception. *Science* **240**:740–749, 1988.
30. Hubel, D.H. and Wiesel, T.N. Receptive fields of single neurons in the cat's striate cortex. *J Physiol* (London) **148**:574–591, 1959.
31. deAngelis, G.D., Ohzawa, I., and Freeman, R.D. Receptive field dynamics in the central visual pathways. *Trends Neurosci* **18**:451–458, 1995.
32. Kuffler, S.W. and Nichols, J.G. *From Neuron to Brain: A Cellular Approach to the Function of the Nervous System*, Sinauer Associates, Sunderland, MA, 1976.
33. Michael, C.R. Color vision mechanisms in monkey striate cortex: dual-opponent cells with concentric receptive fields. *J Neurophysiol* **41**:572–588, 1978.
34. Tootel, R.B.H., Silverman, M.S., Switkes, E., and DeValois, R.L. Deoxyglucose analysis of retinotopic organization of the second visual are in primates. *Science* **220**:737–739, 1982.
35. LeVay, S., Hubel, D.H., and Wiesel, T.N. The pattern of ocular dominance columns in macaque visual cortex revealed by a reduced silver stain. *J Comp Neurol* **159**:559–575, 1975.
36. Horton, J.C. and Adams, D.L. The cortical column: a structure without a function. *Phil Trans Biol Sci* **360**:837–862, 2005.
37. Hubel, D.H. and Wiesel, T.N. Receptive Fields of cells in striate cortex of very young, visually inexperienced kittens. *J Neurophysiol* **26**:994–1002, 1963.
38. Hubel, D.H. and Wiesel, T.N. The period of susceptibility to the physiological effects of unilateral eye closure in kittens. *J Physiol* **206**:419–436, 1970.
39. Shomstein, S. and Behrmann, M. Cortical systems mediating visual attention to both objects and spatial locations. *Proc Natl Acad Sci USA* **103**:11387–11392, 2006.
40. Le Bihan, D., Urayama, S., Aso, T., Hanakawa, T., and Fukuyama, H. Direct and fast detection of neuronal activation in the human brain with diffusion MRI. *Proc Natl Acad Sci USA* **103**:8263–8268, 2006.
41. Gattass, R., Nascimento-Silva, S., Soares, J.G.M., Lima, B., Jansen, A.K., Diogo, A.C.M., Farins, M.F., Marcondes, M., Bothelho, E.P., Mariani, O.S., Azzi, J., and Fiorani, M. Cortical visual areas in monkeys: location, topography, connections, columns, plasticity and cortical dynamics. *Philosophical Transactions of the Royal Society* (London) B 709–731. Online publication 29 April, 2005.
42. Marr, D. *Vision*. Freeman and Co., New York, 1982.
43. Roberts, L.G. Machine perception of three-dimensional solids. In: Tippett, J.T., Berkowitz, D.A., Clapp, L.C., *et al.* (eds.) *Optical and Electro-optical Information Processing*, MIT Press, Cambridge, MA, 1965.
44. Marr, D. and Hildreth, E.C. Theory of edge detection. *Proc R Soc Lond Biol Sci* **207**:187–217, 1980.
45. Canny, J.F. A computational approach to edge detection. *IEEE Trans Pattern Anal Mach Intell* **8**:769–798, 1986.
46. Deriche, R. Using Canny's criteria to derive an optimum edge detector recursively implemented. *Int J Comput Vis* **2**:167–187, 1987.

47. DeValois, R.L. and DeValois, K.K. *Spatial Vision*, Oxford University Press, New York, 1988.
48. Campbell, F.W. and Robson, J.G. Application of Fourier analysis to the visibility of gratings. *J Physiol* **197**:551–556, 1968.
49. Blakemore, C. and Campbell, F.W. On the existence of neurons in the human visual system selectively sensitivity to the orientation and size of retinal images. *J Physiol* **203**:237–260, 1969.
50. Marcelja, S. Mathematical description of the responses of simple cortical cells. *J Opt Soc Am* **70**:1297–1300, 1980.
51. Jones, J. and Palmer, L. An evaluation of the two-dimensional Gabor filter model of simple receptive in cat striate cortex. *J Neurophysiol* **58**:1233–1258, 1987.
52. Stork, D.G. and Wilson, H.R. Do Gabor functions provide appropriate descriptions of the visual cortical receptive fields? *J Opt Soc Am* **A7**:1362–1373, 1990.
53. Field, D.J. Wavelets, vision and the statistics of natural scenes. *Philosophical Trans R Soc Lond A* **357**:2527–2542, 1999.
54. Field, D.J. What is the goal of sensory coding? *Neural Comput* **6**:559–601, 1994.
55. Lee, T.S. and Mumford, D. Hierarchical Bayesian inference in the visual cortex. *J Opt Soc Am Opt Image Sci Vis* **20**:1434–1448, 2003.
56. Friston, K. A theory of cortical responses. *Philosophical Trans Biolo Sci* **360**:815–836, 2005.
57. Kourtze, Z. and DiCarlo, J.J. Learning and neural plasticity in visual object recognition. *Curr Opinion Neurobiol* **16**:152–158, 2006.
58. Friston, K. Hierarchical models in the brain. *PLoS Comput Biol* **4**:1–24, 2008.

Chapter 2
Computational Methods for Feature Detection in Optical Images

Michael Dessauer* and Sumeet Dua*

2.1. Introduction to Computational Methods for Feature Detection

Digital imaging provides medical researchers and computer scientists the opportunity to collaborate in solving medical diagnostic challenges with automated computational methods. The medical researchers in the pathological domain of interest are necessary to define clearly where, how, and why a subject is symptomatic of a specific disease or condition. For retinal images, these domain experts play a significant role in creating automated decision support by grading the severity of pathologies such as diabetic retinopathy and macular degeneration. Computational scientists, leveraging domain expert knowledge, work to solve the pattern recognition task of differentiating retinal images into specific classes of disease severity to provide clinical decision support.

The creation of an automated algorithm that takes retinal fundus images as input and classifies the image into levels of disease severity is a multi-staged process, in which each step provides input into the next. Although retinal-imaging hardware preprocesses the acquired image, more low-level processing steps are necessary to condition an image into a form acceptable for higher-level algorithms. These preprocessing steps include image

*Department of Computer Science, College of Engineering and Science Louisiana Tech University, Ruston, LA, U.S.A.

brightness normalization, non-uniformity correction, noise reduction, and noise artifacts elimination. In longitudinal studies, image registration is an important preprocessing step that allows the retina axis to remain oriented for comparison. After a retinal image has been properly preprocessed, the algorithm typically requires specific anatomical constituents to be identified to a level of specificity somewhere between finding its general area (detection), its central location (localization), and its boundary (segmentation). These retinal regions include the blood vessels, the optic disk, or pathological phenomena such as microaneurysms, hemorrhages, and cotton wool spots. After the identification stage is completed, features are extracted from the different regions to provide the discriminate characteristics for accurate pathological classification. In supervised classification schemes, classifiers are trained with features taken from annotated training sets, creating a data-dependent decision model. *A priori* methods use the algorithm model parameters to make decisions for classification.

Before discussing the computational methods for accomplishing these algorithm steps, we will describe the digital retinal image representation as a discrete function, $f(x, y)$, of pixel size $M \times N$. The RGB color images (typical retinal fundus image format) will give

$$f(x, y) = \{r, g, b\} \tag{2.1}$$

for each (x, y) location (note in matrix form, x refers to rows and y refers to columns). These intensities compose three two-dimensional (2D) matrices, with ranges for eight-bit images between [0, 255], or, if normalized, between [0, 1]. We can also think of these intensities as magnitudes at a particular location (Fig. 2.1), to visualize many of the computational methods discussed in this chapter more easily. As shown in Fig. 2.1, the green channel (G) will contain most of the interesting contrast in retinal images; thus, it will be used frequently in the following sections as the representation of the image, f.

For clarity, this chapter is divided into three sections: in Sec. 2.2, we describe preprocessing methods; in Sec. 2.3, we explain segmentation/localization, and in Sec. 2.4, we explore feature extraction techniques. Methods will overlap in several sections. As you will see in the discussions below, computational methods can both enhance and extract descriptive features that provide the necessary specificity for successful retinal pathology classification. The process of selecting which methods to implement can be a

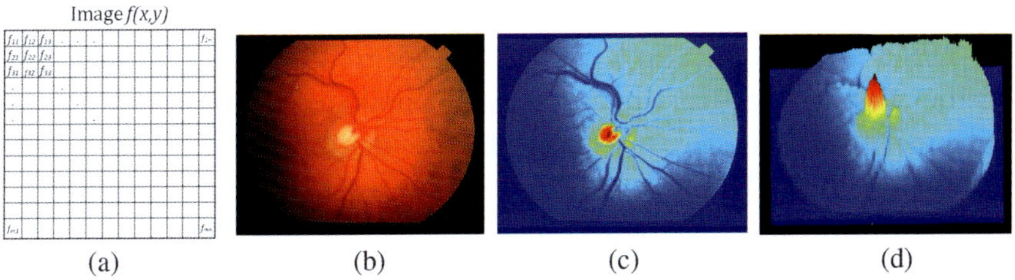

Fig. 2.1. (a) Matrix representation of an image, (b) RGB retinal image, (c) green channel of RGB image, and (d) 3D visualization of green channel intensities.

difficult, trial and error undertaking, with best results typically generated after multiple attempts. In this chapter, we aim to provide an overview of multiple methods to achieve automatic retinal image pathology classification, providing a starting point for a medical researcher or computational scientist to understand image processing used in this field of research.

2.2. Preprocessing Methods for Retinal Images

Image preprocessing is the initial stage of image analysis in which low-level operations are performed on either global or local image areas to reduce noise and enhance contrast. These enhancement steps attribute to significant gains in the quality and accuracy of object detection, segmentation, and feature extraction for classification by removing anomalous image data caused by illumination effects or camera acquisition noise and increasing intra-image contrast from one object to another. Inter-image normalization can also increase automated retinal imaging research results by attenuating image sequences for differences in camera specifications, illumination, camera angle, and retinal pigmentation.[1] The following section will discuss several methods that provide the necessary preprocessing operations required for successful retinal pathology classification.

2.2.1. *Illumination Effect Reduction*

Illumination effects due to changes in reflectance cause nonuniform variance in pixel intensities across an image. These changes in pixel intensity are due to varying retinal reflectivity and background fluorescence from the retinal capillary network.[2] Other factors contributing to nonuniform illumination include varying degrees of pupil dilation, involuntary eye movement, and the presence of a disease that changes normal reflectively properties.[3] These effects can hinder the segmentation of the optical anatomy due to the illumination causes shading of artifacts and vignetting.[4] These changes in the local image statistics result in characteristic variation from the normal optical anatomy pixel representation, which leads to misclassification through weak segmentation and feature extraction. We describe several methods that decrease illumination effects on retinal image segmentation and feature extraction, which range from simple and local to complex and global operations.

2.2.1.1. *Non-linear brightness transform*

A direct, global method to adjust pixel intensity is the application of a brightness transform function to the image. One such function is a nonlinear point transformation that changes only the darker regions of the retinal image, allowing potential features to be detected in subsequent steps[5]:

$$y = \beta * x^{\alpha}, \tag{2.2}$$

where x is the original pixel intensity, y is the adjusted image pixel, and $0 \leq \alpha \leq 1$, $\beta = \text{inmax}^{1-\alpha}$, inmax is the upper limit of intensity in the image. By selecting appropriate parameters α and inmax, an illumination effects-corrected image can be created (Fig. 2.2).

A drawback for this method is that global point transformations do not take into account the luminance variations that are caused by the anatomical regions of the eye, thus decreasing feature contrast in certain areas.

2.2.1.2. *Background identification methods*

Several methods in the literature require finding a background intensity image that can be used to correct the original image for illumination effects. These methods include shade correction through median filtering and background luminosity correction through sampling. We will explain both methods.

As discussed above, the shading of artifacts in an eye image can lead to inaccurate classification. Thus, shade correction methods have been

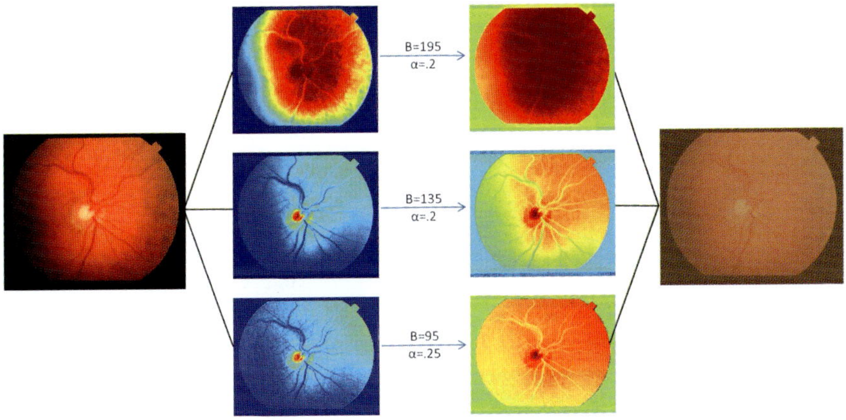

Fig. 2.2. Nonlinear point transform of a color retinal image to correct for illumination effect.

developed in the literature. One such technique smoothes an image using a median filter, with the resulting image treated as the background image.

The median filter belongs to the order-statistic family of smoothing spatial filters. These nonlinear filters order pixels in a determined local area and replace the center pixel with the median pixel value. This type of simple filter is effective in reducing optical image illumination while retaining edge data with minimal blurring. This filter has been implemented in the literature for retinal image illumination reduction by first using a large-scale median filter to create a filtered image and then subtracting from the original.[6] Only anatomical constituents smaller than the filter size remain for further analysis, providing an illumination invariant description (Fig. 2.3).

This scheme uses assumptions based on domain knowledge of retinal fundus imaging techniques and eye geometry to estimate correction variables for recovering the true image without illumination effects. Because the retina is a curved surface with camera illumination occurring near the center, the image region will appear darker as the distance from the eye center increases. Using a linear model, the relation between the true image U and observed image I is:

$$U(x, y) = \frac{I(x, y) - \mathrm{S_A}(x, y)}{\mathrm{S_M}(x, y)}, \tag{2.3}$$

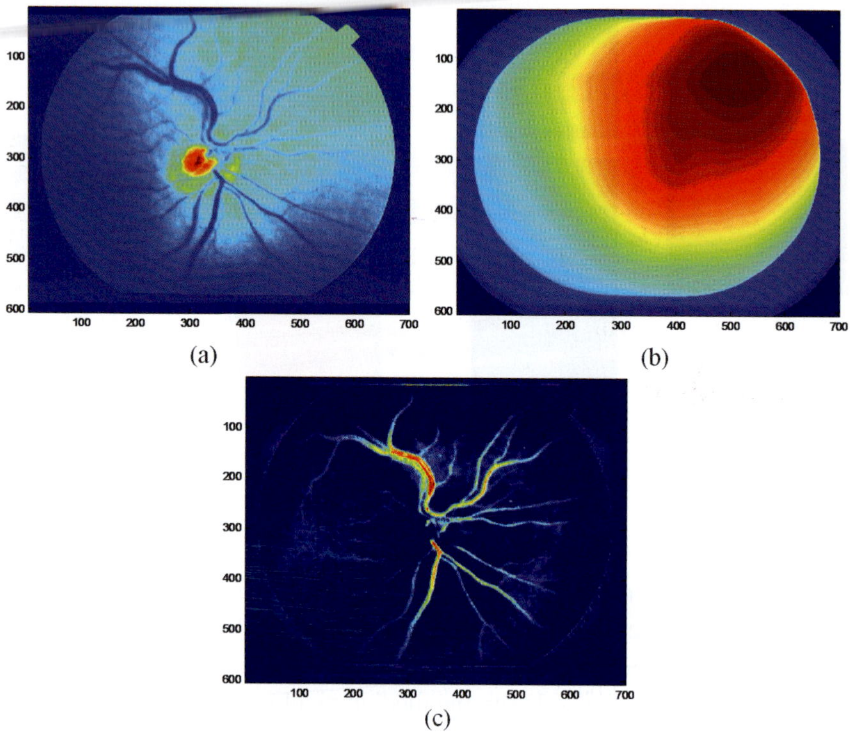

Fig. 2.3. (a) Raw green channel retinal image, (b) median filter response with window size [250×250], and (c) difference image from the previous images.

where S_A is contrast degradation and S_M is the luminosity degradation. To estimate these values correctly, it is necessary, first, to extract the background pixels. Several assumptions are made for a pixel of the image in neighborhood N: S_A and S_M are constant, at least 50% of pixels are background pixels, and background pixels differ from foreground pixels.[7]

Since eye features have either very high or very low pixel intensities in the green color channel, background pixels (not contained within retinal regions of interest) are automatically chosen by interpolating the sampling points to obtain local neighborhood N mean and variance. Computing the Mahalanobis distance $D(x, y)$ determines if this value is above or below a threshold t. All pixels below the threshold are determined to be background pixels:

$$D(x, y) = \left| \frac{I(x, y) - \mu_N(x, y)}{\sigma_N(x, y)} \right|. \quad (2.4)$$

The S_A and S_M values are the standard deviation and mean intensity for location (x, y) in neighborhood N, which is deemed part of the background. This method has also been recently implemented using a nonuniform sampling grid.[5] The sampling points are chosen at nonuniform locations in the r and θ directions, with point density increasing with r.

2.2.2. *Image Normalization and Enhancement*

Like many other fields of imaging research, most retinal images must undergo transformations of color and/or intensity channels to enhance contrast for increased discrimination for further processing stages. Several methods achieve image enhancement, ranging from simple nonlinear spatial transformations as discussed in Sec. 2.2.1 to spectral analysis in the Fourier domain.[8] The typical result of image normalization and enhancement operations is a change in the distribution of the image color/intensity probability density function (PDF) to increase the area of useful pixel values without significantly degrading the image. Several such methods that have been applied to retinal image research are described in detail below.

2.2.2.1. *Color channel transformations*

Due to the inter- and intra-image variability of retinal color brought on by factors such as age, camera, skin pigmentation, and iris color, color normalization is a necessary step in retinal image studies.[1] A typical retinal fundus image will produce three $M \times N$ pixel matrices contributing to the red, green, and blue (RGB) color model based on the Cartesian coordinate system (Fig. 2.4).

A more intuitive model for human interpretation, HSI (hue, saturation, and intensity), decouples the intensity from color-carrying information. Transforming retinal images from the RGB model allow for the normalization of the intensity channel without changing color value information in hue and saturation levels, which has proven effective as a preprocessing step in the localization of the several retinal features.[9] To convert an RGB

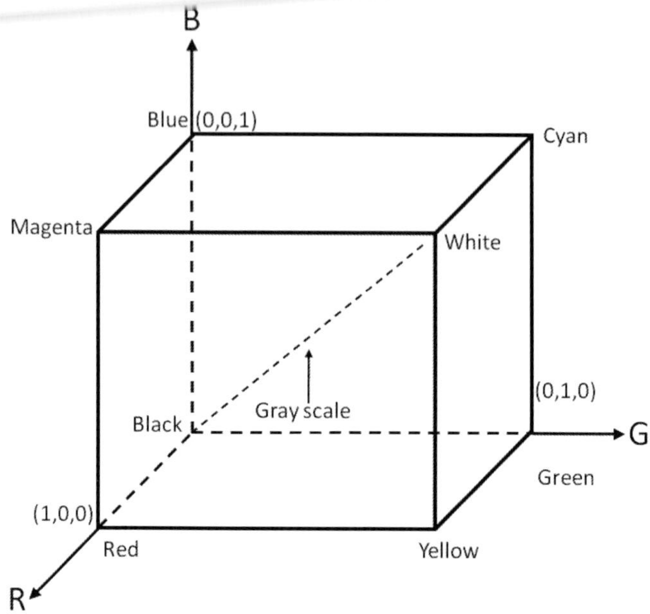

Fig. 2.4. RGB color cube used in three-channel retinal fundus imaging.

image (normalized to range [0, 1]) to HSI, the following series of equations employed each pixel location[8]:

$$\text{Hue component: H} = \begin{cases} \theta & \text{if B} \leq \text{G} \\ 360 - \theta & \text{if B} > \text{G} \end{cases} \quad \text{with}$$

$$\theta = \cos^{-1}\left\{\frac{\frac{1}{2}[(\text{R}-\text{G})+(\text{R}-\text{B})]}{[(\text{R}-\text{G})^2+(\text{R}-\text{B})(\text{G}-\text{B})]^{\frac{1}{2}}}\right\},$$

$$\text{Saturation component: S} = 1 - \frac{3}{(\text{R}+\text{G}+\text{B})}[\min(\text{R},\text{G},\text{B})], \quad \text{and}$$

$$\text{Intensity component: I} = \frac{1}{3}(\text{R}+\text{G}+\text{B}).$$

Using these equations, converted retinal intensity (I) images can be further enhanced to provide sharp retinal feature contrast for better segmentation and classification results (Fig. 2.6).

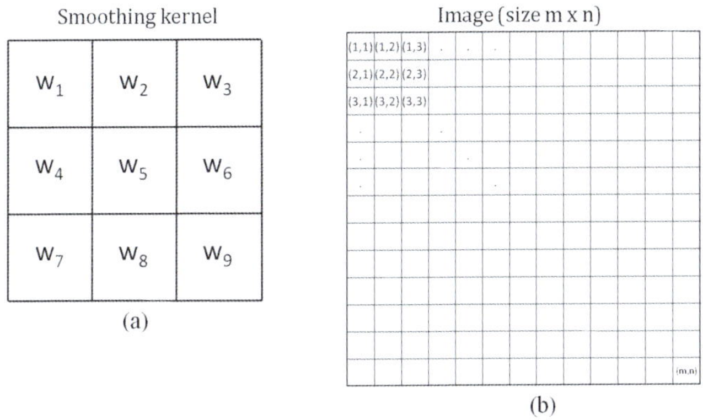

Fig. 2.5. (a) Example smoothing kernel, (b) image convolution at location *I* (2,2).

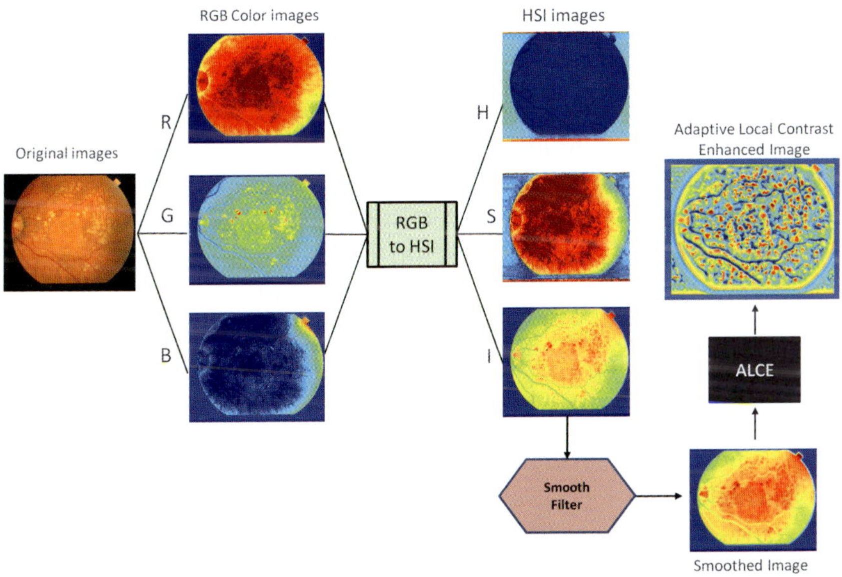

Fig. 2.6. Preprocessing steps of converting RGB to HSI, intensity smoothing, and adaptive local contrast enhancement outputs.

2.2.2.2. *Image smoothing through spatial filtering*

Spatial filters are used in the preprocessing stages to reduce noise and help connect edge and object regions by combining neighborhood pixels to transform central pixel intensity or color value. The Gaussian kernel $g(x, y)$ is formalized as[10]:

$$g(x, y) = \frac{1}{2\pi\sigma^2} e^{-\left(\frac{x^2+y^2}{2\sigma^2}\right)}, \tag{2.5}$$

where σ determines the width of the kernel. A smoothing mask or kernel is a matrix of coefficients (usually rectangular or circular) that is convolved with an image. The coefficients increase toward the center of the kernel to give larger weight to the pixel locations closest to the central pixel of the passing image window. An example of a smoothing filter is provided below along with the convolution summation.

$$G = w_1 * I_{(1,1)} + w_2 * I_{(1,2)} + \cdots + w_9 * I_{(3,3)} \tag{2.6}$$

$$= \sum_{i=1}^{3*3} w_i * I_i. \tag{2.7}$$

Performing smoothing operations on an image will reduce sharp transitions, which can lead to loss of edge information. Experimentation with best size and distribution of smoothing coefficients is required to find the best results for a particular image. A Gaussian smoothing function has been implemented to reduce noise in Ref. [8] to ensure that subsequent contrast enhancement does not amplify noisy data (Fig. 2.6).

2.2.2.3. *Local adaptive contrast enhancement*

In the image processing community, "contrast" describes the amount of variation that occurs within segments of an image. Although noisy images can contain high contrast, images with high levels of contrast between anatomical regions allow for more discrimination in the segmentation stage. Locally adaptive methods can provide enhanced contrast using statistics in a local neighborhood, which use information from pixels in similar regions. We describe a locally adaptive contrast enhancement method that has been successfully implemented on retinal images.[11] For an image f, consider a sub-image W of size $M \times M$ centered at about (i, j). Denote the mean and standard deviation of the intensity image within W by $\langle f \rangle_W$ and $\sigma_W(f)$. We will find a point transformation dependent on W, so that a local intensity distribution will spread across the intensity range. W is found to be large enough to represent the local distribution, but small enough so that gradual spatial changes from illumination effects will not cause a perturbation:

$$f(i, j) \rightarrow g(i, j) = 255 \frac{[\Psi_W(f) - \Psi_W(f_{\min})]}{[\Psi_W(f_{\max}) - \Psi_W(f_{\min})]}, \tag{2.8}$$

where the sigmoidal function is:

$$\Psi_W(f) = \left[1 + \exp\left(\frac{\langle f \rangle_W - f}{\sigma_W}\right)\right]^{-1}, \tag{2.9}$$

while $f_{\min}$ and $f_{\max}$ are the minimum and maximum values of intensity within the image I.

$$\langle f \rangle_{W(i,j)} = \frac{1}{M^2} \sum_{(k,l) \in W(i,j)} f(k, l) \tag{2.10}$$

$$\sigma_W^2(f) = \frac{1}{M^2} \sum_{(k,l) \in W(i,j)} (f(k, l) - \langle f \rangle_W)^2 \tag{2.11}$$

Prior to running the adaptive local contrast enhancement algorithm, retinal images are first converted from RGB to HIS using methods described in Sec. 2.2.1.1, then a Gaussian smoothing kernel (32×32, $\sigma = 1$) is passed over the intensity (I) channel to reduce noise. The results of $M = 49$ window can be seen in Fig. 2.6.

2.2.2.4. *Histogram transformations*

Histograms can be used to describe image color or intensity density distribution by discretely counting pixels in the bins of a specified width (range). The user-defined width of bins is the range in which a pixel must fall for inclusion, with more bins reducing the width of each bin. The histogram is a useful image statistic, providing insight into necessary further contrast enhancement by fitting a histogram into a desired shape, ultimately transforming image pixels from the original image. We will discuss several types of histogram transformations that have been used as retinal image preprocessing steps in the literature, histogram equalization, histogram specification, and multi-level histogram equalization.

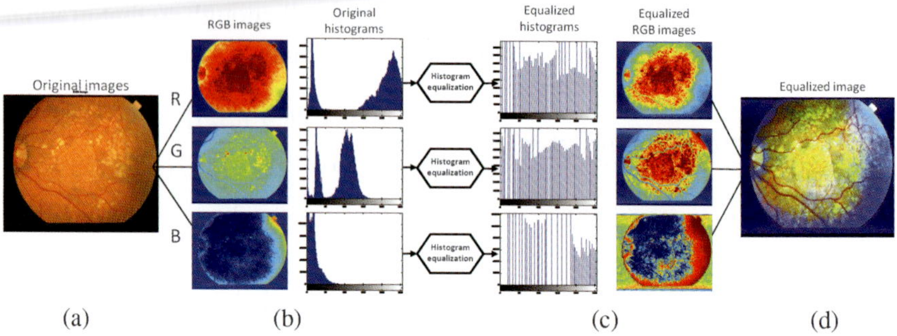

Fig. 2.7. (a) Original image, (b) RGB images and histograms, (c) equalized RGB images and histograms, and (d) equalized image.

An image made up of pixels that occupy the entire range of intensity values with an approximately uniform distribution can be considered to exhibit high contrast.[8] High contrast with minimal noise, as stated earlier, is a goal of image processing methods. Most digital retinal images do not initially display a highly distributed histogram (Fig. 2.7). Histogram equalization can be applied to transform an image with compact pixel intensity distributions. For discrete pixel values, we will represent the probability of the occurrence of intensity level r_k in an image as approximated by:

$$p_r(r_k) = \frac{n_k}{n} \quad k = 0, 1, 2, \ldots, L-1. \tag{2.12}$$

Here, n is the total pixel count of the entire image, n_k is the number of pixels that have intensity r_k, and L is the range of intensity values (0–255 for eight-bit images). The histogram-equalization transform function is then:

$$s_k = T(r_k) = \sum_{j=0}^{k} p_r(r_j), \tag{2.13}$$

$$= \sum_{j=0}^{k} \frac{n_j}{n} \quad k = 0, 1, 2, \ldots, L-1. \tag{2.14}$$

Each image pixel r_k is mapped to s_k through transform T, resulting in an image with a histogram that is much more evenly spread across the range $[0 \cdots L-1]$ of pixel intensities. Histogram equalization has been used as a preprocessing step for retinal images, giving retinal anatomical constituents stronger contrast (Fig. 2.7). Although enhanced contrast is typically valuable for image segmentation, analysis issues in using histogram-equalized images arise with inter-image variations, due to the uniqueness of intra-image transformations.

A second method of enhancing contrast through histogram manipulation is histogram specification. This method uses a reference model histogram (usually from a retinal image considered optimal in contrast for algorithm success) to transform an image's histogram to closely match that of the model. This method provides a baseline histogram shape that decreases inter-image variations that are prevalent in using histogram equalization.

The method begins similarly to histogram equalization where we first determine s_k from the PDF of $p_r(r)$. Next, we specify a probability density function $p_z(z)$ that we wish the PDF of $p_r(r)$ to closely approximate. This discrete formulation has the form[8]:

$$v_k = G(z_k) = \sum_{i=0}^{k} p_z(z_i) = s_k \quad k = 0, 1, 2, \ldots, L-1. \tag{2.15}$$

In Fig. 2.8, we display the output of histogram specification between two retinal images. Histogram specification has been shown to achieve superior results in increasing separation between lesion type clusters when used for color normalization.[12]

A third method of histogram transformation, multi-level histogram equalization (MLE), first equalizes the entire image (global), then performs local window equalization. This method will enhance local areas larger than features that are being detected (drusen, cotton wool spots, etc ...), due to sometimes close similarity to the background. This method has been applied to lesion detection using sequential, non-overlapping windows, which decrease in size until a lesion is detected.[13] The calculation is performed using the same equations as typical histogram equalization, with the addition of the iterative windows of equalized histograms (Fig. 2.9). Because the window size is dependent on detection results, varying levels of equalization occur for inter-image analysis.

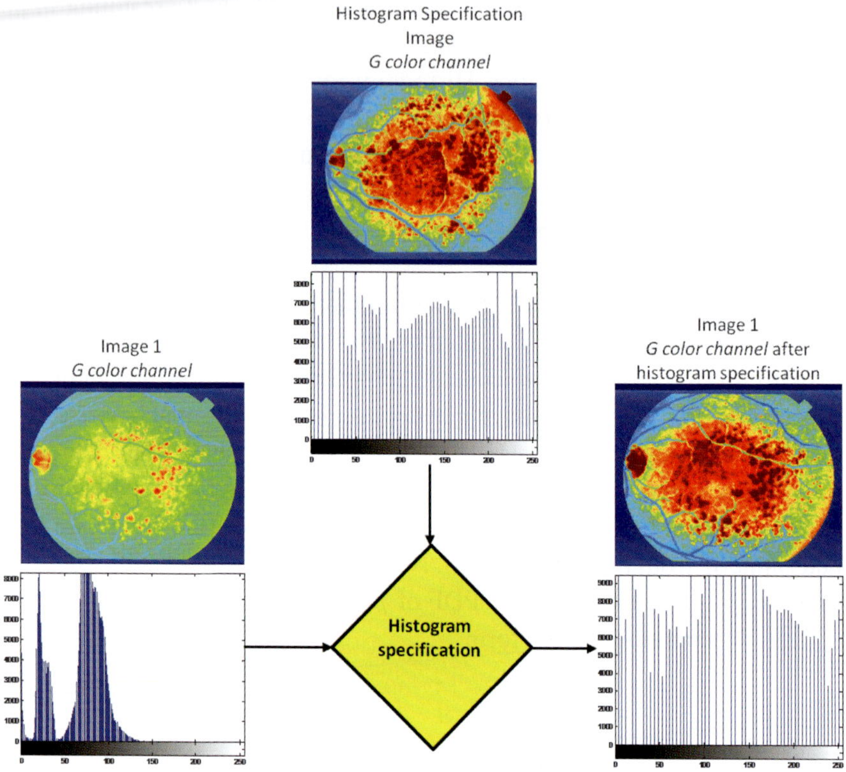

Fig. 2.8. Histogram specification of green intensity channel performed on Image 1 (left) with model histogram (top) to produce an image with similar intensity distribution (right).

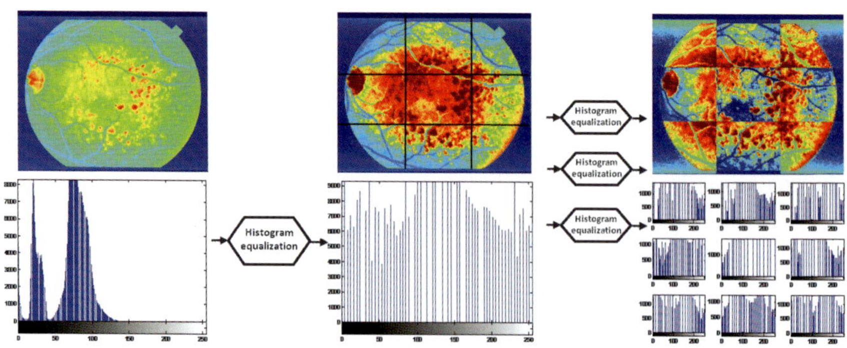

Fig. 2.9. Example of MLE performing two levels of enhancement, using a 3 × 3 non-overlapping window.

2.3. Segmentation Methods for Retinal Anatomy Detection and Localization

Regions of interest (ROIs) can differ in retinal images depending on the researcher's purpose for classification. The detection of particular anatomical features is required in both the preprocessing and classification steps. We will use the term "localization" when referring to an anatomical feature, such as blood vessels or an optic disk, that is assumed to be present in every image. We will use the term "detection" to find retinal features that are not assumed to be present always, like pathological indicators such as lesions, cotton wool spots, and drusen. We will discuss the methods and examples of retinal image segmentation, and note the strengths and weaknesses of these methods when gauging which method to implement. The discussion will begin with initial operations for finding local ROI boundaries. Next, we will discuss more advanced methods for image segmentation and detection, and provide example algorithms for finding the specific anatomical features of the retina.

2.3.1. *A Boundary Detection Methods*

Visually distinguishable region boundaries within an image can be mathematically described as a spatial discontinuity in the image pixel values. The size, length, and variance of this discontinuity depend on the reflectance properties of the region, background, image resolution, illumination effects, and noise. We will use the term gradient to describe these discontinuities. It is important to note that gradients, with both magnitude and direction, have vector qualities. A gradient can be divided into orthogonal components to allow for the combining of discontinuities in multiple dimensions. For example, a 2D image can have a gradient, Δf, with magnitudes in the x and y directions, with their total magnitude being the norm

$$\Delta f = \mathrm{mag}(\Delta f) = \left[G_x^2 + G_y^2\right]^{\frac{1}{2}}. \tag{2.16}$$

The direction of the gradient vector can be calculated from the orthogonal gradient magnitudes as:

$$\alpha(x, y) = \tan^{-1}\left(\frac{G_y}{G_x}\right), \tag{2.17}$$

where this angle is measured from the x-axis. We will discuss several classes of algorithms that find and exploit these discontinuities for retinal image anatomy segmentation.

2.3.1.1. *First-order difference operators*

We will describe the gradient of an image f at a point (x, y) as the magnitude of the first-order derivatives in both the x and y directions. Matrix kernels are convolved with an image f at each (x, y), which represent derivatives, $\nabla f/dx$ and $\nabla f/dy$. These values are then used to find the gradient. The Prewitt and Sobel operators use two 3×3 masks with the coefficients of opposite values to calculate gradients in the x and y directions. The gradient magnitude is approximated using the following,

$$\nabla f = |G_x| + |G_y|. \tag{2.18}$$

A threshold, t, is then used to determine if an edge exists at location (x, y) if $\nabla f > t$, which results in binary image, BW, at $(x, y) = 1$ (Fig. 2.10). Sobel differs from Prewitt by using a larger coefficient in the center of the row or column to help smooth the calculation by giving more weight to the four-connected pixel neighbors. These methods can provide useful information at a small computational expense, but, in practice, are typically only a part of a chain of methods used to obtain reliable results.

2.3.1.2. *Second-order boundary detection*

Finding second-order approximations of the image in spatial directions can be helpful when combined with smoothing functions and first-order derivatives for detecting corners and localizing boundaries. The Laplacian is an isotropic derivative operator that linearly combines the second-order derivative in the x and y directions as shown

$$\nabla^2 f = \frac{\nabla^2 f}{\nabla x^2} + \frac{\nabla^2 f}{\nabla y^2}. \tag{2.19}$$

The Laplacian value at (x, y) will have large values for locations with discontinuities in both x and y directions, as well as de-emphasize areas with close-to-linear intensity changes. The discrete approximation of the Laplacian can use either four- or eight-connected neighborhoods (Fig. 2.11).

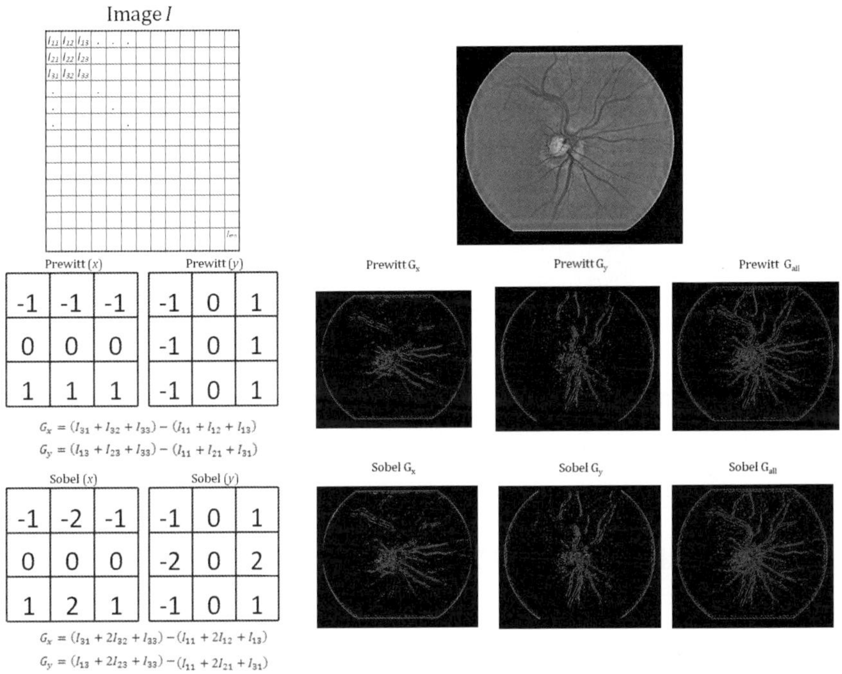

Fig. 2.10. First-order difference operators for edge detection. Top: original grayscale matrix and image; center: Prewitt kernels and edge response binary images; and bottom: Sobel kernels and edge response binary images.

Laplacian 4-connected

0	-1	0
-1	4	-1
0	-1	0

Laplacian 8-connected

-1	-1	-1
-1	8	-1
-1	-1	-1

Fig. 2.11. Matrices of four- and eight-connected Laplacian kernels.

This method does not provide much useful boundary information by itself, but when used with a Gaussian smoothing function or first-order derivatives, can provide useful information with boundary detection (Fig. 2.12).

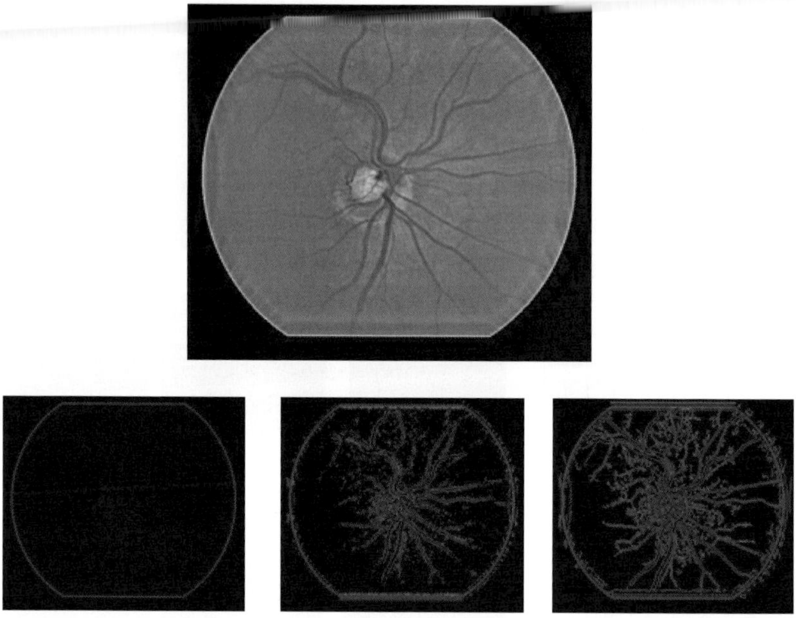

Fig. 2.12. Top: original grayscale image; bottom left: Laplacian kernel response; bottom center: LoG edge detection; and bottom right: Canny edge detector.

The Laplacian response provides little useful information due to its sensitivity to noise, but when combined with a smoothing function (described in Sec. 2.2.2.2), edges are found at zero-crossing locations. The initial step is to smooth the image with a Gaussian smoothing kernel to reduce noise, formalized as

$$h(r) = -e^{-\frac{r^2}{2\sigma^2}}, \tag{2.20}$$

where $r^2 = x^2 + y^2$ and σ is the standard deviation to control the smoothing scale. The convolution of this function blurs an image, thus reducing noise and boundary clarity. The Laplacian of h is then

$$\nabla^2 h(r) = -\left[\frac{r^2 - \sigma^2}{\sigma^4}\right] e^{-\frac{r^2}{2\sigma^2}}. \tag{2.21}$$

The kernel that approximates the LoG (referred to as the Mexican hat function) has a large positive central term (similar to first- and second-order difference operators), then is surrounded by negative values, with zero values filling in the outermost areas. The kernel must have a total value of zero to ensure that the areas of constant gray values will return a zero edge value. The edge response of a typical LoG kernel is given in Fig. 2.12.

2.3.1.3. *Canny edge detection*

As with the LoG method, Canny edge detection begins by smoothing the image with a Gaussian function to reduce false edge detection from noise artifacts. The method then uses edge detection kernels (Prewitt or Sobel) in four-directions (0°, 45°, 90°, and 135°, for example) and finds a gradient magnitude and direction using the method explained in Sec. 2.3.1.1. A nonmaximal suppression step is then employed to determine whether the gradient magnitude is a local maximum in the gradient direction. A high and low threshold is also used to reduce large gradients due to noise. A high threshold is used to ensure the found edges are real, then using directional information from the gradients, edges are traced throughout the image, applying the lower threshold. This method produces a binary image of edge locations, with typically thin edge width due the non-maximal suppression step (Fig. 2.12).

2.3.2. *Edge Linkage Methods for Boundary Detection*

The boundary detection methods described in Secs. 2.3.1.1 through 2.3.1.3 provide useful initial edge information, but are not robust enough to segment retinal features successfully. The edge pixels are typically corrupted by image noise, illumination effects, and occlusion discontinuities, requiring additional methods to both disregard false edge data and link true feature edges to improve segmentation results. Edge linking methods can also provide the initial steps for segmentation by labeling autonomously linked edges together, thus separating edges from one another so that their structures can be compared to possible features. We will discuss several methods that use edge detection input data to enhance edge linkages for the increasing accuracy of retinal feature segmentation.

2.3.2.1. *Local neighborhood gradient thresholding*

A simple method to link similar gradients into a single-edge unit is accomplished by comparing magnitude and orientation values. With the domain knowledge that retinal region boundary intensity is locally constant and the shape can be considered piecewise linear, we can use simple thresholding of the magnitude and orientation in a local neighborhood to connect like gradients to one another. We will use a gradient magnitude threshold equation

$$|\nabla f(x, y) - \nabla f(x_0, y_0)| \leq E, \tag{2.22}$$

where (x, y) and (x_0, y_0) are located within some defined local neighborhood, N, and E is a nonnegative threshold value. The gradient angle, α, is found using the equation presented in Sec. 2.3.1, is also compared to the neighborhood

$$|\alpha(x, y) - \alpha(x_0, y_0)| < A, \tag{2.23}$$

where A is a nonnegative angle threshold. Before implementing the gradient similarity thresholds above, we assume that an initial edge threshold similar to those used in Secs. 2.3.1.1 through 2.3.1.3 has been used. This procedure is recursively repeated throughout the image at every pixel location, with an indexing step used to keep the track of edge labels. Each new edge linkage can be indexed by an integer, which can then be used to find the overall size of each edge. A final minimum/maximum size threshold can be used to erase smaller edges or extremely large edges.

Figure 2.13 displays the multiple steps in the local neighborhood-edge linkage algorithm.

As seen in Fig. 2.13, some edges remain unlinked and some nonedge artifacts remain linked. Additional linking methods can add additional enhancement for more accurate segmentation.

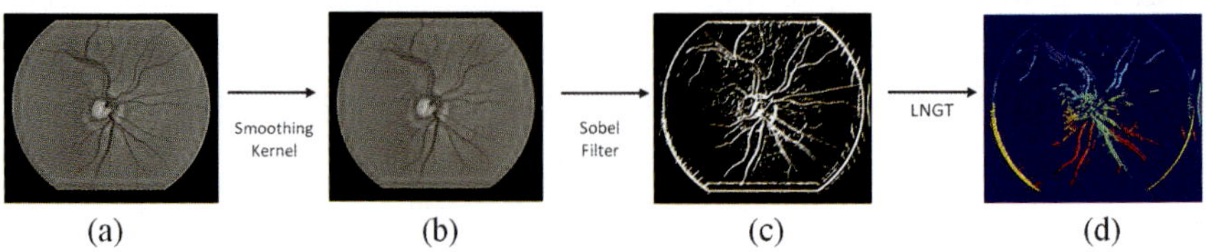

Fig. 2.13. (a) Original grayscale image, (b) image smoothed by Gaussian kernel, (c) edge response using Sobel edge detector, and (d) linked and labeled edges using local neighborhood gradient thresholding.

2.3.2.2. *Morphological operations for edge link enhancement*

Morphological steps that can increase edge linkages artificially dilate and erode edges using structuring elements defined by the user are commonly used for retinal image feature segmentation.[14] A binary image created from an edge detection method will contain broken edges due to soft edges, noise, illumination, or occlusion. The broken edge location will not have similar gradient values to its edge members, thus, methods such as neighborhood gradient thresholding will fail to link the edge. We can use a dilation operation, followed with an erosion step to first link, then thin edges to refine edge structures.

A dilation operation can be defined in set theory as

$$A \oplus B = \{z | [(\hat{B})_z \cap A] \subseteq A\}, \tag{2.24}$$

where A is the set of edge locations and B is a structuring element. The dilation of A by B is the set of all displacements, z, such that $\hat{B}$ and A overlap by at least one element. Although this operation is similar to convolution, dilations are based on set operations, as opposed to the arithmetic operations of a convolution mask. We chose a structuring element intuitively, using domain knowledge of the edge structure we are attempting to link, to close gaps while not adding too many false edges. We show the results of edge dilation in Fig. 2.14 using a simple disk-structuring element.

The counterpart of dilation, erosion, is used to reduce edge elements that are created from noise and illumination effects or from a dilation operation. The definition of erosion that we will use is

$$A \ominus B = \{z | (B)_z \subseteq A\}, \tag{2.25}$$

which indicates that the erosion of A by B is the set of all points z such that B, translated by z, is contained in A. This method is typically used in tandem with dilation operations to either initially reduce edge artifacts or to thin edges. Performing erosion followed by a dilation is referred to as an opening operation, which tends to smooth contours, break narrow edges, and erase thin artifacts. A closing operation is a dilation followed by erosion, which tends to smooth edge contours and fuse narrow edges together. It is also common to perform several erosion and dilation steps using multiple structuring elements to enhance true edges while reducing

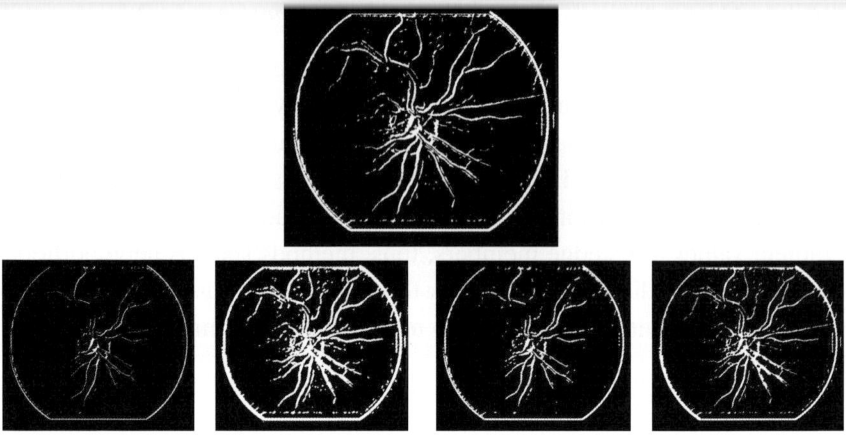

Fig. 2.14. Top: input binary edge image; left: erosion operation; left-center: dilation operation; right-center: opening operation; and right: closing operation.

false edge artifacts. An example of each step of the operation can be seen in Fig. 2.14.

More complex set-theoretic methods can yield better edge linkage and refinement results.[8] Because edges can be described as a group of connected components, morphological operations provide useful enhancements to edge linkage challenges, although the parameter adjustment of structuring elements inhibits these methods from providing robust solutions to retinal image feature segmentation. Opening and closing operations have been used for both vasculature and microaneursym segmentation.[15,16]

2.3.2.3. *Hough transform for edge linking*

Unlike previous methods, which use local neighborhoods to increase true edge linkage and reduce false edge artifacts, the Hough transform uses a global processing method to link points. As the name implies, the points are transformed from spatial coordinates into a shape-space. This predetermined shape should resemble a feature boundary, such as a circle or ellipse for the optic disk.[17] This method can provide boundaries (as shape in the transform) that are robust to noise and breaks in edges. In the case of the line equation transform, we use the normal representation of the line

$$x \cos\theta + y \sin\theta = \rho, \tag{2.26}$$

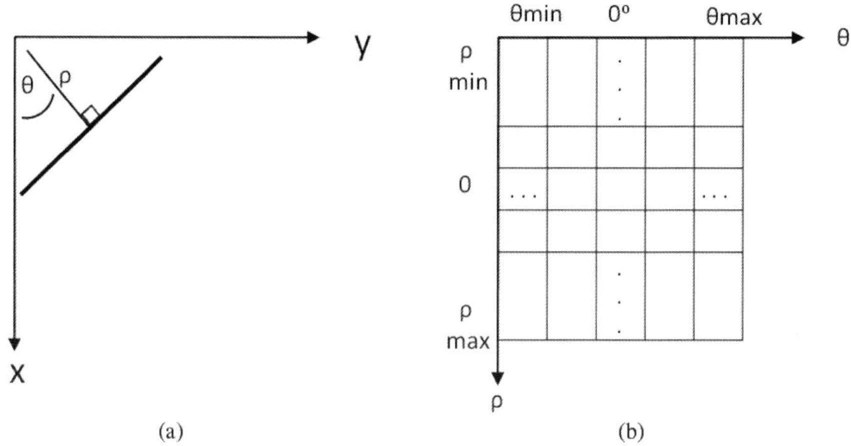

Fig. 2.15. (a) Line representation used in Hough transform and (b) accumulator cells for binning in Hough transform.[8]

to avoid issues of divergence. Figure 2.15 shows the parametric relationship from the transform in line space. A matrix of accumulator cells are then used to bin similar ρ and θ values discretely, with high bins corresponding to line edges in the image (Fig. 2.15).

Line detection is not overly useful in retinal image segmentation, due to the lack of straight retinal feature edges (although some blood vessels can be detected). Finding circular edges is useful for detecting the optic nerve head, and the Hough transform is a popular method because of its robust performance with broken edges (due to blood vessel occlusion).[18] The equation for the circle shape space is

$$(x - c_1)^2 + (y - c_2)^2 = c_3^2, \tag{2.27}$$

which produces a 3D accumulator cell matrix for the three variables. Figure 2.16 shows preliminary results and accumulator cells for both the line and circle Hough transforms.

2.3.3. *Thresholding for Image Segmentation*

We have discussed the use of thresholding in the previous section, providing discrimination on edge point inclusion. In cases where a specific anatomical feature has a distinct intensity range from other features and

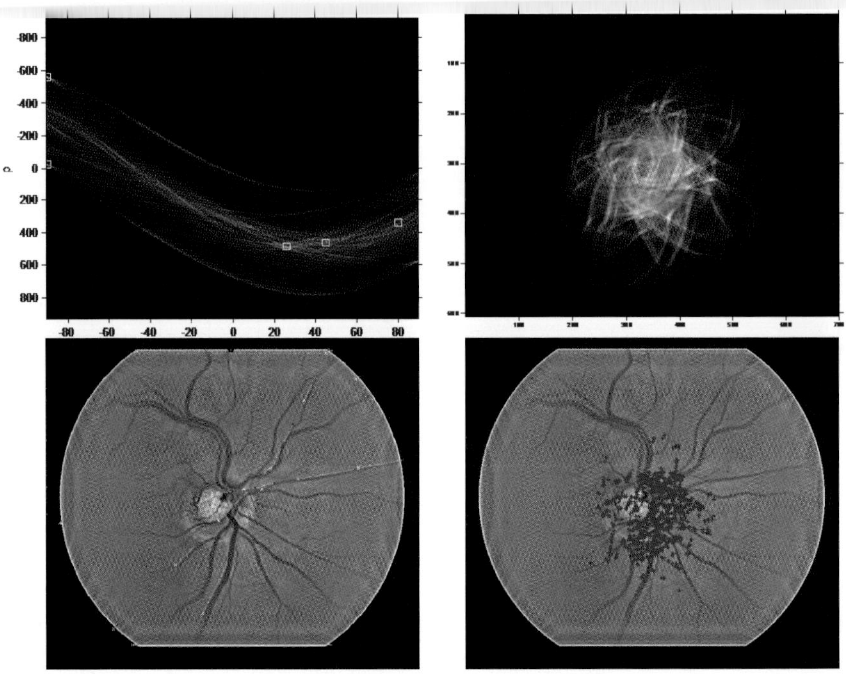

(a) (b)

Fig. 2.16. (a) Accumulator cells and edge detection for line Hough transform and (b) accumulator cells and locations for circle Hough transform.

background, thresholding can provide strong segmentation results with low computation overhead. The optic disk pixels typically have high intensity, low contrast, and discontinuous edges due to blood vessel occlusion, lending itself to thresholding as a viable segmentation approach. We will discuss using intensity thresholds based on global image histograms and domain knowledge to segment a retinal image.

2.3.3.1. *Segmentation with a single threshold*

The simplest method for threshold segmentation is to choose, either manually or at an automatically chosen point, a threshold intensity value, t, that will include all values above (when ROI is bright) or below (when ROI is dark) in a segmentation. For an image, f, the method will determine for each (x, y)

$$g(x, y) = \begin{cases} 1 & \text{if } f(x, y) > t \\ 0 & \text{if } f(x, y) \leq t' \end{cases}, \tag{2.28}$$

where $g(x, y)$ is a binary image labeled with ones at pixels included in the segment. Setting a high threshold for optic disk segmentation and a low threshold for blood vessels, simple thresholding can provide good segmentation results, provided the retinal image has minimal illumination effects (Fig. 2.17).

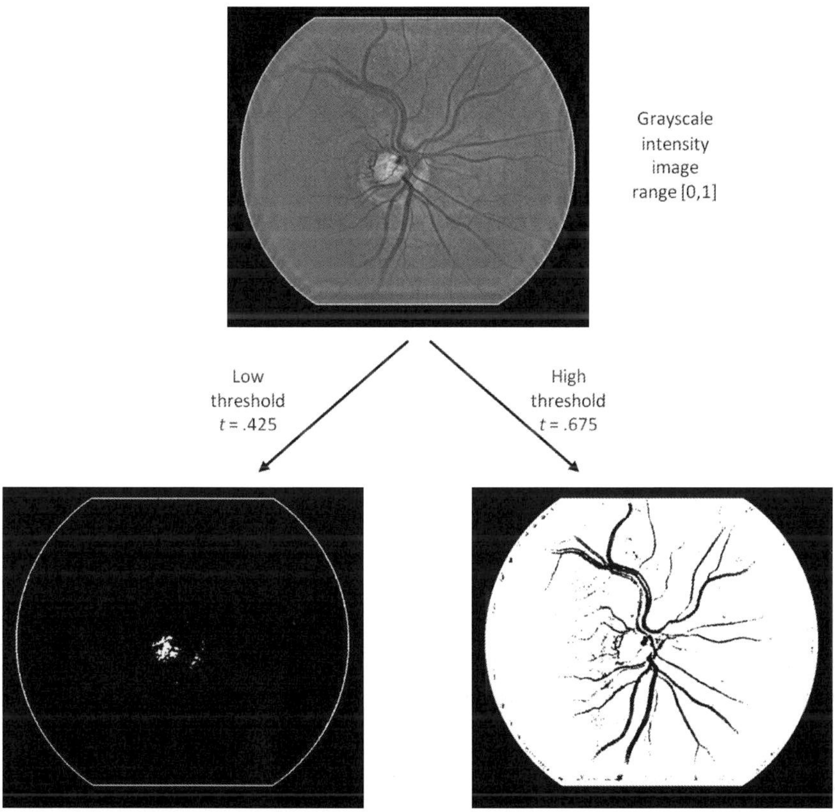

Fig. 2.17. Examples of low and high simple thresholding for the segmentation of grayscale intensity image.

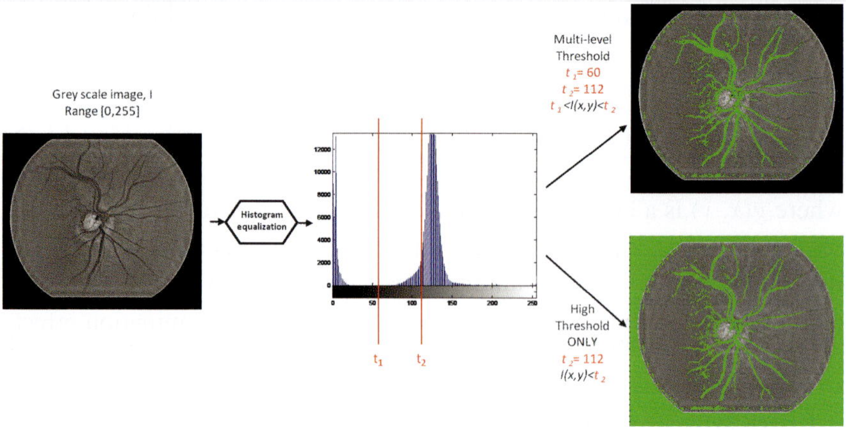

Fig. 2.18. (a) Original grayscale image, (b) image histogram with two thresholds denoted, (c-top) segmentation using high and low thresholds, and (c-bottom) segmentation using only high threshold.

2.3.3.2. *Multi-level thresholding*

For more discrimination than a simple threshold, multi-level threshold allows intensity values to be constrained between a high and a low boundary within the image range. These boundaries can either be predefined or automatically chosen using saddle points in the histogram, which are local minimums formed in curves of the bins (Fig. 2.18). A threshold segmentation equation takes the form

$$g(x, y) = \begin{cases} 1 & \text{if } t_1 < f(x, y) < t_2 \\ 0 & \text{otherwise} \end{cases}, \tag{2.29}$$

where t_1 and t_2 are the low and high thresholds. When a large percentage of a particular anatomical feature can be contained between two value limits, segmentation specificity increases (Fig. 2.18).

2.3.3.3. *Windowed thresholding*

Due to the prevalence of illumination effects and difficulty of obtaining an accurate normalization solution, intra-feature intensity variance will inhibit a robust segmentation through simple and multi-thresholding methods.

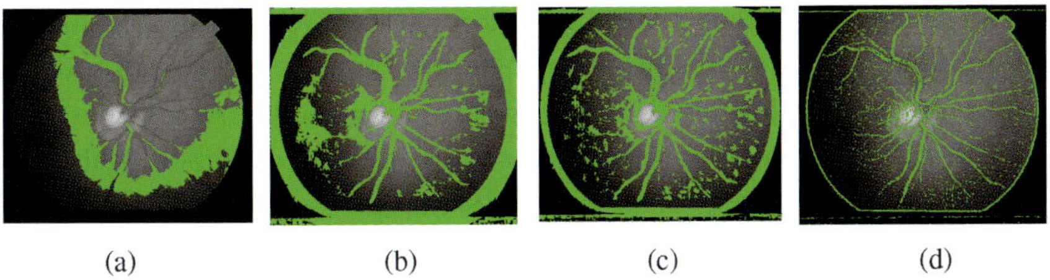

Fig. 2.19. (a) Multi-threshold segmentation, (b) 100×100 pixel windowed segmentation, (c) 50×50 pixel windowed segmentation, (d) 10×10 pixel windowed segmentation.

An effort to segment regions with varying intensities is possible through adaptively setting thresholds in windowed regions of the image. The window size should be small enough so that a given feature should have minimal nonuniform illumination effects. Figure 2.19 demonstrates how window size can alter a simple threshold segmentation of blood vessels for an image with nonuniform illumination effects.

2.3.4. *Region-Based Methods for Image Segmentation*

Although edge detection and linking methods can provide useful information for boundary detection and segmentation, results are not always reliable when boundaries are obscured or noisy. In most cases, region-based segmentation can provide accurate results without depending on linked boundaries to encapsulate an anatomical retinal feature. We will discuss methods that use discriminatory intra-retinal feature statistics to segment regions from within an image.

2.3.4.1. *Region growing*

This method takes provided seed point locations and groups surrounding pixels (four or eight-connected neighborhood, for example) together based on predefined statistical similarity. The basic formulation is

$$\bigcup_{i=1}^{n} R_i = R, \tag{2.30}$$

where R_i is a connected region and $i = 1, 2, \ldots, n$, $R_i \cap R_j = \emptyset$, for all i and j, $i \neq j$, $P(R_i) = \text{TRUE}$ for $i = 1, 2, \ldots, n$, and $P(R_i \cup R_j) = \text{FALSE}$ for adjacent R_i and R_j.

Similarity measurement, $P(R_i)$, and seed point location decisions are based on domain knowledge of the anatomical feature of interest, usually based on intensity, local mean, standard deviations, or higher level textural statistics. Another necessary parameter is the stopping condition. Although a stopping condition can occur when the similarity measure ceases to find similar pixels, region shape statistics can also be used to improve results when prior feature models are known. A recursive region growing method has been used for the segmentation of yellow lesions, due to their homogenous gray-scale intensity.[19] In Fig. 2.20, we show the results of recursive region-growing segmentation on a processed gray-level retinal image when different seed points are chosen. This simple, recursive method can attain powerful segmentation results when anatomical features have continuous regions (as in the image), but fail when regions have large statistical discontinuities from either occlusion or illumination.

2.3.4.2. *Watershed segmentation*

In this segmentation approach, an intensity image is represented in a 3D space where the intensity at each pixel location (x, y) denotes height (Fig. 2.21). Although this approach combines operations from both edge detection and morphological operations, the watershed uses regions peaks and valleys (regional maximums and minimums) that would act as

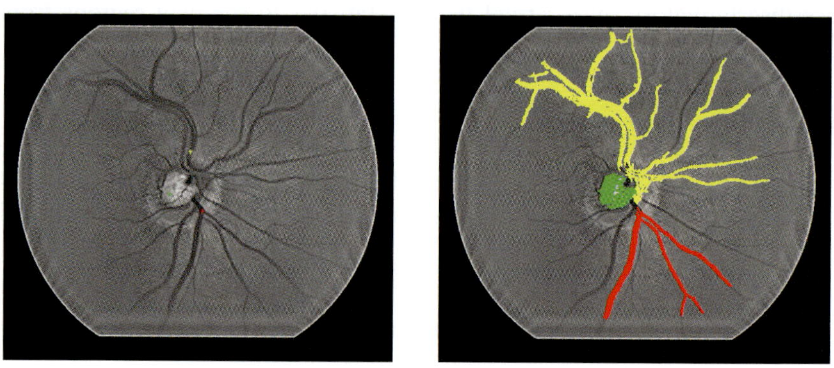

(a) (b)

Fig. 2.20. (a) Seed points used in region growing and (b) region growing using seed points and intensity similarity predicate.

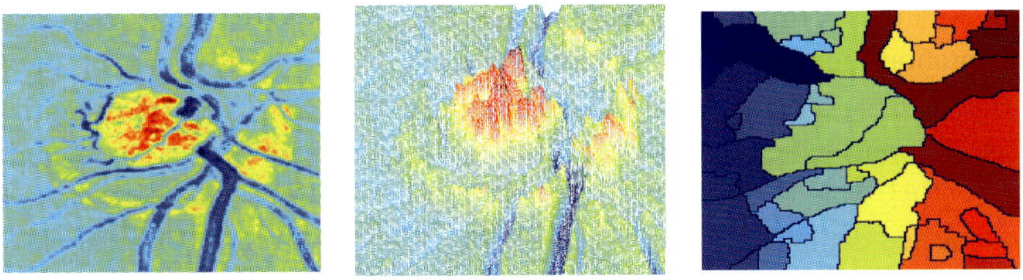

(a) (b) (c)

Fig. 2.21. (a) Localized optic disk image, (b) 3D visualization with intensity magnitude in z direction, and (c) sample watershed segmentation.

catchment basins for liquid. Segments are created from this topographical representation, which are connected components lying within a regional minimum and surrounded by a connected regional maximum.

The watershed segmentation algorithm can be conceptualized as follows.[20] First, let $M_1, M_2, \ldots, M_r$ be the sets of coordinates of the locations in the regional minima of an image $f(x, y)$, which will be a gradient image calculated from using any of the previous methods. Let $C(M_i)$ be a set of locations of the points in the catchment basin of regional minimum M_i. Let $T[n]$ represent the set of locations (s, t) for which $f(s, t) < n$, written as:

$$T[n] = \{(s, t)|g(s, t) < n\}. \tag{2.31}$$

$T[n]$ is the set of locations in $f(x, y)$ below the plane $f(s, t) < n$. The regional area will then be "flooded" in integer increments, from $n = \min +1$ to $n = \max +1$. Now, $C_n(M_i)$ represents the points located in the basin that are below n, which is written as:

$$C_n(M_i) = C(M_i) \cap T[n]. \tag{2.32}$$

This equation gives binary values of one if a location (x, y) belongs to both $C_n(M_i)$ and $T[n]$. Next, we find

$$C[n] = \bigcup_{i=1}^{R} C_n(M_i). \tag{2.33}$$

Then, $C[\max +1]$ is the union of all catchment basins:

$$C[\max +1] = \bigcup_{i=1}^{R} C(M_i). \tag{2.34}$$

As the algorithm begins at $C[\min +1] = T[\min +1]$, watershed segmentation members are detected recursively, with step n occurring only after $C[n - 1]$ has been found. To find $C[n - 1]$ and $C[n]$, let Q be the set of connected components in $T[n]$, so that, each connected component $q \in Q[n]$, which is that $q \cap C[n - 1]$ is empty, contains a single instance of connected components, or contains multiple connected components. $C[n]$ depends on which of these conditions is satisfied:

- Empty set occurring when a new minimum is found, adding q into $C[n - 1]$ to form $C[n]$,
- One connected component, which means q lies within a basin, adding q into $C[n - 1]$ to form $C[n]$, and
- All or part of a peak ridge separating two or more catchment basins is found; thus, a "damn" is built (one-pixel thick connecting ridges by dilating $q \cap C[n - 1]$ with a 3×3 structuring element of ones, and constraining the dilation to q.

Watershed transform has been used in retinal image analysis to segment the optic disk, using the red channel of the RGB color space.[21] A watershed algorithm result is provided. This result shows how the method uses regional minimums and maximums for segmentation (Fig. 2.21).

2.3.4.3. *Matched filter segmentation*

In the cases where prior models of retinal anatomy are available, we can create kernels (templates) that are then convolved with the image, finding maximal responses at locations of high template matching. Blood vessels in retinal images have intensity characteristics that allow for successful template modeling, considering their piece-wise linear segments and Gaussian intensity profile.[22] Using these assumptions, we can formulate the Gaussian curve as:

$$h(x, y) = A\{1 - k * e^{-d^2/2\sigma^2}\}, \tag{2.35}$$

where d is the perpendicular distance between point (x, y) and a straight line passing through the center of the blood vessel along its length, σ gives the intensity profile, and A is the local background intensity. One required step in matched filters that adds to the complexity of this method is that the kernel must be rotated to find objects in various orientations. A rotation matrix is used, given by:

$$\bar{r}_i = \begin{bmatrix} \cos\theta_i & -\sin\theta_i \\ \sin\theta_i & \cos\theta_i \end{bmatrix}, \tag{2.36}$$

where θ_i is the orientation with a corresponding point in the rotated coordinate system given by:

$$\bar{p}_t = [u\ v] = \bar{p}\overline{r_i^T}. \tag{2.37}$$

If we divide the orientations into 15° increments, 12 kernels are necessary to convolve with the image to find all orientations, with the maximum value chosen for each location (x, y). The Gaussian curve tail is truncated at $u = \pm 3\sigma$, and a neighborhood N is defined as being within u. The weights in the ith kernel are given as:

$$K_i(x, y) = -e^{-u^2/2\sigma^2} \forall \bar{p}_i \in N. \tag{2.38}$$

A will denote the number of points in N, with a mean value, m_i, of the kernel given as:

$$m_i = \sum_{\bar{p}_i \in N} \frac{K_i(x, y)}{A}. \tag{2.39}$$

The convolution mask is then given by:

$$K_i'(x, y) = K_i(x, y) - m_i \overline{\forall p_i} \in N. \tag{2.40}$$

In Fig. 2.22, we present the above example of matched filter kernels and segmentation result using a simple magnitude threshold of the maximum

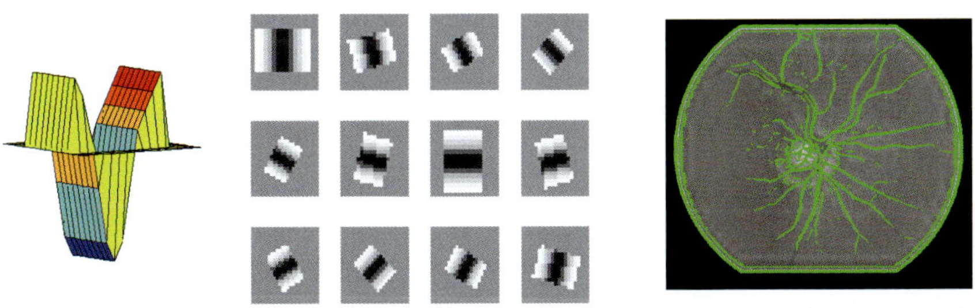

(a) (b) (c)

Fig. 2.22. (a) 3D representation of a matched filter used for blood vessel detection, (b) filter bank for 15° orientation increments, and (c) vessel segmentation results using low bound threshold.

filter response at each location (x, y). Although matched filters require high computation, the method can be parallelized for faster computation time.

We have detailed several methods for retinal image anatomy segmentation that have proven successful when coupled with proper pre- and post-processing steps. Choosing a proper segmentation method should depend on specific domain knowledge, the level of accuracy required, and computational resources available, with no one specific method standing out as producing the best results for every situation. The retinal anatomy creates unique challenges for successful segmentation, due to "distracters" that include blood vessel occlusion, spatially varying albedo, the presence of pathologies such as lesions and cotton wool spots, and residual non-uniform illumination effects.[1]

2.4. Feature Representation Methods for Classification

Along with segmentation, discriminative feature extraction is a prerequisite for accurate pathology classification. In this section, the term feature will describe a quantitative representation of the detected, localized, or segmented regions of the image, which are often specific parts of the retinal anatomy. We chose features that can best classify a retinal image for the specific problem, whether it is binary such as "diseased" or "nondiseased," or varying degrees of a specific pathology. Most of these methods provide input into pattern recognition algorithms that use machine-learning methods for classification. Segmentation and image registration problems can also use feature extraction to localize or detect parts of the retinal anatomy. We will discuss several feature extraction methods that convert low-level information (e.g. pixel intensities) into robust numerical descriptors that can classify retinal pathologies and/or segment anatomy with high levels of specificity.

2.4.1. *Statistical Features*

After an image has been successfully segmented into its constituent parts, these regions of interest's size, shape, color, and texture can provide simple and powerful values for discriminative classification. These descriptors are scale and rotationally invariant, which increases the robustness and accuracy of classifying non-registered or orientation-corrected images. Microaneursym detection and classification using these types of features has been used to classify the degrees of diabetic retinopathy.[23] We will provide the details on how to compute the three most common methods of statistical descriptors.

2.4.1.1. *Geometric descriptors*

First, we will describe geometric descriptors. Assuming that a region of connected components has been established through segmentation, characteristics of the geometry can provide discriminative classification specificity. These values only take into account the overall shape from linked boundary pixels, providing illumination invariant features sets. These scalar descriptors, listed below, also make entry into a classification algorithm straightforward.

Area is a simple sum of pixels contained within a bounded region. Although not overly useful alone, it is typically needed for finding more interesting shape statistics.

Centroid is the center of mass or the geometric center of a region, which provides insight on region localization at a single point (x, y). This location can be used as an axis for orientation adjustment for registration or feature extraction that is not rotationally invariant.

Eccentricity measures how circular a region's shape is. Enclosing a region in an ellipse, the eccentricity is measured as the scalar value of the ratio between the distance from the center to the focus of the ellipse, c, and its major axis length, a:

$$\text{Eccentricity} = \frac{c}{a}.$$

This value will be zero if the region is a line and one if the region is a circle. This shape descriptor is useful when classifying hemorrhages.[22]

Euler number is a scalar value, E, that tabulates the number of objects, O, in a region subtracted from the number of holes, $H : E = O - H$. Knowing the number of holes in a region provides insight that can be used to decide how to extract descriptors.

Extent is a scalar ratio of pixels between the bounding box and region area. This value gives an idea of the overall density of the region.

Axis lengths are the major and minor axis lengths, in pixels, of the elliptical representation of the region, used in computing orientation.

Orientation is the angle between the x-axis and the major axis.

Perimeter gives the length of the boundary of the region.

Compactness is the scalar measure of the ratio of the square of the perimeter, p, and the area, a, given as: compactness $= p^2/a$, which provides a minimal value for circular regions.

We use the shape of boundary segments to create a boundary histogram, $p(v_i)$, by representing the curve of a boundary as a 1D function, $s(x)$, connecting its end-points and rotating to a horizontal access (Fig. 2.23). The amplitude at each boundary point is the distance from the x-axis to the boundary. We select an arbitrary number, A, of amplitude bins for the histogram and use $p(v_i)$ as the estimate of the probability of amplitude bin value v_i occurring in the boundary segment. The nth moment of v about its mean is[8]:

$$\mu_n(v) = \sum_{i=0}^{A-1} (v_i - m)^n p(v_i), \tag{2.41}$$

where

$$m = \sum_{i=0}^{A-1} v_i * p(v_i), \tag{2.42}$$

showing that m is the mean value of v and the second moment gives us the variance. These rotationally invariant shape descriptors retain physical boundary information, requiring only the first few moments to represent unique shape characteristics. Figure 2.23 shows the creation of 1D boundary histogram.

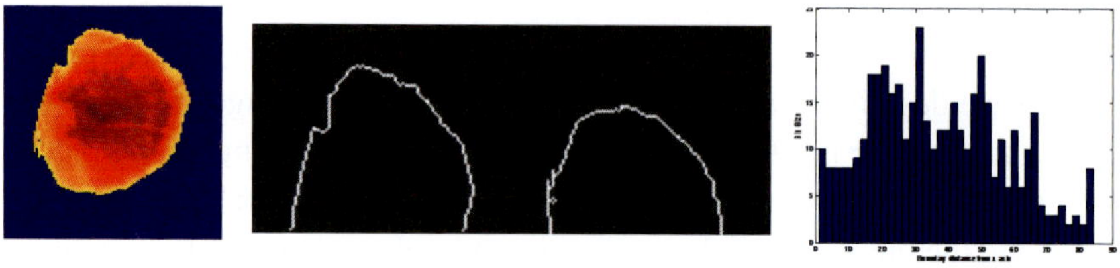

(a) (b) (c)

Fig. 2.23. (a) Segmented region, (b) top and bottom boundaries aligned on the x-axis, and (c) boundary magnitude histogram.

Although most of these geometrical descriptors are straightforward to calculate, they provide insight for refining choices on which regions are suitable for further analysis and as features for classification. These descriptors are not the only methods that can characterize regional information of boundaries and shapes, but do provide a good introduction into how scalar values can represent complex shapes.

2.4.1.2. *Texture features*

Another important statistical description of a segmented region of interest is the intra-region intensity distribution, which is loosely referred to as the texture. The texture of an image region can have multiple values, depending on how the internal pixels are represented. We will discuss two sets of texture feature descriptors that provide discrimination for retinal pathology classification: histogram-based texture features and spatial textures through co-occurrence matrices.

Using the gray-level histogram (explained in Sec. 2.2.2.3), we can calculate n-level moments, each of which provides insight into the characteristics of the intensity distribution, often a discriminate descriptor for classification. First, let z be a random variable with an ROI histogram $p(z_i)$, $i = 0, 1, 2, \ldots, L-1$. We use the moment equation to find the nth moment,[8]

$$\mu_n(z) \sum_{i=0}^{L-1} (z_i - m)^n p(z_i), \tag{2.43}$$

where m is the mean value of z:

$$m = \sum_{i=0}^{L-1} z_i * p(z_i). \tag{2.44}$$

The most straightforward moment is the second moment, better known as the standard deviation. This value measures the intensity contrast, characterizing the relative smoothness of the ROI. The third moment is a measure of skewness, and the fourth moment describes the relative flatness. Other histogram-based texture descriptors include "uniformity," which is formulated as

$$U = \sum_{i=0}^{L-1} p^2(z_i), \tag{2.45}$$

which has a maximum value when all pixels are the same (or "uniform"). Entropy, on the other hand, is a calculation of variability and is zero for a constant intensity, given as

$$e = -\sum_{i=0}^{L-1} p(z_i)\log_2 p(z_i). \tag{2.46}$$

Histogram-based texture features are useful due to their rotational invariability, but lack discrimination for instances where the spatial textures (pixel intensity positions) provide characteristic class information.

Representing a segmented region as pixels with location and intensity $I(x, y)$, we create a set of co-occurrence matrices that tabulate how often a pair of pixels with similar intensity values "occur" in a specific orientation from one another. The user can define an orientation of 0° (pixels $I(x, y)$ and $I(x+1, y)$), 45° (pixels $I(x, y)$ and $I(x+1, y-1)$), 90° (pixels $I(x, y)$ and $I(x, y-1)$), and −45° (pixels $I(x, y)$ and $I(x+1, y+1)$), that will produce a square matrix, G, the size of intensity bins, k, in each dimension. The value at each location $G_{i,j}$ will be the total number of times two pixels of a chosen orientation having similar intensity values "occur" within the region. After constructing co-occurrence matrices in the principal directions (typically 0°, 45°, 90°, and −45°), we normalize the matrices to give a joint probability occurrence of pixel pairs with the corresponding orientation and intensity range. We can find a set of spatially dependent texture descriptors termed Harilick features[24]:

$$\text{Contrast:} \quad \sum_{i,j} |i-j|^2 p(i,j),$$

$$\text{Correlation:} \quad \sum_{i,j} \frac{(i-\mu i)(j-\mu j)\vec{p}(i,j)}{\sigma_i \sigma_j},$$

$$\text{Energy:} \quad \sum_{i,j} p(i,j)^2 \quad \text{and}$$

$$\text{Homogeneity:} \quad \sum_{i,j} \frac{p(i,j)}{1+|i-j|}.$$

Contrast measures intensity differences between neighboring pixels over the image, where correlation returns a measure of how correlated a pixel is to its neighbor. Energy returns a sum of the squared elements in the co-occurrence matrix, returning a value of one for a constant intensity. Homogeneity returns a value of the closeness of the member distribution in the co-occurrence matrix.

A popular method for addressing rotational invariance is to take an average of over the feature values for all the principal orientations, so that co-occurrences that move to a different matrix due to rotational changes will still be measurable.

2.4.1.3. *Invariant moments*

Invariant moments combine the overall shape of the region and intensity distribution to create a set of values that are considered invariant to rotation, scaling, and translation changes. We will refer to these moment invariants as Hu moments,[25] which we attain by first using the discrete version of the moment:

$$M_{ij} = \sum_{x}\sum_{y} x^i y^j I(x, y) \tag{2.47}$$

to find centroids $\bar{x} = M_{10}/M_{00}$, $\bar{y} = M_{01}/M_{00}$. We then define the central moments as:

$$\mu_{pq} = \sum_{x}\sum_{y} (x - \bar{x})^p (y - \bar{y})^q I(x, y), \tag{2.48}$$

where p and q give the moment order. The central moments used in the Hu moments are written as:

$$\begin{aligned}
\mu_{00} &= M_{00}, \\
\mu_{01} &= 0, \\
\mu_{10} &= 0, \\
\mu_{11} &= M_{11} - \bar{x}M_{01} = M_{11} - \bar{y}M_{10}, \\
\mu_{20} &= M_{20} - \bar{x}M_{10}, \\
\mu_{02} &= M_{02} - \bar{y}M_{01}, \\
\mu_{21} &= M_{21} - 2\bar{x}M_{11} - \bar{y}M_{20} + 2\bar{x}^2 M_{01}, \\
\mu_{12} &= M_{12} - 2\bar{y}M_{11} - \bar{x}M_{02} + 2\bar{y}^2 M_{10},
\end{aligned}$$

$$\mu_{30} = M_{30} - 3\bar{x}M_{20} + 2\bar{x}^2 M_{10}, \quad \text{and}$$
$$\mu_{03} = M_{03} - 3\bar{y}M_{02} + 2\bar{y}^2 M_{01}.$$

From these moments, η_{ij} can be made invariant to translation and scale by dividing by the scaled 00th moment, written as:

$$\eta_{ij} = \frac{\mu_{ij}}{\mu_{00}^{\left(1+\frac{i+j}{2}\right)}}. \tag{2.49}$$

Finally, the seven Hu moments, φ_i

$$\varphi_1 = \eta_{20} + \eta_{02},$$
$$\varphi_2 = (\eta_{20} + \eta_{02})^2 + (2\eta_{11})^2,$$
$$\varphi_3 = (\eta_{30} + 3\eta_{12})^2 + (3\eta_{21} + \eta_{03})^2,$$
$$\varphi_4 = (\eta_{30} + \eta_{12})^2 + (\eta_{21} + \eta_{03})^2,$$
$$\varphi_5 = (\eta_{30} - 3\eta_{12}) + (\eta_{30} + \eta_{12})[(\eta_{30} + \eta_{12})^2 - 3(\eta_{21} + \eta_{03})^2]$$
$$+ (3\eta_{21} - 3\eta_{03})(\eta_{21} + \eta_{03})[3(\eta_{30} + \eta_{12})^2 - (\eta_{21} + \eta_{03})^2],$$
$$\varphi_6 = (\eta_{20} - \eta_{02}) + [(\eta_{30} + \eta_{12})^2 - (\eta_{21} + \eta_{03})^2]$$
$$+ 4\eta_{11}(\eta_{30} - \eta_{12})(\eta_{21} - \eta_{03}), \quad \text{and}$$
$$\varphi_7 = (3\eta_{21} - \eta_{03})(\eta_{30} + \eta_{12})[(\eta_{30} + \eta_{12})^2 - 3(\eta_{21} + \eta_{03})^2]$$
$$+ (\eta_{30} - 3\eta_{12})(\eta_{21} + \eta_{03})[3(\eta_{30} + \eta_{12})^2 - (\eta_{21} + \eta_{03})^2].$$

In Fig. 2.24, we show that Hu moments produce almost identical values for rotated, scaled, and translated versions of a segmented region. Invariant moments work well in situations in which both the orientation of a segment can be unreliable, and characteristics of the region's spatial intensity provide important information for classification.

2.4.2. *Data Transformations*

These next feature descriptor sets use methods to transform the data from the spatial domain, or Euclidean space, to coordinate systems that provide different types of characterization for an ROI that is not obvious or directly calculable from the image matrix. We will briefly describe and discuss

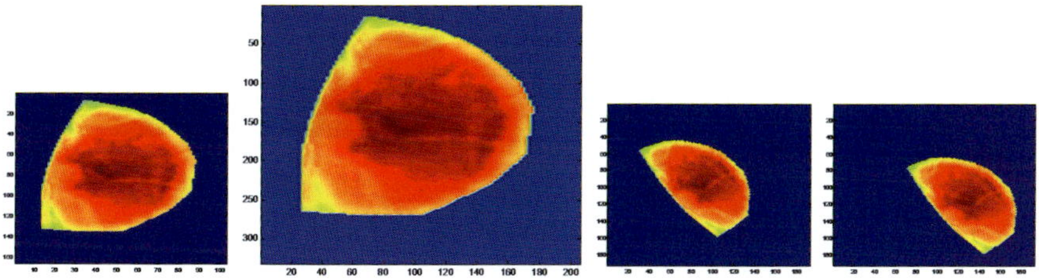

Image	Hu Moments (abs(log(ϕ)))						
	ϕ_1	ϕ_2	ϕ_3	ϕ_4	ϕ_5	ϕ_6	ϕ_7
Original	0.0009	0.0000	4.4625	4.5401	5.0126	4.6777	5.0479
Scale 2x	0.0009	0.0000	4.4625	4.5401	5.0126	4.6777	5.0479
rotated	0.0009	0.0000	4.4627	4.5402	5.0126	4.6776	5.0480
rotated & translated	0.0009	0.0000	4.4627	4.5402	5.0126	4.6776	5.0480

Fig. 2.24. (Top: left–right): an original segmented optic disk, scaled 2x, rotated 45°, rotated 45° and translated, (bottom) table of Hu moments for each image.

two methods that can both contribute useful discriminative information to a classification algorithm and reduce data dimensionality by retaining only data of interest. As you will see, computational methods for image transformations use classic linear algebraic operations to produce refined sets of data that can better represent image regions in a reduced, descriptive domain.

2.4.2.1. *Fourier descriptors*

The frequency domain describes an image not by its intensities at specific (x, y) locations on a 2D matrix, but as magnitudes of periodic functions of varying wavelengths. Although the functions' coefficients no longer retain any spatial information, they are divided into low, middle, and high frequencies with low frequencies providing overall structural information and high frequencies containing the detail. Feature descriptors can be localized from this set of Fourier coefficients, with low frequency coefficients retaining region structure (Fig. 2.25). To obtain the 2D discrete Fourier transform coefficients of a region $f(x, y)$ of size $M \times N$, we use the equation[26]

$$F(u, v) = \frac{1}{MN} \sum_{x=0}^{M-1} \sum_{y=0}^{N-1} f(x, y) e^{-f2\pi\left(\frac{ux}{M} + \frac{vy}{N}\right)}, \tag{2.50}$$

with $F(u, v)$ being calculated for $u = 0, 1, 2, \ldots, M - 1$ and $v = 0, 1, 2, \ldots, N - 1$. These values for $F(u, v)$ can be directly used as

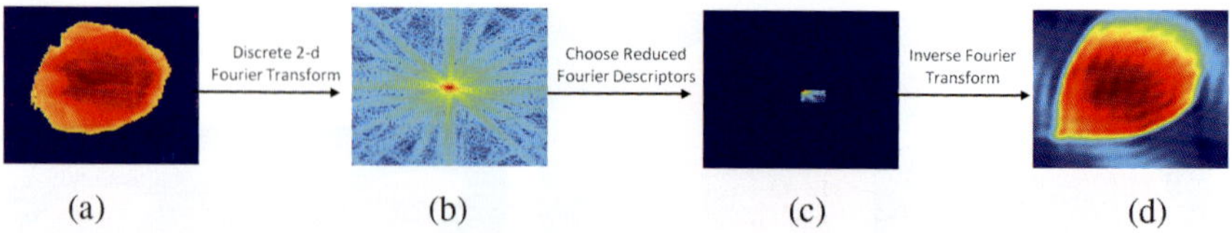

Fig. 2.25. (a) Original ROI, (b) visualization of a center-shifted 2D discrete Fourier transform, (c) reduced set of Fourier coefficients, and (d) 2D inverse transform of the reduced coefficients.

descriptors, or we can set values of F to zero and transform back to Euclidean space using the inverse transform equation:

$$F(x, y) = \sum_{u=0}^{M-1} \sum_{v=0}^{N-1} F(u, v) e^{j2\pi\left(\frac{ux}{M}+\frac{vy}{N}\right)}. \tag{2.51}$$

for and $x = 0, 1, 2, \ldots, M-1$ and $y = 0, 1, 2, \ldots, N-1$. In Fig. 2.25, we demonstrate how a Fourier transform can retain the structural information of a 166×133 region using a reduced set of 13×13 Fourier descriptors. A subset of frequency-domain descriptors has previously been used to extract blood vessels details from retinal images successfully.[27] Although not rotation, scale, or translation invariant, a linear operation with constant values in the frequency domain can correct for the identified spatial changes.[8]

2.4.2.2. *Principal component analysis (PCA)*

Another linear transformation that creates a useful feature set is principal component analysis, PCA. This method is defined as an orthogonal linear transformation of data projected onto a new coordinate system, where the greatest variance of any projection of the data lies on the line that passes through the first coordinate (principal component), with the second greatest variance on the second coordinate, and so on.[28] This method reduces a set of variables by calculating the eigenvalues of the covariance matrix, after a mean normalization occurs for each attribute.

We show how to use PCA with image regions by first vectorizing an $M \times N$ image region into a vector, v, with $M * N$ dimensions. We then take a set, K, of vectorized images, which creates $M * N$ sets of 1D vectors, $\mathbf{x}$, of size k:

$$x = \begin{bmatrix} x_1 \\ x_2 \\ \vdots \\ x_k \end{bmatrix}, \quad i = 1, 2, \ldots, M * N. \tag{2.52}$$

We then construct a vector of mean values, $\boldsymbol{m}_x$, calculated for each i as

$$\boldsymbol{m}_x = \frac{1}{K} \sum_{k=1}^{K} \boldsymbol{x}_k, \tag{2.53}$$

which we can then use to form a covariance matrix, $\boldsymbol{C}_x$, which is written as

$$\boldsymbol{C}_x = \frac{1}{K} \sum_{k=1}^{K} \boldsymbol{x}_k \boldsymbol{x}_k^T - \boldsymbol{m}_k \boldsymbol{m}_k^T. \tag{2.54}$$

Next, let $\boldsymbol{A}$ be a 2D matrix with rows that are filled with eigenvectors of $\boldsymbol{C}_x$ sorted so that eigenvector rows correspond to eigenvalues of descending size. We use $\boldsymbol{A}$ to map $\boldsymbol{x}$ values into vectors, $\boldsymbol{y}$, using the following:

$$\boldsymbol{y} = \boldsymbol{A}(\boldsymbol{x} - \boldsymbol{m}_x). \tag{2.55}$$

Using matrix algebra, we can find the covariance matrix $\boldsymbol{C}_y$ by

$$\boldsymbol{C}_y = \boldsymbol{A}\boldsymbol{C}_x\boldsymbol{A}^T, \tag{2.56}$$

which has diagonal terms that are the eigenvalues of $\boldsymbol{C}_x$. We can then recover any $\boldsymbol{x}$ from a corresponding $\boldsymbol{y}$ without using the entire matrix $\boldsymbol{A}$. We can use only the k large eigenvectors, using a transform matrix, $\boldsymbol{A}_k$, of size $k \times n$ by using the following:

$$\hat{x} = \boldsymbol{A}_k^T \boldsymbol{y} + \boldsymbol{m}_x. \tag{2.57}$$

This method is used in optic disk localization by normalizing and registering a set of well-defined $M \times N$ optic disk regions, then vectoring each image to form a set of $1 - \mathrm{D}\ M * N$ vectors. The PCA transform is performed first on a set of training image vectors, and the top six eigenvalues are kept as features to represent the training set. Tested ROIs are projected onto these eigenvectors, and a Euclidean distance metric is used to measure optic disk "likeness." We provide an example of how these lower dimensional eigenvectors reconstruct an optic disk using a training image convolved with differing Gaussian filters in Fig. 2.26.

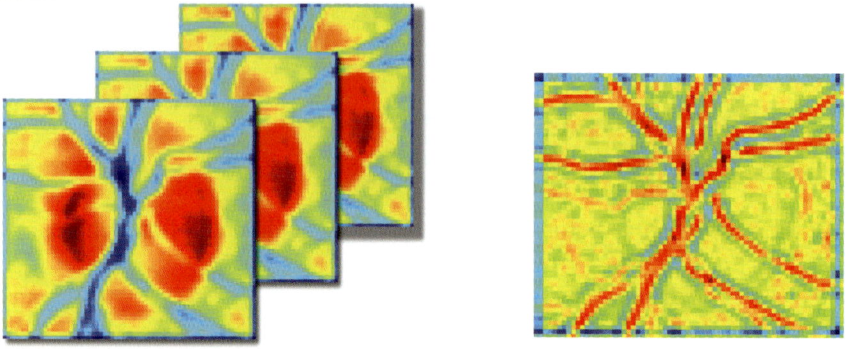

(a) (b)

Fig. 2.26. (a) Three optic disk images smoothed with Gaussian filters of increasing scale and (b) first component from PCA transform, containing the largest variability in the image.

2.4.3. *Multiscale Features*

We will next discuss representing a 2D image with varying degrees of scale through computational operations that reduce the resolution, yet retain multiple levels of detail that can be used as feature descriptors. The texture or shape of an image region at its native resolution can provide an excessive amount of unnecessary detail or noise artifacts, which can reduce classification accuracy. In addition, discriminative characteristics of an image region may occur only at certain reduced scales, which would go unnoticed at a lower scale. We will describe two methods that reduce an image's dimensionality at multiple levels (scales), representing the image as 3D pyramid structure of linearly decreasing resolutions to more easily extract discriminative features. We will give examples of how such methods can be used to extract features in retinal images.

2.4.3.1. *Wavelet transform*

We will discuss an image transformation that creates a multi-scale pyramid of coefficients that can each represent the ROI at different resolutions. Unlike the Fourier transform, wavelet transformations retain spatial and texture information, which can then be used as input in other feature extraction methods, such as those described above. Wavelets retain this information because their basis functions (or wavelets) are localized in the image, where a Fourier basis function spans the entire image. One such method is the discrete wavelet transform.

Although a full explanation of wavelet theory will not be provided here, we will give the mathematical formulation of the 2D discrete wavelet transform,[29] which is typically the wavelet transform used with images. We first need to chose basis functions, which includes a wavelet, $\psi(x, y)$, and a scaling function, $\varphi(x, y)$. Both functions are linearly separable, so that the 2D combinations give information for a horizontal (H), vertical (V), and diagonal (D) direction:

$$\psi^H(x, y) = \psi(x)\varphi(y); \tag{2.58}$$

$$\psi^V(x, y) = \varphi(x)\psi(y); \tag{2.59}$$

$$\psi^D(x, y) = \psi(x)\psi(y). \tag{2.60}$$

Each of these values describes the variations at a particular point along the specified direction. The linear combination of the scaling function produces: $\varphi(x, y) = \varphi(x)\varphi(y)$.

We use scaling and translation variables with the functions above to position the function correctly before convolving with the image of size $M \times N$ using the following formula:

$$\varphi_{j,m \cdot n}(x, y) = 2^{\frac{j}{2}}\varphi(2^j x - m, 2^j y - n), \tag{2.61}$$

$$\psi^i_{j,m \cdot n}(x, y) = 2^{\frac{j}{2}}\psi^i(2^j x - m, 2^j y - n), \quad i = \{H, V, D\}.$$

The 2D discrete wavelet transform of the function $f(x, y)$ of an image of size $M \times N$ is

$$W_\varphi(j_0, m, n) = \frac{1}{\sqrt{MN}} \sum_{x=0}^{M-1} \sum_{y=0}^{N-1} f(x, y)\varphi_{j_0,m \cdot n}(x, y) \tag{2.62}$$

$$W^i_\psi(j, m, n) = \frac{1}{\sqrt{MN}} \sum_{x=0}^{M-1} \sum_{y=0}^{N-1} f(x, y)\psi^i_{j,m \cdot n}(x, y), \quad i = \{H, V, D\}.$$

j_0 is the starting scale, which is typically set to $j_0 = 0$, then set to $N = M = 2^j$, $j = 0, 1, 2, \ldots, J - 1$, and $m, n = 0, 1, 2, \ldots, 2^j - 1$. We can

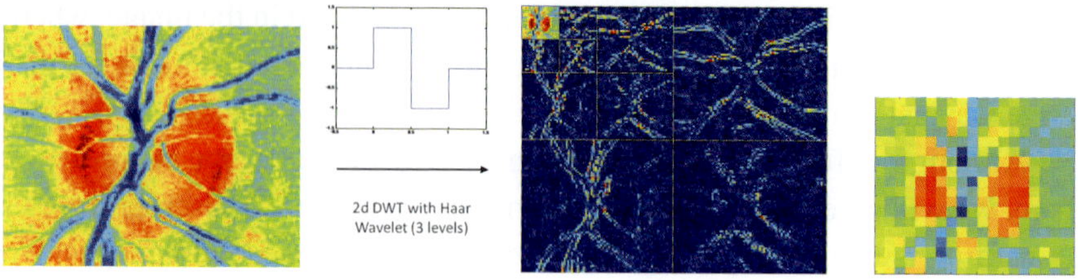

(a) (b) (c)

Fig. 2.27. (a) Original ROI, (b) 2D discrete wavelet-transform image pyramid, and (c) third-level reconstruction from approximation coefficients.

then perform the inverse discrete wavelet transform to find $f(x, y)$ using

$$f(x, y) = \frac{1}{\sqrt{MN}} \sum_{m} \sum_{n} W_{\varphi}(j_0, m, n)\varphi_{j_0,m\cdot n}(x, y), \tag{2.63}$$

$$+ \frac{1}{\sqrt{MN}} \sum_{i=H,V,D} \sum_{j=j_0}^{\infty} \sum_{m} \sum_{n} W_{\psi}^{i}(j, m, n)\psi_{i,m\cdot n}^{i}(x, y). \tag{2.64}$$

We use the simplest wavelet function, the Haar wavelet (Fig. 2.27), to construct a wavelet decomposition pyramid, containing details in the horizontal, vertical, and diagonal directions at different sizes. We can use these details to extract shape, texture, and moments at multiple scales, thus, increasing available ways to describe a ROI. The approximation of the image at varying resolutions also provides a compact representation of a region, reducing dimensionality in cases where fine details of are not necessary.

A method similar to that illustrated in Fig. 2.26 is used to localize the optic disc by performing a four- and five-level Haar wavelet decomposition, with the optic disc reduced to a small cluster of coefficients.[30]

2.4.3.2. *Scale-space methods for feature extraction*

As discussed in the previous section, we are interested in finding discriminative characteristics of a ROI, many of which occur over varying spatial scales. Several methods other than wavelet decomposition can provide a scale-space representation by convolving an image with kernels of varying size, typically suppressing details of varying scales. We will discuss two sets of scale space representations that can be used to extract multi-scale features from the retinal anatomy: difference of Gaussians and Hessian determinants.

As discussed in Sec. 2.2.2.2, Gaussian kernels smooth an image by varying amounts based upon the overall size and magnitude of the discrete 2D Gaussian curve. We can use subsequently smoothed images, $L(x, y; t)$, convolved with increasing kernel sizes ($t = \sigma^2$) to create a set of difference images, written as

$$G(x, y; t_1, t_2) = \frac{1}{2\pi t_1} e^{-\left(\frac{x^2+y^2}{2t_1}\right)} - \frac{1}{2\pi t_2} e^{-\left(\frac{x^2+y^2}{2t_2}\right)}, \tag{2.65}$$

which reveal scale-specific features of the ROI (Fig. 2.28). This method can also be used to approximate the Laplacian of the Gaussian (LoG as discussed in Sec. 2.3.1.3) for edge detection. We can now perform shape and texture feature extraction operations to the new set of multi-scale images to find a scale-specific features.

We can derive a set of 2D matrices that find corners for each smoothed image, $L(x, y; t)$ by first calculating the Hessian matrix, which is the square

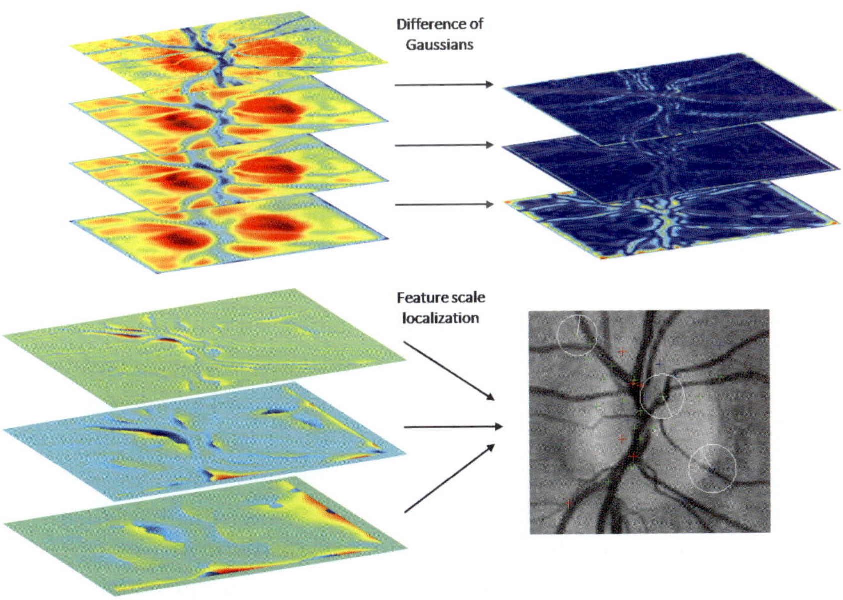

Fig. 2.28. Top-left: Gaussian smoothed image with increasing t; top-right: difference of Gaussian images (DoG); bottom-left: determinant of Hessian matrices with increasing t; and bottom-right: feature location (markers) and scale (radius of circle) from Hessian scale space.

matrix of second-order derivatives of $L(x, y; t)$. For each scale t and location (x, y), we find:

$$\mathcal{H}(x, y, t) = \begin{vmatrix} L_{xx}(x, y, t) & L_{xy}(x, y, t) \\ L_{xy}(x, y, t) & L_{yy}(x, y, t) \end{vmatrix}. \tag{2.66}$$

We then use these values to calculate the determinant of the Hessian using

$$\det HL(x, y; t) = L_{xx}L_{yy} - L_{xy}^2, \tag{2.67}$$

which can then be searched using a nonmaximal suppression algorithm to find scale-space localized, rotationally invariant features.[31] These locations in scale-space represents locations that have "corners," or large gradients in both x and y directions.

Retinal mosaicing, using the above methods to find location and scale, extracts a feature vector of binned gradients from the 2D Haar wavelet responses that are then used to match images for registration (Fig. 2.28).[32] Scale space provides a unique multi-level view of images, which exploit scale-specific features.

2.5. Summary

We have presented many of the most popular methods of retinal image pre-processing, segmentation/localization, and feature extraction for automated clinical decision support. Methods that are more complex have been introduced and can provide further guidance of how advanced methods, such as steerable filters,[33] active contours,[34] and neural networks,[35] are constructed. A retinal image classification algorithm can be assessed and observed as a series of steps consisting of many of the methods described in this chapter. It is important for the researcher/scientist to attempt multiple methods at each stage in the algorithm to best assess how the method affects the overall performance of the algorithm. Although the ultimate goal of a fully automated decision support system for retinal pathology detection has not been fully realized, there continue to be great strides toward reaching this challenge through computational methods.

References

1. Winder, R.J., Morrow, P.J., McRitchie, I.N., Bailie, J.R., and Hart, P.M. Algorithms for digital image processing in diabetic retinopathy. *Comput Med Imaging Graph* **33**:608–622, 2009.
2. Frame, A.J., Undrill, P.E., Cree, M.J., Olson, J.A., McHardy, K.C., Sharp, P.F., and Forrester, J.V. A comparison of computer based classification methods applied to the detection of microaneurysms in ophthalmic fluorescein angiograms. *Comput Biol Med* **28**:225–238, 1998.
3. Joshi, G.D. and Sivaswamy, J. Colour retinal image enhancement based on domain knowledge. *ICVGIP'08: Proceedings of the 2008 Sixth Indian Conference on Computer Vision, Graphics & Image Processing*. Washington, DC, USA: IEEE Computer Society, pp. 591–598, 2008.
4. Hoover, A. and Goldbaum, M. Locating the optic nerve in a retinal image using the fuzzy convergence of the blood vessels. *IEEE Trans Med Imaging* **22**:951–958, 2003.
5. Wang, H., Hsu, W., Goh, K.G., and Lee, M.L. An effective approach to detect lesions in color retinal images. *Proc IEEE Comput Soc Conf Comput Vis Pattern Recognit*, pp. 1–6, 2000.
6. Spencer, T., Olson, J.A., McHardy, K.C., Sharp, P.F., and Forrester, J.V. An image-processing strategy for the segmentation and quantification of microaneurysms in fluorescein angiograms of the ocular fundus. *Comp Biomed Res* **29**:284–302, 1996.
7. Foracchia, M., Grisan, E., and Ruggeri, A. Luminosity and contrast normalization in retinal images. *Med Image Anal* **9**:179–190, 2005.
8. Gonzalez, R.C. and Woods, R.E. *Digital Image Processing*, Prentice Hall, 2002.
9. Sinthanayothin, C., Boyce, J.F., Cook, H.L., and Williamson, T.H. Automated localization of the optic disc, fovea, and retinal blood vessels from digital color fundus images. *Br J Ophthalmol* **83**:902–910, 1999.
10. Shapiro, L.G. and Stockman, G.C. *Computer Vision*, pp. 137–150, Prentice Hall, 2001.
11. Newsom, R.S.B., Sinthanayothin, C., and Boyce, J. Contrast enhancement for oral fluorescein angiograms. *BJOP* **1**:29–32, 1998.
12. Goatman, K.A., Whitwam, A.D., Manivannan, A., Olson, J.A., and Sharp, P.F. Color normalization of retinal images. *Proceedings Medical Image Understanding and Analysis*, pp. 49–52, 2003.
13. Rapantzikos, K. Detection and segmentation of drusen deposits on human retina: potential in the diagnosis of age-related macular degeneration. *Med Image Anal* **7**:95–108, 2003.
14. Yulong, M. and Dingru, X. Recognizing the glaucoma from ocular fundus image by image analysts. *Engineering in Medicine and Biology Society, Proceedings of the Twelfth Annual International Conference of the IEEE*, pp. 178–179, 1990.
15. Gregson, P., Shen, Z., Scott, R., Kozousek, V. Automated grading of venous beading. *Comput Biomed Res* **28**:291–304, 1995.
16. Spencer, T., Olson, J.A., McHardy, K.C., Sharp, P.F., and Forrester, J.V. An image-processing strategy for the segmentation and quantification of microaneurysms in fluorescein angiograms of the ocular fundus. *Comp Biomed Res* **29**:284–302, 1996.

17. Pinz, A., Bernogger, S., Datlinger, P., and Kruger, A. Mapping the human retina. *IEEE Trans Med Imag* **17**:606–619, 2002.
18. Yulong, M. and Dingru, X. Recognizing the glaucoma from ocular fundus image by image analysts. *Engineering in Medicine and Biology Society Proceedings of the Twelfth Annual International Conference of the IEEE* pp. 178–179, 1990.
19. Sinthanayothin, C., Boyce, J.F., Williamson, T.H., Cook, H.L., Mensah, E., Lal, S., and Usher, D. Automated detection of diabetic retinopathy on digital fundus images. *Diabetic Med* **19**:105–112, 2002.
20. Meyer, F. Topographic distance and watershed lines. *Signal Processing* **38**:113–125, 1994.
21. Walter, T. and Klein, J.-C. Segmentation of color fundus images of the human retina: detection of the optic disc and the vascular tree using morphological techniques. *Medical Data Analysis*. Crespo, J., Maojo, V., and Martin, F. (Ed.), *Lecture Notes in Computer Science* **2199**:282–287, 2001.
22. Chaudhuri, S., Chatterjee, S., Katz, N., Nelson, M., and Goldbaum, M. Detection of blood vessels in retinal images using two-dimensional matched filters. *IEEE Trans Med Imag* **8**:263–269, 1989.
23. Hipwell, J.H., Strachan, F., Olson, J.A., McHardy, K.C., Sharp, P.F., and Forrester, J.V. Automated detection of microaneurysms in digital red-free photographs: a diabetic retinopathy screening tool. *Diabetic Med* **17**:588–594, 2000.
24. Haralick, R.M., Shanmugan, K., and Dinstein, I. Textural features for image classification. *IEEE Trans Syst Man Cybern* **SMC-3**:610–621, 1973.
25. Hu, M.K. Visual pattern recognition by moment invariants. *IRE Trans Info Theory* **IT-8**:179–187, 1962.
26. Cooley, J.W. and Tukey, J.W. An algorithm for the machine computation of the complex Fourier series. *Math Comput* **19**:297–301, 1965.
27. Tan, W., Wang, Y., and Lee, S. Retinal blood vessel detection using frequency analysis and local-mean-interpolation filters. *Proceedings of SPIE Med Imaging: Image Processing* **4322**:1373–1384, 2001.
28. Pearson, K. On lines and planes of closest fit to systems of points in space. *Philosophical Magazine* **2**(6):559–572, 1901.
29. Mallat, S. *A Wavelet Tour of Signal Processing*, Academic Press, New York, 1999.
30. Lalonde, M., Beaulieu, M., and Gagnon, L. Fast and robust optic disc detection using pyramidal decomposition and Hausdorff-based template matching. *IEEE Trans Med Imag* **20**:1193–1200, 2001.
31. Bay, H., Ess, A., Tuytelaars, T., and Vangool, L. Speeded-up robust features (SURF). *Computer Vision and Image Understanding* **110**:346–359, 2008.
32. Cattin, P., Bay, H., Van Gool, L., and Székely, G. Retina mosaicing using local features. *Medical Image Computing and Computer-Assisted Intervention — MICCAI 2006*. Larsen, R., Nielsen, M., and Sporring, J. (Ed.), *Lecture Notes in Computer Science* **4191**:185–192, 2006.
33. Kochner, B., Schuhmann, D., Michaelis, M., Mann, G., and Englmeier, K.-H. Course tracking and contour extraction of retinal vessels from color fundus photographs: most efficient use of steerable filters for model-based image analysis. *Society of*

Photo-Optical Instrumentation Engineers (SPIE) Conference Series. Hanson, K.M. (Ed.), pp. 755–761, 1998.

34. Saradhi, G.V., Balasubramanian, S., and Chandrasekaran, V. Performance enhancement of optic disc boundary detection using active contours via improved homogenization of optic disc region. *International Conference on Information and Automation* pp. 264–269, 2006.
35. Gardner, G., Keating, D., Williamson, T.H., and Ell, A.T. Automatic detection of diabetic retinopathy using an artificial neural network: a screening tool. *Br J Ophthalmol* **80**:940–944, 1996.

Chapter 3
Computational Decision Support Systems and Diagnostic Tools in Ophthalmology: A Schematic Survey

Sumeet Dua and Mohit Jain

Computer decision support systems are computer applications that can help clinicians make diagnostic decisions for patient treatment. Computer decision support systems provide easy access to patient file repositories. By accessing the data from the computational system, clinicians can decide whether to opt for surgery, therapy, or another form of treatment. Moreover, the system can alert clinicians to new patterns in patient data. Because of these benefits, a computer decision support system can save a clinician's time and can be more effective than the clinician can in making treatment decisions. Designing a computational system is not easy because the software and hardware infrastructure requirements are complicated and may negatively affect the computer's user friendliness. Additionally, training clinicians for this computational system can be costly. Despite these difficulties, the benefit and convenience that this system provides surpasses those offered by the old methods. This automatic and interactive system can help clinicians make better clinical decisions, which will help them to treat patients more effectively by helping clinicians avoid risks.

Department of Computer Science, College of Engineering and Science, Louisiana Tech University, Ruston, LA, U.S.A.

3.1. Evidence- and Value-Based Medicine

The two most prevalent forms of clinical diagnostics are evidence- and value-based medicine (VBM).[1] Evidence-based medicine (EBM) is a medical application in which individual expertise, such as experience, education, and skills, are integrated with the reliable, unambiguous, and thoughtful use of the most current, best evidence for making decisions about the care of the patient. EBM is usually initiated by a patient who questions the therapy, diagnostic tests, or prognosis provided by a physician.[1] The patient brings his or her own concerns and expectations to the clinician, providing the clinician with the best possible evidence for proper healthcare. Such evidence will not help the clinician make a decision, but will aid the clinician in deciding which treatment is best.

EBM is a learning process in which clinicians use clinically relevant or pertinent information about patient treatment, diagnosis, therapy, or other healthcare problems.[1] Instead of reading papers, clinicians see the evidence with respect to the individual patient problems, develop questions, and search a database that helps them get pertinent information to benefit the patient. Because interactive care includes the active participation of the patient, this system increases optimal clinician outcomes.

3.1.1. *EBM Process*

The EBM process consists of six steps, as listed below.

Step 1: A patient goes to the clinician for the treatment.
Step 2: A clinical question arises from the patient treatment.
Step 3: The clinician selects the most reliable and unambiguous evidence, and uses it to make decisions about patient care.
Step 4: The clinician evaluates this evidence.
Step 5: The clinician returns to the patient and integrates the current evidence with some form of expertise, such as experience, education, and/or skills, for treatment.
Step 6: The clinician evaluates the treatment with the patient.

3.1.2. *Evidence-Based Medical Issues*

Occasionally, problems arise from the use of EBM. The three most common of these problems are listed below.

1. Using EBM, clinicians and computational systems apply results obtained from large groups of people to individual cases, which may have different circumstances or unique characteristics that are not reflected in existing EBM records.
2. Since there is no optimal or standard way to search, it is often difficult to search EBM evidence effectively.
3. EBM does not measure or include quality of life.

3.1.3. *Value-Based Evidence*

VBM circumvents many of these problems.[1] VBM assembles the best EBM from clinical trials, and then converts the data to value-based evidence. To create a database of VBM, a clinician must conduct a cost utility analysis. A cost utility analysis measures the health and quality of life of a patient with values ranging from 0 to 1, where 0 is death and 1 is perfect health.[1] These values are also referred to as preferences. For example, if vision decreases, then the value of the utility analysis also decreases. VBM is composed of three essential components, as listed below:

1. collection of evidence-based data,
2. conversion of EBM to VBM, and
3. integration of associated costs with the value.

VBM provides a more accurate measure of patient healthcare than EBM. For example, evidence-based data from a clinical trial for diabetes might show a result, such as the improvement of the average life expectancy from eight to nine months, while the value-based data also shows this increased life expectancy, but also shows severe vomiting or pain during this time that decreases the patient's overall quality of health.

3.2. Economic Evaluation of the Prevention and Treatment of Vision-Related Diseases

Over the last three decades, many countries (such as Australia, Canada, and Germany) shifted large portions of their GDP toward healthcare.[2] This shift indicates an awareness of the economic importance of the maintenance of health among a nation's residents. The economic importance of the maintenance of health can be determined through economic evaluation.

3.2.1. *Economic Evaluation*

An economic evaluation of healthcare can be accomplished by comparing the cost of healthcare programs, such as medical therapy, health education, diagnostic tests, etc. with the advantage or profit that such programs can provide the physicians, patients, and healthcare providers. Economic evaluation is composed of three essential components, as listed below:

1. cost minimization,
2. cost benefit, and
3. cost effectiveness.

Each of these components should be tested to determine whether a healthcare program is economically sound. Economically successful programs should cost only a minimal amount, should save money in lost productivity, and should provide good results. The most common form of economic evaluation in healthcare is cost-effectiveness analysis (CEA).[2] The decision analysis method described below is used to perform CEA.

3.2.2. *Decision Analysis Method*

Decision analysis is a method of solving complex clinical problems in an easy and transparent manner, identifying all potential solutions, weighing all possible solutions, and selecting the optimal course of action. The decision analysis method consists of four elements: a decision tree, probabilities, costs, and benefits. A decision tree is a graphical representation that aids clinicians in making a decision based on the history of a patient's disease, the success of care as it was delivered in the past, and the success of care as it was delivered to patients with similar problems. Probability is the risk of an event occurring. Probability must be calculated from the strongest evidence available. An example of probability is the likelihood of curing cataracts versus the likelihood of preventing the condition from worsening. For example, if there is 50% probability of having a cataract, then the remaining 50% probability is of not having a cataract. Costs are the monetary values, which include the care or treatment provided to the patient and the costs of traveling to a particular medical center, including time off from work. Benefits are the outcomes that measure clinical trials, vision, life expectancy, quality-adjusted life year (QALY), and other improvements to the health of the patient.

3.2.3. *Advantages of Decision Analysis*

There are three advantages to decision analysis: structure, evidence, and evaluation. The structure of the decision tree model allows the clinician to make an informed decision based on the history of the disease and to evaluate the effect of treatment based on other information in the model.[3] Decision analysis helps to integrate evidence from a variety of data sources by introducing heterogeneity. Decision analysis can aid a clinician or other healthcare professional in translating the available evidence into estimates of cost and benefits. However, because each clinician may have had different experiences, the final decision will be affected by his or her goals and perspective.

3.2.4. *Perspective in Decision Analysis*

The perspective of the decision maker must be considered when applying decision analysis. The perspective may be different for a physician, an insurance representative, or a patient. Because each of these individuals may have different goals, an analytical report should clearly state who provided the opinion. For example, if the perspective originates from an insurance company, then it is likely that the costs incurred by the patient and physician are not included. In addition to considering the decision maker's perspective, a decision tree can help the clinician decide on the best course of action for a particular patient.

3.2.5. *Decision Tree in Decision Analysis*

A decision tree can be especially helpful to ophthalmologists, who need to manage difficult clinical problems.[3] For example, if a student has an examination in the morning and have to decide whether to watch a soccer match or not, then a decision tree containing actions, events, and outcomes can be useful. In this case, the decision tree would contain two actions:

1. "Watch Soccer Match" or 2. "Do Not Watch Soccer Match."

The events that influence the outcome of the actions are:

"Easy Questions" and "Difficult Questions," and reflect the ease with which the student was able to complete the examination. Consider the decision tree in Fig. 3.1.

Once a student or ophthalmologist makes the decision tree, then he or she can add information, such as the cost of outcome and event probability. The cost will depend on the rank of the outcome or on which outcome is more important and can vary from person to person.

The outcome of the ranking system for our student example is shown in Fig. 3.2, in the rightmost column. Our first ranked outcome is "Fortunate," for which we assigned a weight of 1.0; our second ranked outcome is "Right Decision," for which we assigned a weight of 0.75. Our third ranked outcome is "Should Have Watched," for which we assigned a weight of 0.5, and our

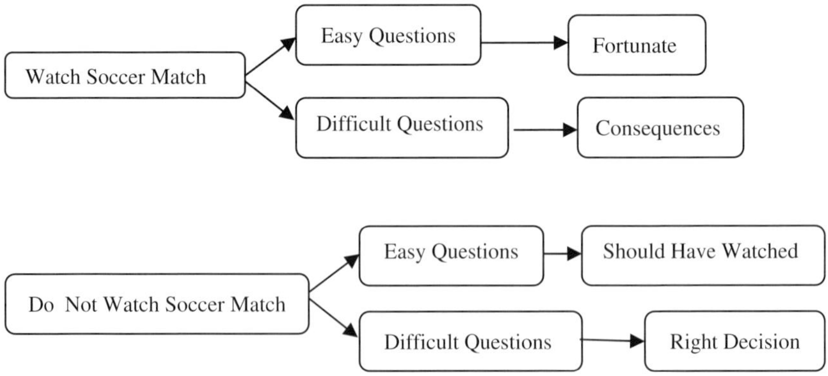

Fig. 3.1. Example of a decision tree.

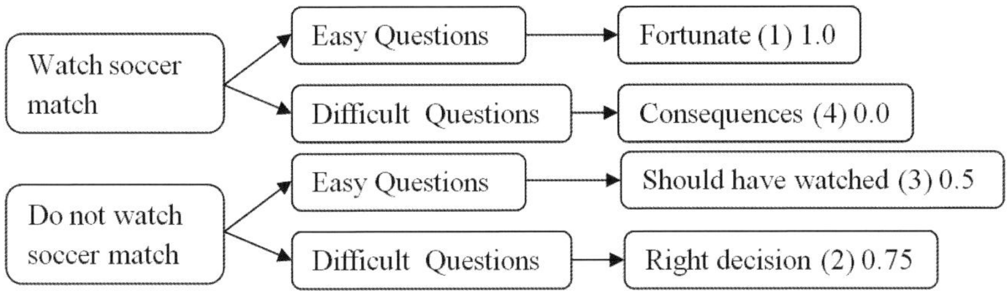

Fig. 3.2. Example of a decision tree after assigning weight and rank to outcomes. Outcomes rank (1), (2), (3), and (4), based on weights (the cost of the outcome) of 1.0, 0.75, 0.5, and 0.0.

fourth ranked outcome is "Consequences," for which we assigned a weight of 0.0. The weighting system is described below.

As the name indicates, event probability denotes the chance or probability that a particular event will occur. Let us continue with our example, and assume that if student watches soccer, then there is a 35% chance that the student will find the questions easy and a 65% chance that the student will find the questions difficult. If the student does not watch the soccer match, then there is a 60% chance that he or she will find the questions easy, and a 40% chance that the student will find the questions difficult. These event probabilities are shown below in Fig. 3.3.

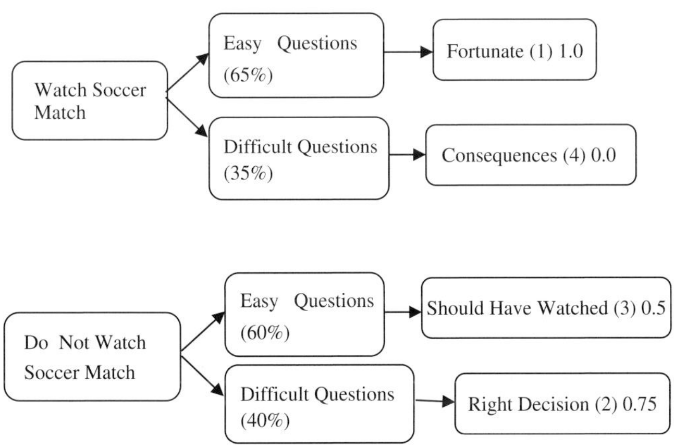

Fig. 3.3. Example of decision tree after assigning event probabilities.

The final probabilities are calculated as:

"Watch Soccer Match": $0.35*1 + 0.65*0 = 0.35$ and
"Do Not Watch Soccer Match": $0.6*0.5 + 0.4*0.75 = 0.3 + 0.3 = 0.6$.

In this example, the student should choose "Do Not Watch Soccer Match," since it has the highest weight.

For another example, consider a team of ophthalmologists that go to a city to perform cataract operations on several people. They must determine whether to include all the patients that need laser treatment to prevent blindness, because treating only the necessary people will save resources, time, and money. Fig. 3.4 shows a decision tree based on their actions, events, and outcomes.

First, a diagram of actions will be drawn (for example, in the above figure "Everyone" and "Not Everyone"), events (for example, in the above figure "Cataract" and "Not Cataract"), and outcomes (for example, in the above figure "Prevent Blindness," "Waste of Time," "Probable Blindness," and "No Blindness or Waste of Time"). Based on the events and on personal perception, we determine how to rank the outcomes.

The ranks are described in Step 2.

Second, the outcomes of the events will be ranked, as shown in Fig. 3.5. In our example, the ophthalmologists rank the outcomes of the events as:

Rank 1. Not Everyone -> No Cataract -> No blindness, not examining people who do not have cataracts saves time and money. Rank 2. Prevention of Blindness -> The patient had cataracts removed after laser surgery.

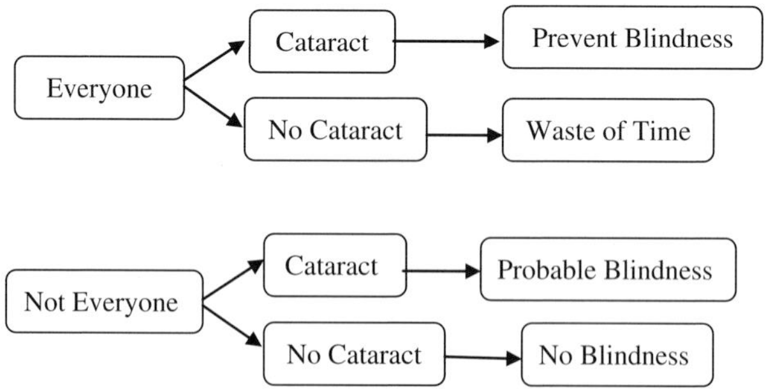

Fig. 3.4. Example of decision tree with actions and events. The methodology for creating this tree consists of four steps, as defined below.

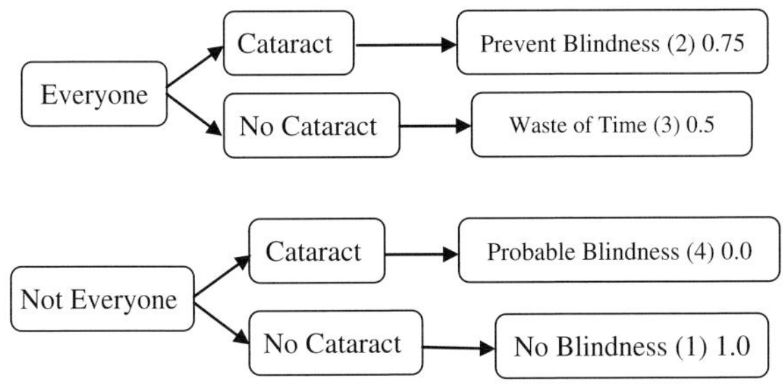

Fig. 3.5. Example of decision tree after assigning weight and rank to outcomes. The ranks are (1), (2), (3), and (4), and are based on weights (the cost of the outcomes) of 1.0, 0.75, 0.5, and 0.0.

Rank 3. Everyone -> No Cataract -> Waste of time. -> Since we have included everyone for examination, some people who do not have cataracts will be examined, which results in loss of time and money. Rank 4. Not Everyone -> Cataract -> Probable blindness, i.e. a person who has cataracts is not examined and may become blind because of not being included. "Not Everyone" ranks fourth since we do not want these cases to be missed.

The above ranking will vary from surgeon to surgeon. For a team, the surgeons will need to sit together and discuss which cases to give the best rank and which the least.

Our outcome ranking system, as explained above, is shown in Fig. 3.5 in the rightmost column. Our first ranked outcome is "No Blindness or Waste of Time," which we assigned a weight of 1.0; our second ranked outcome is "Prevent Blindness," which we assigned a weight of 0.75. Our third ranked outcome is "Waste of Time," which we assigned a weight of 0.5, and our fourth ranked outcome is "Probable Blindness," which we assigned a weight of 0.0.

Third, event probability will be determined, as shown in Fig. 3.6. Let us assume that if everyone is examined, the chance of an examined individual having a cataract is 50% and the chance of an examined individual not having a cataract is 50%.

Fourth, a numerical calculation will be made. The numerical calculation for the weighting is as described below:

"Everyone": $0.5^*0.75 + 0.5^*0.5 = 0.375 + 0.25 = 0.625$ and
"Not Everyone": $0.5^*0 + 0.5^*1 = 0.5$.

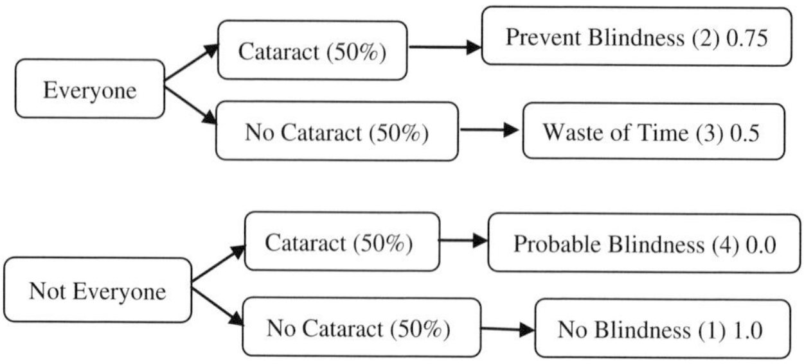

Fig. 3.6. Example of decision tree after assigning event probability.

Since the weight of "Everyone" is higher, everyone is included for the diagnostic examination. The above example illustrates how decision trees can help an ophthalmologist make complex clinical decisions and diagnose patients.

3.3. Use of Information Technologies for Diagnosis in Ophthalmology

Information technology helps healthcare professionals to build decision support systems, such as diagnosis, therapy planning, and monitoring.[4] The components of decision support systems are image acquisition, data mining, and graphical user interface (GUI). There are four ways in which images can be acquired, as explained below.

1. Slit-lamp cameras are used to capture the details of the eye structure by changing the parameters or width of the beam.
2. P2, a digital, integrated platform for ocular imaging, combines the functions of the slit lamp and BIO with a CCD camera and acquires 3D images from the front and back of the eye without dilation. The 3D images accurately depict the current eye condition and will help improve diagnostic accuracy.
3. A RetCam can be used by nurses in neonatal intensive care units (NICUs) to take images.
4. Optical coherence tomography (OCT) creates high quality, micrometer-resolution, 3D images from within optical scattering media.[5]

Data mining, or knowledge discovery, can be used to gather and utilize information in healthcare. Because it handles multiple data entries, it is ideal for ophthalmology, where clinics may have a variety of patients with similar diseases but wide-ranging symptoms and personal circumstances.

3.3.1. *Data Mining in Ophthalmology*

Data mining consists of five steps: preprocessing, feature extraction, feature selection, clustering, and classification.[6] Image preprocessing consists of three steps. If images have different brightness or intensity, the clinician must normalize all the images to the same brightness level, or it will be difficult to differentiate between images of healthy and/or diseased eyes. This process can be completed by using an image histogram. An image histogram is a pictorial representation of the distribution of intensities of the image through which one can increase or decrease the intensity or brightness of the image as per the requirement for better viewing. An image histogram helps to create uniform brightness for each image and improve the quality of the image.

As with brightness, every image should be the same size. It is difficult to perform matrix operations, such as addition or subtraction, on images that are not the same size. Images that are not the same size can be normalized by padding, i.e. adding 0s that do not affect picture information to lengthen the shorter images. Padding does not distort the image by adding external information. Noise is undesirable information that has to be removed from the image. Gaussian noise,[7] quantization noise,[8] and film grain[9] are methods that can be used to remove noise.

Feature extraction is a form of dimensionality reduction. When the input data is large and redundant, then the feature extraction algorithm will divide large datasets into smaller sections or into feature vectors. For feature extraction, a system first reads an image and extracts the features or key points of the image that will be represented as unique features. The feature vectors help to distinguish between healthy and diseased eyes. Feature extraction can be performed by extracting the features of the image using global or regional feature extraction. For global feature extraction, the system will take the entire image as input and extract the feature vectors. The global feature extraction can be performed using algorithms, such as scale invariant feature transform[10] and principal component analysis.[11] For regional feature extraction, the system will divide the image into several equal-size parts. For each part, the system will extract the features and then merge the extracted features into one feature vector. From the vectors, the system can extract the regional features by using algorithms, such as gray-scale co-occurrence matrix[12] and grid-based segmentation.[13] The feature vector will become the output of these algorithms for the next phase.

Feature selection is also known as attribute selection or as feature reduction. Feature selection aids in selecting important or relevant features. By selecting important features, the clinician can speed up the learning process or remove the effect of the curse of dimensionality. Feature selection can be performed by feature ranking, where the clinician will get the rank of the features or the attributes, and, based on the rank, the clinician can select the features. Methods by which the clinician can rank our features include the ChiSquared attribute,[14] the GainRatio attribute,[15] and the info-gain attribute.[16] The clinician can use tools such as Weka, Orange, and Rapidminer, as well for feature ranking.[17]

Clustering is the process of grouping together objects that are similar to one another but are dissimilar to objects in other clusters or groups. Once the clinician has selected the features, he or she can give a class label to the features or to the category in which they belong, i.e. healthy eyes or the diseased eyes. However, if the clinician does not know the class labels or the category title, then we use clustering to learn the class labels or the category to which group the feature belongs. The quality of a cluster will depend on the following three factors: high intra-class similarity and low inter-class similarity, the similarity measure used by the method and its implementation, and the cluster's ability to discover hidden patterns.

Once the clinician has the class labels, then he or she can begin classification. A classification algorithm is used to predict categorical or discrete values based on the training data or the data the clinician gets after feature selection or clustering. There are two steps for this process: model construction and model usage. In model construction, the clinician builds or trains a model according to the training data, and in model usage, the clinician classifies future or unknown objects or estimates the accuracy of the classification. In ophthalmology, the clinician determines how many images have been correctly classified to the healthy eyes and diseased eyes classes.

Before starting this step, the clinician preprocesses the images by removing missing or NAN values or normalizing the data to a particular range. The classification algorithms that the clinician uses to obtain these values are Naïve Bayes, Neural Network, SVM, and Multi-Class Classifier.[18] Once the data mining has been completed, then the clinician will input the values into the GUI.

3.3.2. *Graphical User Interface*

A GUI allows users to interact with the system, i.e. the user can click on the screen, retrieve values, or give commands. Examples of GUIs include computer games and interactive mp3 player functions. Software with GUI functions includes the following. NeoSoniX is the first hand piece to combine sonic, nonlinear oscillation with linear ultrasound.[19] Together, with AdvanTec software, NeoSoniX decreases the time and energy required for cataract removal.[20] The software improves the sensory feedback system by increasing the accuracy and consistency of ultrasound energy, and maximizes ultrasonic performance. Intraocular lens (IOL) power revolutionizes the measurement of the ocular axis and sets new standards for speed and accuracy, even with eyes that are not ideally formed. IOLs have an accelerated workflow, produce user-independent measurement results, and are easy on patients with no ocular contact. The automatic and interactive system provides healthcare professionals with decision support. Once these steps are completed, the clinician will have the tools to diagnose specific diseases and the capabilities to avoid the risks of visual disability.

3.4. Role of Computational System in Curing Disease of an Eye

Computational systems can help detect, diagnose, and treat eye diseases such as diabetic retinopathy, cataracts, glaucoma, blepharitis, rosacea, Sjögren, dry eyes, retinal detachment (RD), and amblyopia. Our computational system will use eye images, taken by using slit-lamp, p2, or a similar method, as input and will classify the images into classes such as cataract, diabetic retinopathy, etc. This system will not only save the clinician time, but it will also provide the clinician with precise information about the eye disease, thus aiding in quick treatment and reducing the risks of visual disability.

3.4.1. *Computational Decision Support System: Diabetic Retinopathy*

Diabetic retinopathy occurs when neurons stop transmitting signals from the retina to the brain, damaging the blood cells in the retina, and can lead to severe vision loss or blindness, and a diabetic retinopathy patient may not notice changes to his or her vision at the beginning of the disease. Late diagnoses may cause lasting problems, because, if this disease is not cured early, treatment may be unsuccessful. If the disease is detected early, the worst-case can be anticipated, and the patient and ophthalmologist can prepare appropriately. A computerized interactive system can screen a large number of retinal images and classify abnormal and normal images, saving the physician time, which can then be spent on surgery and treatment. Computer decision support system methodologies for ophthalmology are wavelet-based neural network and content-based image retrieval (CBIR) based on wavelet transform coefficients distribution.[21,22] Steps to implement the wavelet-based network are shown in Fig. 3.7.

3.4.1.1. *Wavelet-based neural network*[23]

Applying wavelet-based neural networks to diagnose diabetic retinopathy requires four steps, as described below.

First, color retinal images can be acquired using Nikon, Sony, or fundus cameras and can be used by an ophthalmologist for diagnosing a patient. Once these images are acquired, they are preprocessed using the steps shown in Fig. 3.8. The RGB color system is converted to HIS as an optic disc and is one of the brightest regions in the image. Then, the disc is converted to a gray-scale image. Next, an image histogram is applied to adjust the brightness or intensity level of the image. Last, noise is removed. To reduce noise from the image, a median filter or Gaussian filter is applied.

Second, the discrete wavelet transform is used to extract features, such as an optic disc, which acts as the reference for other features.[23] Blood vessels help ophthalmologists to analyze diabetic retinopathy over the course of the disease. These blood vessels can be extracted through segmentation with

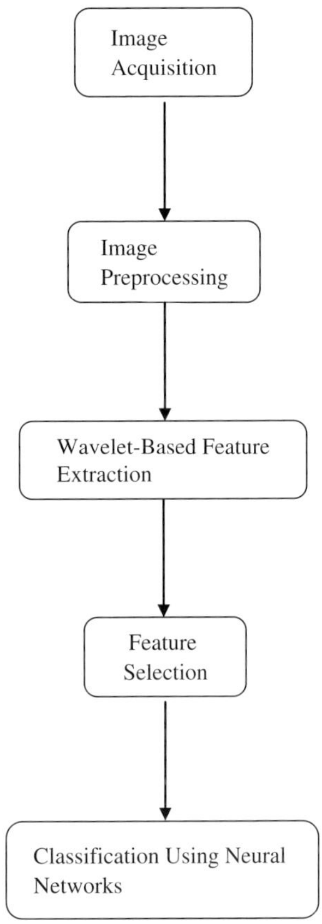

Fig. 3.7. Computational steps for wavelet-based neural network.

wavelet transform. Exudates are lesions that commonly occur in diabetic retinopathy. The shape and brightness of exudates vary from patient to patient and can be detected using a thresholding approach. The identification of the changes in blood vessels and exudates in the retina over time can aid physicians in making early detections of diabetic retinopathy.

Third, a feature selection method, such as the ChiSquared attribute, the GainRatio attribute, and the infogain attribute, is used to select the important or relevant features of an image. By selecting important features, a clinician can speed up the learning process or remove the effect of the curse of dimensionality.

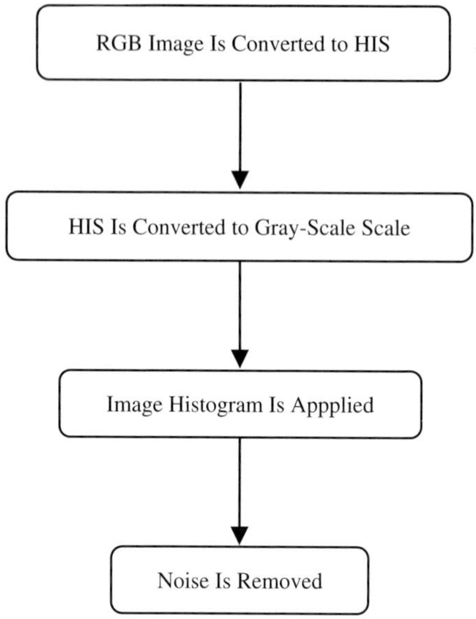

Fig. 3.8. Computational steps for image preprocessing.

Fourth, the features that are extracted during feature selection become input for the neural network. In this step, a classifier helps the clinician to classify images into healthy or diseased eyes categories. First, to construct a model using input, the features selected to build or train the model can distinguish images of different classes. This automated system can be used to analyze the retinal images of different classes using wavelet-based extraction, and an image classifier based on artificial neural network can classify the images according to disease conditions.

3.4.1.2. *Content-based image retrieval*

CBIR is the application of computer vision in which images are retrieved according to the image content, for example, color, shape, textures, or any other information from the large database. The steps to retrieve images using CBIR are shown in Fig. 3.9.

CBIR[22] is based on retrieving the K nearest neighbor (KNN) using similarity measures based on the input image means to find all the nearest matches of the input image and its procedure or steps are drawn in Fig. 3.10.

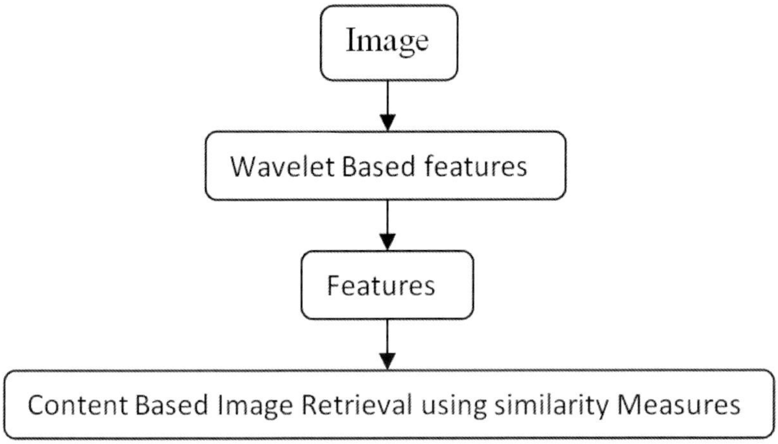

Fig. 3.9. Computational steps for CBIR.

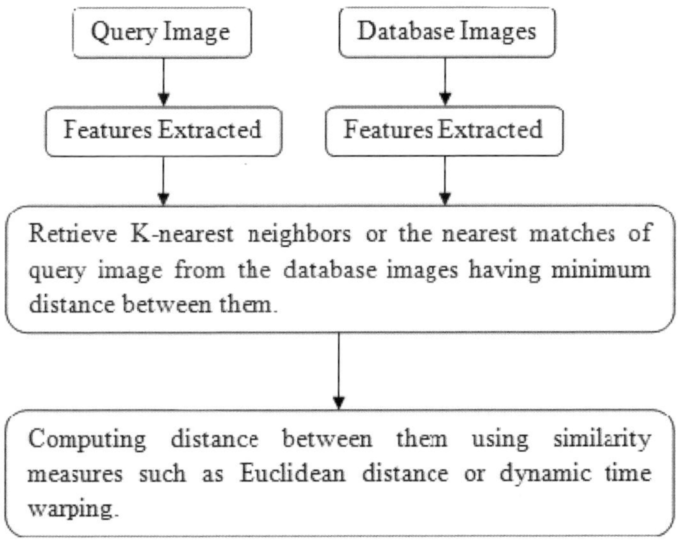

Fig. 3.10. Steps for retrieving images using similarity measures.

To explain this methodology, let us say that a clinician has 500 retinal images in his or her database and one input or query image. First, the clinician should extract the features of all the images using feature-extracted algorithms (explained in the Sec. 3.3.1). Once the clinician extracts the features, then he or she computes the distance between the input image and the database images one by one, using similarity measures, such as Euclidean distance and the dynamic time warping algorithm. In this way, the clinician gets 500 values, and out of those 500 values, the clinician retrieves KNNs or the nearest matches of those query images having a minimum distance between them. Using the KNN method of CBIR, the clinician can better distinguish between the healthy and diseased eyes by retrieving the nearest matches of query image.

3.4.2. *Computational Decision Support System: Cataracts*

Cataract is an eye disease that is developed in the crystalline lens of the eye and is the leading cause of blindness worldwide.[24] Cataracts develop slowly, and the patient is normally not aware of the gradual loss of eyesight. If untreated, this disease can result in vision loss in both eyes. An early diagnosis of the disease has a greater chance of preventing blindness and curing the disease. A cataract can be age-related (further classified as a nuclear or cortical cataract), congenital, or trauma-related.[25] The causes of cataracts include exposure to the sun and high radiation, diseases such as diabetes and hypertension, trauma, aging, or genetics.

Surgery is the most effective and common way to treat cataracts; phacoemulsification is the most popular form of cataract surgery. During phacoemulsification, surgeons classify the eye into the normal or diseased eye categories. Due to the advancement of image processing and machine-learning techniques, computer aided classification can be used to diagnose and classify the images and to aid decision support.

As with diagnosing diabetic retinopathy, to diagnose a cataract, the computer decision support system must perform image acquisition, image preprocessing (equalization and binarizing), feature extraction, feature selection, and image classification.[4,26,27] These steps are described below.

First, in image acquisition, lit-lamp cameras are used to capture the details of the eye structure by changing the parameters or width of the beam. Then, images are divided into training and testing images. Training images are used to build the model and to aid in the classification of the images.

Second, image preprocessing for computer decision support is a three-step process. The images are equalized, binarized, and converted to gray-scale. Equalization is performed to make the intensity, brightness, or contrast of all images the same. Equalization can be performed by using image histograms to detect the images of healthy and diseased eyes. Binarization is performed with a carefully chosen threshold. The RGB image is converted into gray-scale, and it carries only intensity information. Gray-scale images vary from black at the weakest intensity to white at the strongest and are easier to handle than RGB images.

Third, in feature extraction, features, the key or unique points of an image, are extracted from the raw images, which are based on the lens structure. Features help to extract the important information from the image, which helps to distinguish between the healthy and diseased eyes. Examples of features are a big ring area, a small ring area, homogeneity, and BW morph. A big ring area can result when the color at the outer surface of the cornea is not the same in all classes; some colors are brighter and some are dimmer. A small ring area can result when the color brightness at the inner surface of the cornea is not same in all the classes. Homogeneity measures the closeness of the distribution of the elements. BW morph is a morphical operation performed on the binary images.

Normal cataract images have too many sudden changes in the gray levels. After binarization, the image is converted to black and white to find the area between the images. Other features, such as mean intensity inside the lens, color on the posterior reflex, mean intensity of the sulcus, and the intensity ratio between the anterior and posterior lentils, can be extracted using principal component analysis and regional features using a gray level co-occurrence matrix such as energy, homogeneity, correlation, or contrast.

Fourth, feature selection is also known as attribute selection or feature reduction. During feature selection, the computational system will select the important or relevant features. A clinician can perform a feature selection method such the ChiSquared attribute, the GainRatio attribute, or the infogain attribute.

Fifth, once features have been selected, the clinician can classify the images in two ways. He or she can use a classifier, such as a support vector machine, a neural network, or a fuzzy classifier, or can use KNNs.

3.4.2.1. *Using classifiers*

In order to choose to use a classifier, a clinician must first construct a model. In model construction, clinicians build or train a model according to the training data; in model usage, clinicians classify future or unknown objects or estimate accuracy. In ophthalmology, clinicians classify how many images have been correctly classified.

3.4.2.2. *K nearest neighbors*

KNNs are used to find the nearest neighbors of the input image or those images that have the highest probability of being matched with the input image. K can be any integer value. To find the nearest matches, clinicians can use similarity measures such as Euclidean distance or dynamic time warping, which gives the distance between the two images. If the distance between the two images is zero, then they are matched exactly.

3.4.2.3. *GUI of the system*

The main menu of the GUI will include options, for example image acquisition, feature extraction, and classification, which the clinician can choose. This automatic and interactive system aids healthcare professionals in using the decision support system and gives them important information to avoid the risk of a patient developing a visual disability.

3.4.3. *Computational Decision Support System: Glaucoma*

Glaucoma is an eye disease that affects the optic nerve, and, if untreated, can lead to blindness. There are two categories of glaucoma: open-angle glaucoma and closed-angle glaucoma. Closed-angle glaucoma is painful and can lead to blindness quickly. Open-angle glaucoma progresses slowly, and has often developed significantly, before the patient learns he or she has it.

Open-angle glaucoma is generally caused by increased pressure within the eye, and it is difficult to find symptoms at the early stage. Thus, regular checks are necessary to avoid this disease. Damage to the optic nerve, a side effect of glaucoma, cannot be reversed; thus, early treatment is important to the maintenance of good eye health. Early detection can be maintained by reducing intraocular pressure (IOP). Once the disease is detected, medications can help to stop the progress of the diseases. Therefore, it is important to diagnose glaucoma early to minimize the risk of blindness. Factors to consider when performing glaucoma screening are sufficient sensitivity and specificity, cost effectiveness, the quickness of the diagnosis, and high quality equipment.

Sensitivity and specificity are statistical measures of the performance of a test. Sensitivity measures whether a person with glaucoma is correctly identified. Specificity measures whether a person with healthy eyes is correctly identified. Glaucoma screening should be economical to avoid stress or patient reluctance to do a screening. Diagnosis should be efficient and effective, as glaucoma treatment and medication is expensive. Equipment should be easy to use and should give precise results. Therefore, an automatic and interactive computational system that can help avoid visual loss caused by glaucoma is necessary.

Fuzzy logic and classifiers are computational system methodologies for efficiently diagnosing glaucoma.[28–30]

3.4.3.1. *Using fuzzy logic*

The Heidelberg retina tomograph (HRT), which allows 3D images of the retina and the topography of the optical nerve head to be obtained, can aid in the acquisition of images that can be analyzed over the time. Once images are acquired, fuzzy logic can be applied in three steps: image preprocessing, visual field examination, and intra ocular pressure (IOP).

First, in glaucoma screening, appropriate image processing is applied to enhance the visibility of retinal nerve fiber layer defect (RNFLD) features that occur during early-stage glaucoma. The image processing shows the images of the optic disc. Image enhancement methods include loading the images, converting the RGB images into gray-scale, resizing the images, applying an image histogram so that all the images have uniform brightness or intensity, and removing the noise from the images using filtering techniques, such as Gaussian or adaptive filtering.

Second, a visual field examination is performed. In a visual field examination, different areas of different colors and shapes are scattered over the visual field. The technician uses the computational unit to make an area flicker, and the patient is asked to close one eye and press a key on the computer keyboard indicating which area flickered. This system is designed by professionals to detect the common glaucomatous defects.

Third, inter ocular pressure and fuzzy logic are applied. Increased IOP may damage the optic nerve and be responsible for visual loss. By using a set of fuzzy "IF-THEN" rules, a decision can be make as to whether a patient has glaucoma or not. For example,[28,29,31]

1. **IF** both the IOP and the *Abnormal visual field test* scores are "High," **THEN** the *Risk* is "High";
2. **IF** IOP is "High," and the *Abnormal visual field test* scores are low, **THEN** the *Risk* is "Moderate";
3. **IF** *Family history* is "Bad," *Age* is "Old," and the *Abnormal visual field test* scores are "High," **THEN** the *Risk* is "High";
4. **IF** *Family history* is "Bad," *Diabetes* is "High," and *Abnormal visual field test* scores are "High," **THEN** the *Risk* is "High."

This system is cost effective and appropriate for detecting early-stage glaucoma.

3.4.3.2. *Computational system using different classifiers*

The computational system for diagnosing glaucoma performs as it does when diagnosing diabetic retinopathy and cataracts. Figure 3.11 reiterates these steps.

3.4.4. *Computational Decision Support System: Blepharitis, Rosacea, Sjögren, and Dry Eyes*

Blepharitis is an eye disease similar to dry eye that is caused by inflammation in the eyelid margins. The inflammation causes eye redness, itching, and irritation. Dry eye is caused by decreased tear production or increased tear film evaporation.[32,33] The symptoms, itchy, tired, painful, and red eyes, are similar to those caused by blepharitis. However, treatment for dry eye and blepharitis may differ. Thus, it is important for clinicians to differentiate between the diseases.

A computational system can aid in the classification of healthy, blepharitis, rosacea, Sjögren, and dry eyes through clustering analysis.[33] When a clinician constructs the model for the classification of the images, he or

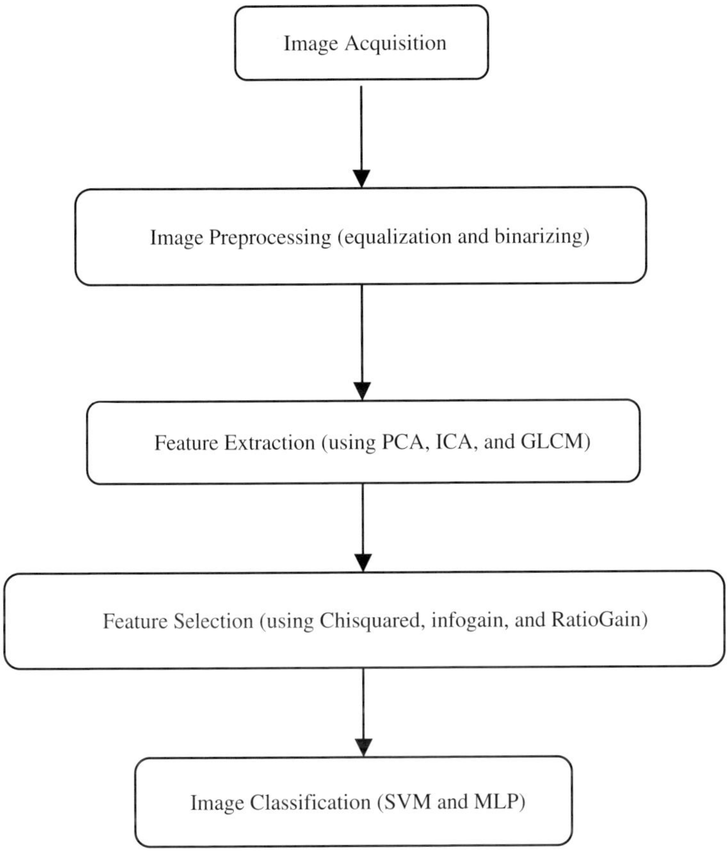

Fig. 3.11. Computational steps for diagnosing glaucoma using classifiers.

she classes image labels to their respective categories (healthy or diseased eyes). Once the clinician determines the correct category, he or she will program the data to the classifiers to build a model. The model can then be used for the future prediction of images. The computational steps for this process are described in Fig. 3.12.

The first five steps (image acquisition through classification) have been explained in previous sections. Once the relevant features are selected or feature selection is performed, then the clinician will classify images. Classification can be performed using 10-fold cross-validation. In 10-fold cross-validation, the dataset images are partitioned into 10 sub datasets. Of the

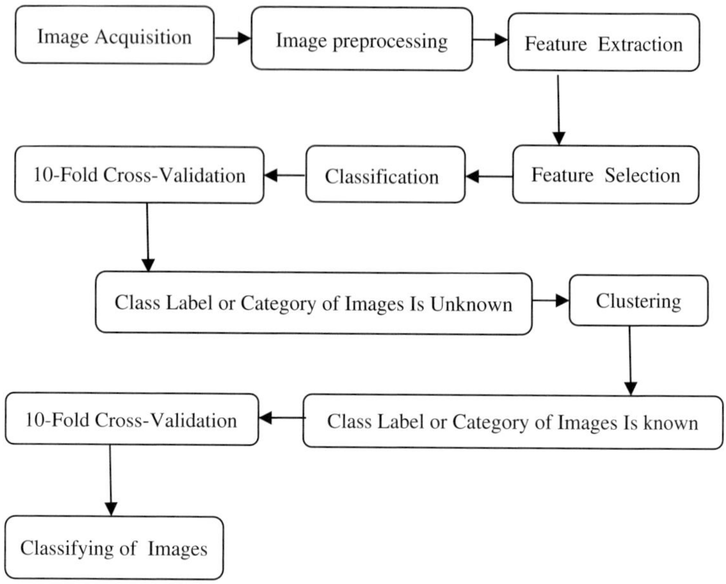

Fig. 3.12. Computational steps for diagnosing glaucoma using clustering analysis.

10 datasets, a single dataset is used for testing the model, and the remaining nine datasets are used as training data. Once the algorithm is trained, the cross-validation process is repeated once with each of the 10 datasets as the new testing data. The 10 results from the folds then can be averaged to produce a single result.

A two-step process model construction and model usage predicts categorical class labels (discrete or nominal). Model construction classifies data based on the training set of images and uses this model to classify new images. In model construction, each sample or image is assumed to belong to a predefined class, and its procedure is described in Fig. 3.13. However, if the class label is unknown, then the clinician will perform clustering to learn the class labels before constructing a model as mention in Fig. 3.14.

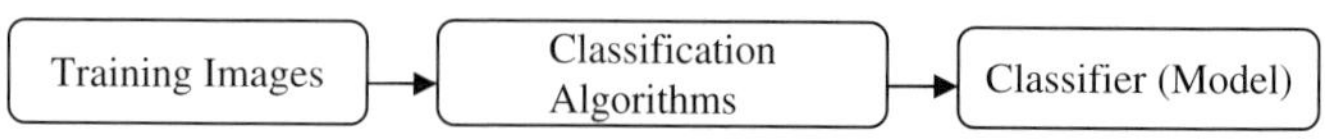

Fig. 3.13. Model construction.

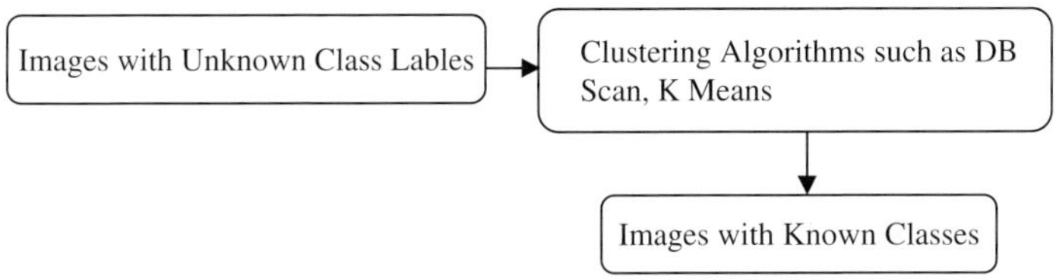

Fig. 3.14. Class labeling.

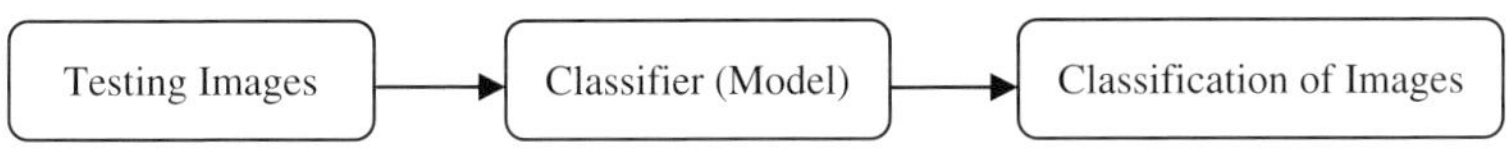

Fig. 3.15. Model usage.

Various clustering algorithms, such as K means, hierarchal, and DB scan, can be used in this step. By using a classification algorithm, the images of eyes that are healthy, that are contaminated with blepharitis, and that are dry can be classified into relevant groups that have common features or characteristics.

The classification model is used to classify future or unknown images. The known label of a test sample is compared with the classified result from the model, and an accuracy rate is calculated as mentioned in Fig. 3.15. The accuracy rate is the percentage of test images that are correctly classified by the model. To avoid over-fitting, the test set must be independent of the training set.

Model construction, class labeling, and model usage will create an automatic and interactive system that can classify the images of different categories using clustering and the 10-fold cross-validation system. This system will help the ophthalmologist to make a diagnostic decision.

3.4.4.1. *Utility of bleb imaging with anterior segment OCT in clinical decision making*

Recent studies have shown that bleb imaging with anterior segment OCT (ASOCT) can provide relevant information to an ophthalmologist that is not present in clinical evaluation.[34] This information is especially useful for determining whether laser suture lysis (LSL) should be performed.

LSL is used after glaucoma surgery to decrease the resistance to outflow under the scleral flap. Yu *et al.* determined the effect of ASOCT in decision making by comparing two observer decisions and seven eyes, each from a different patient. Yu *et al.* determined that all the patients had poorly controlled IOP. ASOCT produces images of the sclera flap, the sclera, the internal ostium, the bleb cavity, and the bleb wall. Using clinical examination alone, LSL was recommended to all the patients due to the presence of a deep AC and a poorly formed bleb, whereas, using ASOCT imaging, LSL was recommended to only five of the seven patients. The two cases that did not required LSL indicated good aqueous flow through the bleb. ASOCT showed the position of sclera flap relative to the sclera, the presence or absence of a patent hyporeflective sub flap space, and bleb wall thickening, none of which were seen with the clinical examination at the slit lamp. As shown in this case study, images obtained using ASOCT impact clinician decisions of whether or not to undertake LSL. Therefore, ASOCT imaging can save the resources, money, and time of patients and ophthalmologist.

3.4.4.2. *Computational decision support system: RD*

RD is an eye disease that occurs when the sensory and pigment layers of the retina separate. If untreated, RD can lead to vision loss or blindness, and is considered an ocular emergency that requires immediate medical attention. RD generally occurs in middle age and elderly people.

The types of RD include rhegmatogenous RD, exudative RD, and tractional RD. The symptoms of RD include heaviness in the eye, an increased number of floaters, a sudden decrease in vision, and central visual loss. Treatment can take many forms, including cryopexy and laser photocoagulation, scleral buckle surgery, and pneumatic retinopexy. RD can be prevented by avoiding direct trauma to the eye, cataract surgery, and a sudden increase or decrease in eye pressure.

3.4.4.3. *Role of computational system*

A computational system can predict the development of proliferative vitreoretinopathy (PVR) in patients who are experiencing rhegmatogenous RD. PVR is the predominant cause of failure of RD surgery.[35] Therefore, by identifying the high risk for PVR, a computational system will help

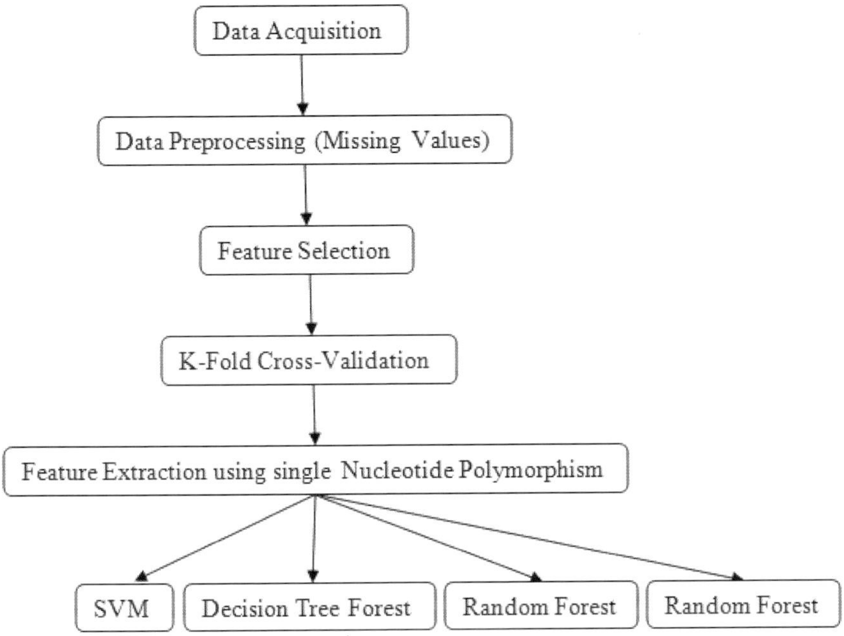

Fig. 3.16. Computational steps for diagnosing RD.

surgeons select specific treatments. Even with advancements in RD surgery, the risk of RD is high in patients with established PVR. A computational system will aid clinicians in identifying PVR and avoiding risks involved in surgery. The computational procedure for diagnosing PVR is described in Fig. 3.16.

First, data are obtained from patient gene expressions or genomic data from those patients who have undergone rhegmatogenous and have a higher PVR and from those patients who have healthy eyes.

Second, missing values in the gene dataset are determined using the KNN impute function. KNN replaces the missing values from the nearest neighbor column, which is computed using Euclidean distance. If the nearest neighbor column also has missing values, then the next neighbor will be used. Once all missing values in a gene are filled, then the feature extraction is performed.

Third, a training set will be developed. The complete gene dataset will be used to train the model for classifiers or to construct the model, so that it can be used to test the data. The training set can also be used to classify the healthy eyes and the diseased eyes.

Fourth, feature extraction will be performed to balance the large number of features present in the genome dataset with the small number of samples. Feature extraction can be performed by SNP measurement, so that SNP helps in dimensionality reduction. Examples of features extracted are CTGF, PDGF, PDGFR_, PI3KCG, EGF, FGF2, MIF, MMP2, and MMP7. The main objective of SNP is to eliminate irrelevant features, which highlights important information about the disease and reduces the cost of storing a large dataset. SNP also helps to extract hidden relationships and correlations.

Fifth, feature selection will be performed. The final aim of SNP is to have a good disease classifier and to minimize the prediction error of the model. The information gain method is used to reduce the uncertainty degree in the data. By selecting important features, the clinician speeds up the learning process or removes the effect of the curse of dimensionality. Feature selection can be performed by ranking features so that the clinician can determine the rank or cost of the features or the attributes. Based on the rank, the clinician can select the features. Methods by which the clinician can rank his or her features are ChiSquared attribute, GainRatio attribute, and infogain attribute.

Sixth, K fold cross-validation using classifiers will be performed. Classification can be performed using 10-fold cross-validation. In 10-fold cross-validation, the dataset or training data are partitioned into 10 sub datasets. From the 10 datasets, a single dataset is selected as the testing the model; the remaining nine datasets are used for training. The cross-validation process is then repeated once with each of the 10 datasets. The 10 results from the folds can then be averaged to produce a single result. The classifiers that can be used for cross-validation are SVM, Naïve Bayes, Random Forest, and Decision Tree. Naïve Bayes uses values of each SNP in patients known to have PVR. This information is used to predict the class of a new patient and assumes that each SNP is independent from every other SNP. SVM constructs hyperplanes in the n-dimensional space of the input data and optimizes the separation between the diseased eye data and the healthy eye data. A decision tree starts as a single node, representing the training samples. If all the samples are from the same class, then the node becomes a leaf and is represented by a category or a class.[36] This category can be the healthy eye class or the diseased eye class. The random forest model is based on a large number of decision trees. To classify a new patient, each tree provides a classification, and the forest chooses the classification with the most samples over all the trees in the forest.

Seventh, the ophthalmologist will use the GUI. The GUI is an automatic and interactive system that aids clinicians in making important clinical decisions.

3.4.5. *Computational Decision Support System: Amblyopia*

Amblyopia is an eye disease that results in poor or indistinct vision. It is caused by either no transmission or poor transmission of the visual image to the brain for a sustained period during early childhood.[37,38] Detecting the disease at an early stage can increase the chances of successful treatment. Amblyopia can be corrected by patching the good eye, or by instilling topical atropine in the eye with better vision. Symptoms of amblyopia include visual disorders, poor depth perception, low sensitivity to contrast and motion, and difficulty seeing in 3D images. Amblyopia may be strabismic, anisometropic, or occlusional.

3.4.5.1. *Role of computational decision support system in amblyopia*

The interactive and automatic system of amblyopia helps surgeons to diagnose and treat amblyopia using a computational system such as multimedia technology, artificial intelligence (fuzzy logic), and neurophysiology in patients, such as babies, who are unable to describe their symptoms.[39] Traditional treatments do not easily check gray vision, stereo vision, or confluent vision, and a 3D computational system graph can fill these diagnostic voids and treat early amblyopia in children. The computational steps for amblyopia are described below in Fig. 3.17.

After diagnosing amblyopia, an ophthalmologist determines the type of amblyopia that the patient has, and adds an optimum treatment into the database. Next, this data is distributed to other repositories through servers. The database and the analytical system in the patient data and

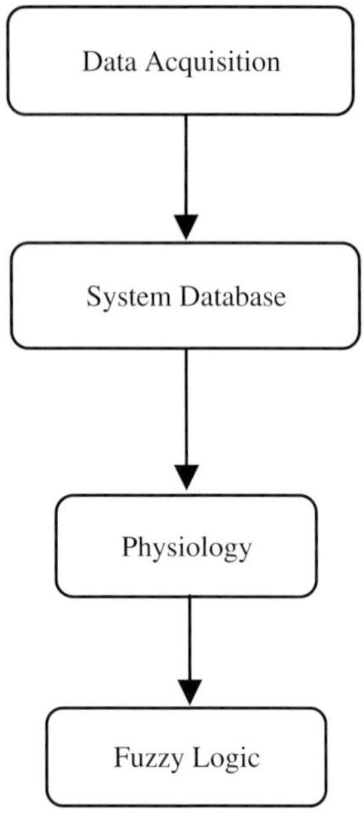

Fig. 3.17. Computational steps for diagnosing amblyopia.

expert system provide the physician with the tools to treat the amblyopia efficiently.

As mentioned above, amblyopia develops during childhood. In healthy eyes, the brain receives images from each eye and merges them into a single image. When a patient develops amblyopia, the brain cannot merge the images it receives from the eyes, and the brain's vision for an eye stops responding. As time progresses, vision loss occurs. Once vision loss occurs, it is difficult to cure the eye with glasses or surgery, because the disease is in the brain, not in the eye. Campbell and Hess presented a pulse method that can improve eyesight using the sensitiveness of cones to red light.[38,40] The parameters that will influence the effectiveness of the treatment are the frequency of light, the intensity of light, and the treatment time. Because amblyopia is physiologically connected to the brain, not limited to the eye, Campbell and Hess stimulated the amblyopia diseased eye from different directions by utilizing black-and-white bars with a different frequency. A thin bar indicated a higher space frequency, which made the center cell strengthen the development and improve eyesight. As shown in this example, the frequency of a bar influences the result of treatment. With a change in eyesight, the frequency of the bar should also change.

Through computer graphics and java, clinicians can develop a GUI, which can simulate the glimmer of red light and the rotation of the black-and-white bars on a computer screen. Different operations on the screen can be performed using the mouse. Computational methods also maintain the refresh rate of the red light and the continuity of the backgrounds.

The main functions of such a computational system in diagnosing amblyopia would be managing patient information, diagnosing the type of amblyopia, recording its symptoms, and determining proper treatment; managing the data of treatment by maintaining the database and the servers; and analyzing the results of the treatment by classifying and analyzing the type of amblyopia and the degree of amblyopia. The computational system would then compare the results with previous results.

This automatic and interactive system can help the clinicians find a case history similar to that of an undiagnosed patient by using the fuzzy logic IF…THEN function. For example, if a patient has amblyopia and a clinician wants to examine the past records of similar diseases, he or she can access them by using the IF…THEN function, specifying the symptoms and degree or type of amblyopia.

3.5. Conclusion

A computational diagnostic system will aid ophthalmologists in developing an efficient and cost-effective computer system that can provide support to clinicians or surgeons by lowering costs, saving time, making shorter waiting lists, and determining painless treatment and better care for patients. This computational decision-support system will provide information about the diseases, the most efficient diagnostic tests, and the most suitable therapies for optimal patient management. The suggested tests can then be performed to reduce the uncertainty about the patient's condition. The suggested therapies can be performed to aid in recovery and help the patient avoid visual loss.

References

1. Brown, C.G., Brown, M.M., Sharma, S., Brown, H., Smithen, L., Leeser, B.D., and Beauchamp, G. Value based medicine and ophthalmology: an appraisal of cost-utility analysis. *Trans Am Ophthalmol Soc* **102**:177–188, 2004.
2. Kymes, S.M. An introduction to decision analysis in the economic evaluation of the prevention and treatment of vision related diseases. *Ophthalmic Epidemiol* **15**:76–83, 2008.
3. Korah, S., Thomas, R., and Muliyil, J. An introduction to clinical decision analysis in ophthalmology. *Indian J Ophthalmol* **47**:41–48, 1999.
4. Acharya, U.R., Kannathal, N., Ng, E.Y.K., Min, L.C., and Suri, J.S. Computer-based classification of eye diseases. *28th Annual International Conference of the IEEE Biology Society and Engineering in Medicine, 2006. EMBS '06*, pp. 6121–6124, 2006.
5. http://www.octnews.org/.
6. Paunksnis, A., Barzdziukas, V., Jegelevicius, D., Kurapkiene, S., and Dzemyda, G. The use of information technology for diagnosis in ophthalmology. *J Telemed Telecare* **12**(Suppl 1):37–40, 2006.
7. Liyu, L. and Amin, M.G. Performance analysis of GPS receivers in non-Gaussian noise incorporating precorrelation filter and sampling rate. *IEEE Transaction on Signal Processing*, 2008.
8. Robertson, M.A. and Stevenson, R.L. DCT quantization noise in compressed images. *IEEE Trans Circuits Syst Video Technol* **15**:27–38, 2005.
9. Sadhar, S.I. and Rajagopalan, A.N. Image estimation in film-grain noise. *IEEE Signal Process Lett*, 2005.
10. Zhang, L., Junwei, C., Yue, L., and Wang, P. Face recognition using scale invariant feature transform and support vector machine. *The 9th International Conference for Young Computer Scientists* (ICYCS 2008), pp. 1766–1770, 2008.
11. Zhao, H., Kwok, P.C., and Yuen, J.T. A novel incremental principal component analysis and its application for face recognition. *IEEE Trans Syst Man Cybern B Cybern* **36**:873–876, 2006.
12. Fan-Hui K. Image retrieval using both color and texture features. *Machine Learning and Cybernetics, 2009 International Conference on*, Vol. 4, pp. 2228–2232, July 12–15, 2009.
13. Katzer, M., Kummert, F., and Sagerer, G. Methods for automatic microarray image segmentation. *IEEE Trans Nanobiosci* **2**:202–214, 2003.
14. Dai, L., Hu, J., and Liu, W.C. Using modified CHI square and rough set for text categorization with many redundant features. *International Symposium on Computational Intelligence and Design (ISCID '08)*, pp. 182–185, 2008.
15. Haung, P.-W. and Liu, C.L. Using genetic algorithms for feature selection in predicting financial distresses with support vector machines. *IEEE International Conference on Systems, Man and Cybernetics SMC '06*, pp. 4892–4897, 2006.
16. Mejia-Lavalle, M., Morales, E.F., and Rodriguez, G. Fast feature selection method for continuous attributes with nominal class. *Fifth Mexican International Conference on Artificial Intelligence (MICAI '06)*, pp. 142–150, 2006.

17. Cossock, D. and Zhang, T. Statistical analysis of Bayes optimal subset ranking. *IEEE Trans Inform Theory* **54**:5140–5154, 2008.
18. Radke, R.J., Andra, S., Al-Kofahi, O., and Roysam, B. Image change detection algorithms: a systematic survey. *IEEE Trans Image Process* **14**:294–307, 2005.
19. http://journals.lww.com/co-ophthalmology/Abstract/2003/02000/AdvanTec_Legacy_System_and_the_NeoSoniX_handpiece.5.aspx.
20. http://www.ophmanagement.com/article.aspx?article=85409.
21. Koreen, S., Gelman, R., Martinez-Perez, M.E., Jiang, L., Berrocal, A.M., Hess, D.J., Flynn, J.T., and Chiang, M.F. Evaluation of a computer-based system for plus disease diagnosis in retinopathy of prematurity. *Ophthalmology* **114**:e59–e67, 2007.
22. Lamard, M., Cazugue, G., Quellec, G., Bekri, L., Roux, C., and Cochener, B. Content based image retrieval based on wavelet transform coefficients distribution. In: *Proceedings of 29th Annual International Conference of the IEEE EMBS*, pp. 4532–4535, 2007.
23. Yagmurl, F., Karlik, B., and Okatani, A. Automatic recognition of retinopathy diseases by using wavelet based neural network. In: *Proceedings of Engineering in Medicine and Biology 27th Annual Conference*, pp. 454–457, 2008.
24. http://www.nlm.nih.gov/medlineplus/cataract.html.
25. Klein, B., Klein, R., Lee, K., and Grady, L. Statin use and incident nuclear cataract. *J Am Med Assoc* **295**:2752–2758, 2006.
26. Li, H., *et al.* Computerized systems for cataract grading. *2nd International Conference on Biomedical Engineering and Informatics BMEI '09*, pp. 1–4, 2009.
27. Moore, J. A cataract decision support system: its requirement is increasing. *Personalised Health Management Systems* **117**:147–153, 2005.
28. Hahn, S. Patient-centered communication to assess and enhance patient adherence to glaucoma medication. *Am Acad Opthalmol* 2009.
29. Song, X., Song, K., and Chen, Y. Computer-based diagnosis system for early glaucoma screening. In: *Proceedings of Engineering in Medicine and Biology 27th Annual Conference*, pp. 6608–6611, 2005.
30. Varachiu, N., Karanicolas, C., and Ulieru, M. Computational intelligence for medical knowledge acquisition with application to glaucoma. *First IEEE International Conference on Cognitive Informatics* (ICCI'02), 2002.
31. Varachiu, N., Karanicolas, C., and Ulieru, M. Computational intelligence for medical knowledge acquisition with application to glaucoma. In: *Proceedings of First IEEE International Conference on Cognitive Informatics (ICCI'02)*, pp. 233–238, 2002.
32. http://www.dryeyeinfo.org/.
33. Mathers, W.D. and Choi, D. Cluster analysis of patients with ocular surface disease, blepharitis and dry eye. *Arch Ophthalmol* **122**:1700–1704, 2004.
34. Singh, M., Aung, T., Aquino, M.C., and Chew, P.T.K. Utility of bleb imaging with anterior segment optical coherence tomography in clinical decision-making after trabeculectomy. *J Glaucoma* **18**:492–495, 2009.
35. Rojas, J., *et al.* Development of predictive models of proliferative vitreoretinopathy based on genetic variables: the retina 4. *Invest Ophthalmol Vis Sci* **50**:2384–2390, 2009.

36. Clemencon, S. and Vayatis, N. Tree-based ranking methods. *IEEE Trans Inform Theory* **55**:4316–4336, 2009.
37. http://en.wikipedia.org/wiki/Amblyopia.
38. Campbel, F.W., Hess, R.F., Watson, P.G. and Banks, R.V. Preliminary results of a physiologically based treatment of amblyopia. *Br J Ophthalmol* **62**:748–755, 1978.
39. Qiu, F., Wang, L., Li, H., and Liu, Y. Research on interactive system of amblyopia treatment. *The 1st International Conference on Bioinformatics and Biomedical Engineering, 2007 (ICBBE 2007)*, pp. 1177–1180, 2007.
40. Banks, R., Campbell, F.W., Hess, R.F., and Watson, P.G.A. New treatment for amblyopia. *Br Orthopt J* **35**:1–12, 1978.
41. Kamarthi, S.V. and Yu, G. Texture classification using wavelets with a cluster-based feature. Systems and Control in Aerospace and Astronautics, ISSCAA 2008. 2nd International Symposium on, 2008.

Chapter 4

Hyperspectral Image Analysis for Oxygen Saturation Automated Localization of the Eye

Bahram Khoobehi and James M. Beach

4.1. Introduction to Oxygen in the Retina

Both the retina and optic nerve head (ONH) have a high demand for oxygen, and changes in the supply of oxygen resulting from vascular disease play an important role in retinal and ONH pathology. The development of a non-invasive means of measuring oxygen saturation in the fundus of the human eye would be useful in the diagnosis and monitoring of numerous disorders. For example, the measurement of retinal and ONH oxygen saturation is essential for a better understanding of the relationship among oxygen consumption, activity, and metabolism, information that is vital to our understanding of diabetic retinopathy. In addition, accurate measurement of vascular efficiency in terms of oxygen saturation could be used to detect the very early onset of glaucoma, a disease in which early detection is crucial for effective treatment. Evidence that vascular inefficiency plays an important role in the pathogenesis of glaucomatous optic neuropathy has been accumulated. Because vascular inefficiency in the ONH resulting from various systemic disorders is directly related to blood flow in the ONH, which, in turn, is related to ONH tissue oxygenation, we propose to develop a practical system to evaluate oxygen saturation in the clinical setting using a recent innovation, hyperspectral imaging (HSI). The hyperspectral technique measures spectral changes within the visible and infrared

LSU Eye Center, Louisiana State University Health Sciences Center, New Orleans, LA, USA.

(IR) spectra and provides information on the molecular state of hemoglobin. We propose to adapt an existing prototype hyperspectral camera to perform clinically useful measurements of the oxygenation status of the retina and ONH to quantitate the role of hypoxia in ocular vascular disease. More specifically, our proposed experimental instrumentation will accommodate parallel studies of the oxygen saturation response to increased intraocular pressure (IOP) in normal monkeys and in monkeys with progressive stages of induced early phase glaucoma. The stimulus–response relationship is a stress–strain response to perturbations from the normal IOP (approximately 15 mmHg). In the absence of glaucomatous disease, autoregulation is expected to maintain normal volumetric blood flow and, hence, normal oxygenation in the ONH tissue up to a homeostatic threshold. It is presently believed that autoregulation is impaired in glaucomatous disease,[1–5] possibly because of anatomical vascular impairment of the retina and the ONH. It would be of considerable interest to determine if the threshold for autoregulation impairment is affected during the pre-onset stages of early phase glaucoma. The ONH has three distinct microcirculations: (1) the surface nerve fiber layer (NFL), which interfaces with the juxtaposed NFL and is predominantly nourished by the retinal arterioles; (2) the prelaminar region of the ONH, which is sandwiched between the surface NFL and the lamina cribrosa; and (3) the lamina cribrosa region, which is generally nourished by centripetal branches arising directly from the short posterior ciliary arteries.[4,6–8] With this new approach, we expect to determine how acute changes in IOP alone or in combination with chronic IOP elevation (glaucoma) affect these circulations independently and/or collectively. The proposed studies are motivated by the potential for the clinical application of this innovative technology in the early diagnosis and monitoring of therapy for ocular vascular diseases in which the associated hypoxia may eventually lead to loss of vision.

Oxygen delivery to tissue can be assessed in part by measuring the oxygen saturation of blood. Although a more complete assessment of tissue oxygen delivery requires blood flow measurements, specific ophthalmic disorders have been linked with abnormal blood oxygen saturation.[9–13] Several methods for evaluating the retinal oxygen saturation in retinal vessels have been reported based on spectrographic techniques in combination with photography, electro-optic detection, and digital imaging.[14–23]

In addition, phosphorescence,[24–27] oxygen microelectrodes,[28–30] and magnetic resonance imaging[31,32] methods have been developed to monitor oxygen tension (PO2) in structures of the retina. More recently, Khoobehi *et al.* describe a method for mapping relative oxygen saturation in the ONH, based on digital imagery obtained over a range of optical wavelengths by HSI methods.[33]

4.1.1. *Microelectrode Methods*

The most effective treatment for diabetic retinopathy is panretinal photocoagulation (PRP), which completely or partially stops neovascularization.[34,35] This treatment applies widespread grids of laser photocoagulation to destroy photoreceptors in the peripheral retina to prevent further damage of the central retina.

One hypothesis[36–38] states that the excess oxygen from the choroid no longer being consumed by the peripheral photoreceptors[39–42] will diffuse into the inner half of the retina. Without this oxygen diffusion, the possible hypoxia in the inner retina may cause capillary loss[43] and regression of neovascularization. Most studies for this hypothesis have only measured preretinal oxygen tension, or preretinal PO2, after PRP. According to studies in the cat retina, PO2 levels only increased after 100% O_2 breathing, not during air breathing.[36,44,45] In rabbits, there was an increase in preretinal PO2 after PRP during air breathing, but the rabbit retina has limited circulation and is metabolically different from the human retina.[46,47] Increased preretinal PO2 was also seen in miniature pigs after photocoagulation with[48] and without[49] venous occlusion.

A second hypothesis is that photocoagulation balances the levels of growth factors promoting or inhibiting angiogenesis.[43] This hypothesis is supported by microarray studies of changes in gene expression in the retina. However, the levels of growth factors may be a direct result of changing PO2 levels.[9]

The long-term effects of PRP on intraretinal PO2 are clinically relevant, yet previous experiments have only made measurements the same day when lesions were made or only measured preretinal PO2, which has more direct access to the oxygen supply. Linsenmeier measured intraretinal PO2 after PRP under normoxic conditions in vascularized retinas using O_2-sensitive microelectrodes inserted into the eye. The understanding of the mechanism of PRP may lead to more clinically appropriate treatments[50] and his measurement will help to understand PRP. However, it is too invasive for clinical measurement of O_2.

4.1.2. *Phosphorescence Dye Method*

R.D. Shonat and C.E. Riva have continued research with phosphorescence. C.E. Riva uses a less invasive phosphorescence method to measure PO2 in the retinal and choroidal vessels, as well as the microvasculature of the ONH rim. Phosphorescence is the emission of light during a transition from a long-lived, spin-forbidden, and excited triplet state to a ground state.[51–55] When a molecule capable of phosphorescence is excited into the triplet state, it may emit phosphorescence or transfer its energy to another molecule without light emission (quenching). In the blood, oxygen is the only significant quenching agent[56] with the degree of quenching dependent on the concentration of oxygen near the phosphorescent molecule. Therefore, by measuring the quenching effect, the intravascular PO2 can be determined.

A Pd complex of *meso-tetra* (4-carboxyphenyl) porphine (Porphyrin Products, Logan, UT) was used as the triplet-state oxygen probe. A Wild Macrozoom microscope with an epifluorescence attachment (Wild-Leitz USA, Malvern, PA) was used for excitation of the probe and collection of the phosphorescence. The phosphorescence resulting from a 45-W xenon flash lamp mounted on a Leitz lamp housing was observed with an intensified CCD camera (Xybion Electronic Systems, San Diego, CA) placed in the image plane of the microscope. The filter block inside the epifluorescence attachment permitted control of the excitation and collection wavelengths. Excitation was through one of two filter combinations, with the bandwidth covering one of the two excitation peaks of the probe. A long-pass cutoff filter with 50% transmission at 630 nm was used for the observation of the phosphorescence.

The phosphorescence lifetime can be calculated from images collected at different delay times after a flash. The microcomputer was programmed to trigger a flash, wait the prescribed delay time, turn on the CCD intensifier for 2.5 ms, digitize the image, and store it. Eight images were averaged for each delay time.

To calculate the lifetime and PO2, each image was passed through a two-dimensional (2D) smoothing filter, and the background image (at 2500 μs) was subtracted. For each pixel element, the integral from the prescribed delay time to infinity of the function of intensity versus delay time was constructed as follows: $I_0\tau\exp(-t_d/\tau)$, where I_0 is the intensity at time zero in arbitrary units, t_d is the delay time in microseconds, and τ is the lifetime in microseconds.[57] This technique requires the injection of a new generation of porphyrin dyes, which have not yet been approved for human use.

4.1.3. *Spectrographic Method*

Schweitzer used all the wavelengths, but the complete spectrum approach did not produce images based on these wavelengths.[58] He used a spectrograph capable of detecting the whole spectrum from a slit-image of the retina. The imaging ophthalmospectrometer was based on a modified fundus camera (RCM 250, Carl Zeiss, Jena, Germany). However, the spectrograph (CP 200, Jobin-Yvon, Longjumeau, France) and intensified CCD matrix camera (576S/RB, Princeton Instruments, Trenton, NJ, USA) were adapted to the ophthalmoscope. A conventional xenon flash lamp illuminated the fundus camera and a slit-like field stop can be switched into the illuminating beam. The illuminated field was confocally placed onto the entrance slit of the spectrograph. The influence of the intraocular scattered light was reduced by the confocal imaging principle. The light coming from the fundus was spectrally dispersed within the spectrograph, and the intensified CCD matrix camera detects a complete reflectance spectrum at the exit plane of the spectrograph in each line. The optical spectrum taken from the vessel blood column was used to determine oxygen saturation using curve fitting methods.[58]

4.1.4. *Three Wavelength Method*

Delori employed a three-wavelength spectrophotometric model to measure oxygen saturation in the artery and vein in the retina. The three-wavelength method developed by Pittman and Duling and advanced by Zheng *et al.*[59–61] to measure the percentage of oxyhemoglobin in the microvascular system was expanded by Delori to use in the retina. The method is based on the difference in the light absorption of oxygenated and deoxygenated hemoglobin. The specific extinction coefficient is given by[16]

$$E_\lambda = E_{o,\lambda}(\mathrm{O_2\,sat}) + E_{r,\lambda}(1 - \mathrm{O_2\,sat}), \tag{4.1}$$

which can be written as:

$$E_\lambda = E_{r,\lambda} + \Delta_\lambda(\mathrm{O_2sat}). \tag{4.2}$$

In these formulas, E_λ is the specific extinction coefficient for a mixture of HbO_2 and Hb (cm^2/μmole), $E_{o,\lambda}$ is the specific extinction coefficient of HbO_2 (cm^2/μmole), $E_{r,\lambda}$ is the specific extinction coefficient of Hb (cm^2/μmole), and $\Delta_\lambda \equiv E_{o,\lambda} - E_{r,\lambda}$. Then, Delori defined the optical density (OD) as follows:

$$D_\lambda = S + sCd(E_{r,\lambda} + \Delta_\lambda(\mathrm{O_2sat})), \tag{4.3}$$

where D_λ is the OD, C the total hemoglobin concentration (μmole/cm^3), d the path length (cm), S and s wavelength independent parameters that describe the light scattering by red blood cells.

D_λ can be measured, and it is equal to the negative logarithm of the intensity of reflection at the vessel divided by the intensity of the reflection of the background using

$$D_\lambda = -\log(I_{\mathrm{in}}/I_{\mathrm{out}}). \tag{4.4}$$

By measuring the OD with at least three wavelengths, Delori solved unknown parameters such as sCd and O_2sat.[16]

The techniques of Delori and Schweitzer are highly innovative and measure the blood oxygen saturation in single retinal vessels. Drawing from these works, a next step was to determine oxygen saturation from all vessels contained in images of the funds. A new method combining spectroscopy and imaging was needed.

4.1.5. *Dual Wavelength Technique Using OD Ratio (ODR)*

Hickam *et al.* first employed two wavelengths using photographic film recordings to measure retinal vessel saturation.[17] Beach pioneered a two-wavelength method for digital oximetry in human subjects, which has been adopted by commercial imaging companies. The ratio of the ODs of blood at two wavelengths was shown to bear an approximately linear relationship to hemoglobin oxygen saturation.[14,62] Laing *et al.* later showed that a linear relationship held over the physiological range of oxygen saturation in their falling rabbit paper.[19] In the digital method, two high-resolution digital images of the retinal vessels were obtained simultaneously with a single pulse of light. The longer of the wavelengths, at 600 nm, was sensitive to oxygen saturation, while the shorter, at 569 nm, was recorded at a wavelength in which oxy- and deoxyhemoglobins had equal light absorbance. Vessel ODs were found for both wavelengths with vessel-tracking algorithms that used the light reflected outside the vessel to estimate incident light on the vessel (I_{out}), and light reflected from inside vessels (I_{in}), as a measure of the amount of light absorption in the blood[63,64]:

$$\text{OD}_{\text{vessel}} = \log_{10}(I_{\text{out}}/I_{\text{in}}).$$

At 600 nm, OD was strongly dependent on the amount of oxyhemoglobin present, while at 569 nm, the OD was independent of saturation. Thus, the ODR provided a determination for blood oxygen saturation that compensated for the differences in vessel diameter and small amounts of misfocus that occurred in retinal images. The ODR was determined from the ratio of the vessel ODs found over the same vessel segments in the image: $\text{ODR} = \text{OD}_{600}/\text{OD}_{569}$.

With this analysis, the ODR varied linearly with blood oxygen saturation measured by pulse oximetry. In Fig. 4.1, the ODR was plotted against systemic blood oxygen saturation for seven subjects. The relationship between the ODR and saturation was offset by different degrees across subjects, and some of the slopes differed significantly from the mean.

The results were partly explained by differing degrees of retinal pigmentation in subjects. The retinal pigment density, which varies across individuals, can affect the light returned from outside the vessels, mostly at the shorter wavelength, and, thus, a pigment correction was added by employing the longer wavelength, where the retinal pigment influence is small, to estimate incident light for vessels in both images. To apply this correction, the relative light throughput for each imaging channel had to be known. This relationship was characterized using the ratio (η) of recorded light intensity at each wavelength that was returned from a diffuse reflecting surface. A pigment-corrected OD was, thus, found for

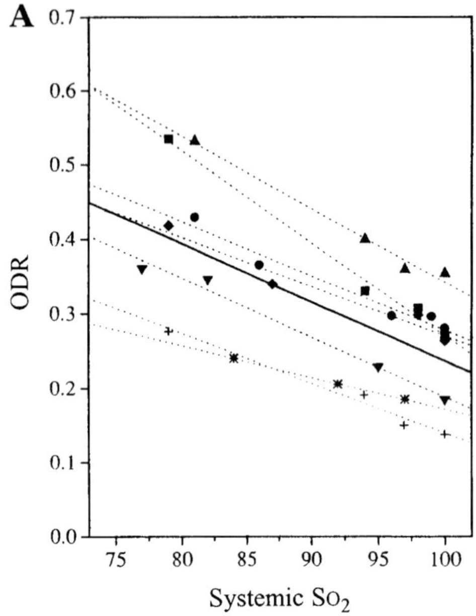

Fig. 4.1. ODRs from retinal arteries vs. systemic SO_2 (seven subjects). ORD obtained from direct calculation of vessel OD by using method I. Dashed lines, line fits through points from each subject: solid line, line obtained by averaging slope and intercept from all subjects: 000787 +/− 0.00098 (SE). Reprinted with permission from Beach, J.M., Schwenzer, K.J., Srinivas, S., Kim, D., and Tiedeman, J.S. Oximetry of retinal vessels by dual-wavelength imaging: calibration and influence of pigmentation. *J Appl Physiol* **86**:748–758, 1999.

the oxygen-insensitive image and used to find the pigment corrected ODR: $OD_{cor,569} = \log 10\,(\eta I\text{out}_{600}/I\text{in}_{569})$, $ODR_{cor} = OD_{600}/OD_{cor,569}$.

The corrected ODR was plotted against the pulse oximeter reading below for the same subjects (Fig. 4.2):

The correction reduced the coefficient of variation for the ODR across subjects by approximately two-fold. The ODR method reproducibly measured saturation in the larger first- and second-order retinal arteries and veins. Calibration was done empirically with linear fits between artery ODRs and pulse oximetry using inhalation of different concentrations of oxygen. Jim Tiedeman and colleagues in a study of diabetic patients without background retinopathy used an early version of the oximeter.[13] This study showed that the retinal vein oxygen saturation was reduced during acute hyperglycemia, and the degree of reduction was greater in

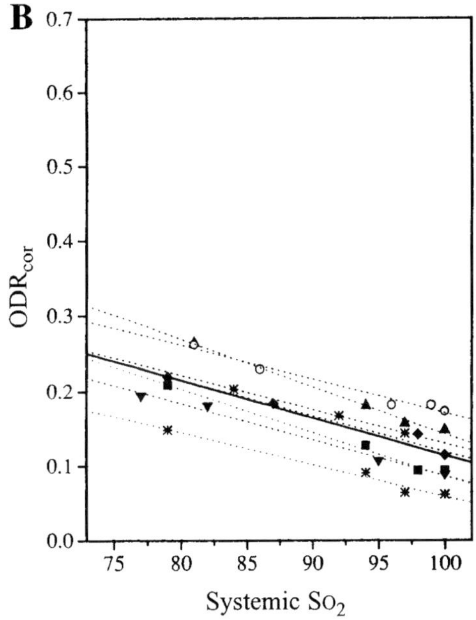

Fig. 4.2. ODRs from retinal arteries vs. systemic SO_2 (seven subjects). ODR corrected for retinal pigmentation ODR_{cor} by using method II. Dashed lines, line fits through points from each subject; solid line, line obtained by averaging slope and intercept from all subjects. 0.00504 +/− 0.00029 (SE) for method II. Reprinted with permission from Beach, J.M., Schwenzer, K.J., Srinivas, S., Kim, D., and Tiedeman, J.S. Oximetry of retinal vessels by dual-wavelength imaging: calibration and influence of pigmentation. *J Appl Physiol* **86**:748–758, 1999.

patients with longer disease duration, suggesting that the autoregulatory response to high sugar becomes impaired. One explanation is that inner retinal oxygen consumption is increased by high levels of sugar, and in the disease condition, autoregulatory responses are not able to increase blood flow to meet demand. Beach *et al.* collaborated in further developments, which produced an automated retinal oximeter that could identify the saturation in individual vessels without user interaction.[64] Specialized software automatically selects measurement points on the oximetry images and calculates the OD (absorbance) of retinal vessels at two wavelengths. Stefansson *et al.* have applied the automated technique to clinical and vision studies.[64,65] Martin Hammer and colleagues have done similar research. Hammer's method is of special significance and he describes it as follows.[66]

"The calculation of the oxygen saturation SO_2 was adopted from Beach *et al.*[14] but slightly modified to:

$$SO_2 = 100\% - (\text{ODR} - \text{ODR}_{a,100\%})/\text{os."} \qquad (4.5)$$

"Here, ODR is the corrected optical density ratio

$$\text{ODR} = \text{Log}(I_{\text{out},610}/I_{\text{in},610})\ \text{Log}((\eta)I_{\text{out},610}/I_{\text{in},548}),\text{"} \qquad (4.6)$$

"where I_{in} and I_{out} are the intensities of reflected light inside and outside a vessel at the indexed wavelengths measured as gray values in the respective images and η is the ratio of the intensities measured at an ideal white reflector (Spectralon, Labsphere Inc., North Sutton, NH) at 548 nm and 610 nm. In Equation [(4.5)], $\text{ODR}_{a,100}$ is an offset which was introduced as the ODR at the arteriole during inhalation of pure oxygen by Beach *et al.*,"[14] "but was set constant and determined experimentally here. OS is the oxygen sensitivity, another constant to be determined. Since we found linear dependences of the SO_2 — value on the vessel diameter as well as on the fundus pigmentation (see results section), we introduced compensation terms into the oximetry equation finally resulting in

$$SO_2 = 100\% - (\text{ODR} - \text{ODR}_{a,100\%})/\text{os} - (a - VD)(b) + (c - \log(I_{\text{out},610}/I_{\text{out},548}))(d)\text{."} \qquad (4.7)$$

"Here, VD is the diameter of the vessel in microns, measured from the image,[67] and

$$\log(I_{\text{out},610}/I_{\text{out},548})\text{"} \qquad (4.8)$$

"represents the fundus pigmentation by melanin, which extinction decreases approximately linear with the wavelength in the considered spectral range".[68] "The constants *a*, *b*, *c*, and *d* were determined experimentally from measurements in twenty healthy volunteers (see below). User-friendly software, easy to operate by clinicians and researchers, comprises the following algorithm: In an image, obtained by the camera and filter assembly described as the oximeter above, the operator has to mark the vessel of interest by a mouse click. The vessel is traced automatically; the vessel walls are located as photometric edges in the green channel image, and the vessel is segmented according to changes of its direction at the fundus. For each vessel segment, the pixels inside and outside the vessel are determined by an iterative threshold procedure starting from the vessel walls. The gray-scale values of the pixels inside and outside the vessel in the green and red camera channels are averaged and used as the respective intensities I in Eq. (4.7) for the calculation of the oxygen saturation. Finally, the SO_2-values are averaged over the vessel. In order to observe changes of the SO_2 along a vessel, the measurement can be restricted to an area of interest defined by the operator".[68]

4.1.6. *HSI Method*

Imaging the retina and ONH with a modified fundus camera can provide a noninvasive method of monitoring oxygenation. The following experiments use this HSI technique (described in detail later) to measure oxygenation in anesthetized cynomolgus monkeys. However, a limitation is that these imaging techniques are not yet quick enough to image the conscious human eye due to the drawback of involuntary movement.

4.2. Experiment One

4.2.1. *Methods and Materials*

4.2.1.1. *Animals*

Two normal cynomolgus monkeys with normal eyes were used. The monkeys were anesthetized and their eyes dilated. We placed a contact lens on the cornea to prevent drying and obtained reflectance HSI measurements in one eye of each monkey.

4.2.1.2. *Systemic oxygen saturation*

An ear oximeter probe (model 3700; Ohmeda, Wallingford, CT) was placed on the monkey's earlobe to measure systemic oxygenation. A tracheal tube was positioned at the trachea and connected to a small-animal breathing chamber (Quantiflex; MDS Matrx Co., New York, NY). The oxygen chamber was supplied through a pressure regulator from an oxygen tank at a rate of 3 L/min at atmospheric pressure. This procedure brought the oximeter reading to 100% saturation, and hyperspectral images were obtained while the monkey breathed room air and during inspiration of pure oxygen.

4.2.1.3. *Intraocular pressure*

To raise IOP, we inserted a 27-gauge needle into the anterior chamber under slit lamp examination. The needle was connected to a 500-mL reservoir containing saline solution with 0.1 mL gentamicin (40 mg/mL), 0.03 mL clindamycin (150 mg/mL), and 4 mL dexamethasone (4 mg/mL). IOP was raised by elevating the reservoir. IOP was monitored by means of a tonometer (Tonopen XL; Medtronic, Jacksonville, FL). Imaging was performed at normal (15 $\pm$ 2 mmHg) and high (60 $\pm$ 2 mmHg) IOPs.

4.2.1.4. *Fundus camera*

Images were acquired with a fundus camera (TRC-50vt; Topcon, Tokyo, Japan), with a lens and a c-mount through the vertical path of the camera. Hyperspectral images were obtained through the vertical viewing port using an imaging spectrograph and digital camera. Figure 4.3 shows the components and position of the hyperspectral imager on the fundus camera. Vertical mounting facilitated image scanning by maintaining the center of gravity of the moving components over the line of travel. A sleeve held the system at the proper height to sample the focused image. The entrance slit of the spectrograph was placed at a conjugated image plane of the eye

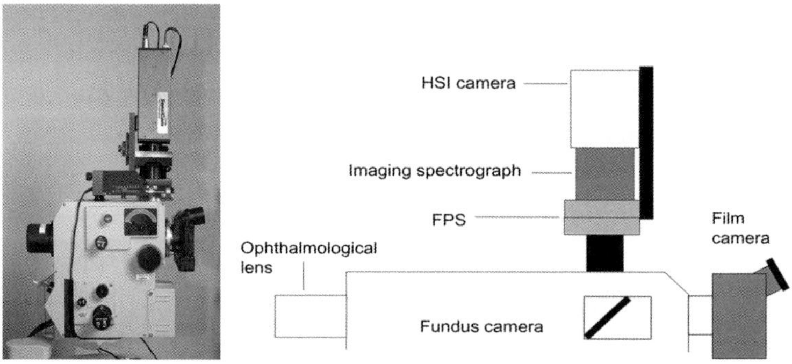

Fig. 4.3. HSI system in relation to the fundus camera. The image normally forms at the film camera port. During HSI, the image is redirected upward by a mirror. The imaging system is translated over the camera port by a linear actuator mounted below the imaging spectrograph and CCD camera, FPS, focal plane scanner. Reprinted with permission from Khoobehi, B., Beach, J., and Kawano, H. Hyperspectral imaging for measurement of oxygen saturation in the optic nerve head. *Invest Ophthalmol Vis Sci* **45**:1464–1472, 2004. ©2004 Association for Research in Vision and Ophthalmology.

fundus with the aid of the lens and c-mount. The tungsten aiming light of the fundus camera provided illumination.

4.2.1.5. *Hyperspectral imaging*

The hyperspectral images were obtained by translating an imaging spectrometer and charged coupled device (CCD) camera (model VNIR 100; Photon Industries Inc., Stennis Space Center, MS) across the fundus image (Fig. 4.4). Images of the back of the eye were acquired using the 35° viewing mode of the fundus camera. The image from the vertical camera port was focused onto the entrance slit of the spectrograph. The output spectrum was in turn focused onto the CCD image sensor. This arrangement caused the spectrum of all points along a line in the fundus image to be recorded in a single CCD frame. Two spatial and four spectral pixels were binned together to give spectral images containing 512 spatial points and 256 spectral bands.

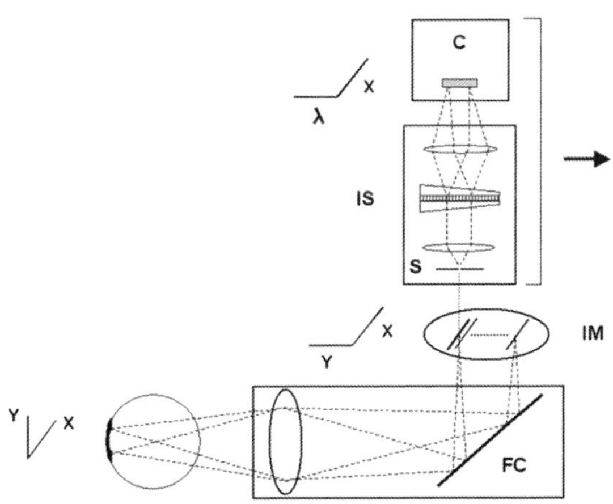

Fig. 4.4. Optical diagram of the retinal hyperspectral imager. The area of interest on the retina is imaged with a fundus camera (FC). Dotted lines represent the light collection path only. The intermediate image (IM) is formed at the slit (S) of an imaging spectrograph (IS). The spectrograph is drawn above the image for clarity. The output spectrum is focused on the sensor of a CCD camera (C). As the spectrograph and camera are translated along the y-axis, the spectrum from points on consecutive lines of the image is recorded in a series of frames. Motion is controlled to create a 1:1 aspect ratio between adjacent pixels in the x direction and lines in the y direction. Reprinted with permission from Khoobehi, B., Beach, J., and Kawano, H. Hyperspectral imaging for measurement of oxygen saturation in the optic nerve head. *Invest Ophthalmol Vis Sci* **45**:1464–1472, 2004. ©2004 Association for Research in Vision and Ophthalmology.

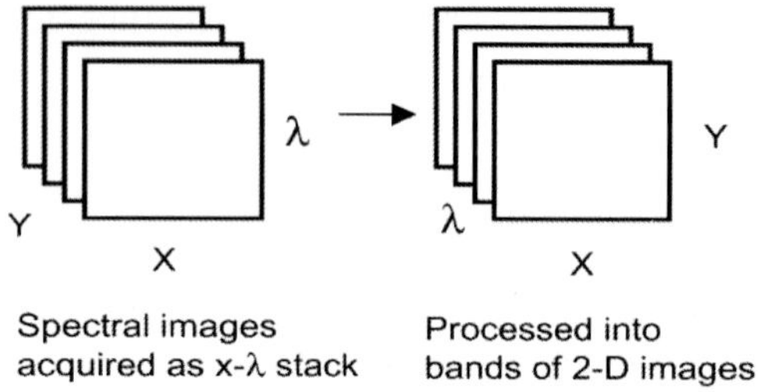

Fig. 4.5. Organization of spatial (x, y) and spectral (λ) information in acquired image frames (left) and after conversion to band-sequential images (right). Reprinted with permission from Khoobehi, B., Beach, J., and Kawano, H. Hyperspectral imaging for measurement of oxygen saturation in the optic nerve head. *Invest Ophthalmol Vis Sci* **45**:1464–1472, 2004. ©2004 Association for Research in Vision and Ophthalmology.

The second spatial dimension was obtained by translating the imaging system at constant velocity in the direction transverse to the orientation of the slit. Each frame of Fig. 4.5 holds the spatial (x) and spectral (λ) axes for each line of the acquired hyperspectral image, with successive lines forming the z-axis in the stack of frames. A "band-sequential" hyperspectral image is obtained by the rotation of the stack of images, interchanging the z- and λ-axes. After rotation, each frame contains a 2D spatial image at a distinct wavelength in which intact structures are recognizable.

4.2.1.6. *Extraction of spectral curves*

Images were corrected for dark values by subtracting an image obtained after blocking illumination. A five-point moving average filter was applied to individual curves of each time point, and the smoothed data were then averaged to obtain final curves that represent the spectral signatures obtained before the application of high oxygen, after the application of high oxygen, and before the high IOP.

4.2.1.7. *Mapping relative oxygen saturation*

Relative saturation was assessed from amplitudes of the hemoglobin spectral signatures that were contained in the reflectance spectra from retinal blood. Spectral minima at 542 and 577 nm from oxygenated hemoglobin (HbO_2 spectral signature) were converted to a single minimum at 555 nm from deoxyhemoglobin (Hb spectral signature). Isosbestic points at 530,

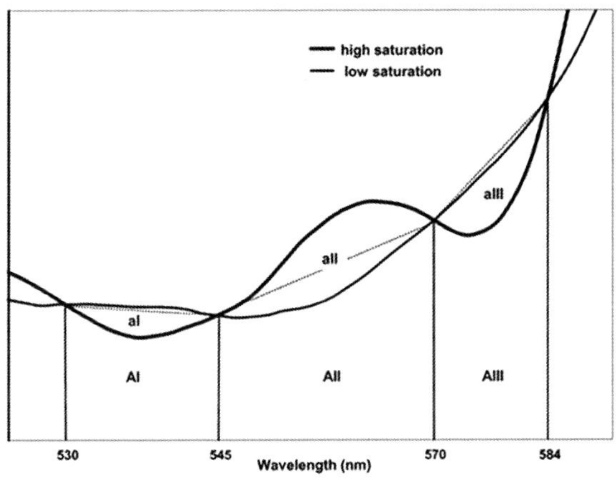

Fig. 4.6. Reflectance spectra of saturated blood (HbO_2 signature, bold curve) and desaturated blood (Hb signature, thin curve) from retinal recordings. The HbO_2 curve contains two minima corresponding to wavelengths of peak light absorption. The Hb curve contains a single broad minimum. The sloping baseline causes minima wavelengths to be shifted slightly to shorter wavelengths. Vertical lines extend from the axis to wavelengths where HbO_2 and Hb have equal reflectance and absorbance (isosbestic points of hemoglobin in distilled water). Dotted lines connect pairs of isosbestic points. Regions I, II, and III are defined between isosbestic points. The areas between the curves and the dotted lines are denoted as aI, aII, and aIII. AI, AII, and AIII represent the areas under the dotted lines. Saturation maps were determined from region II (partial signature) and from the combination of regions I, II, and III (full signature). Reprinted with permission from Khoobehi, B., Beach, J., and Kawano, H. Hyperspectral imaging for measurement of oxygen saturation in the optic nerve head. *Invest Ophthalmol Vis Sci* **45**:1464–1472, 2004. ©2004 Association for Research in Vision and Ophthalmology.

545, 570, and 584 nm were selected from recorded spectra. The curve of saturated blood passed above the line that connects the points at 545 and 570 nm (region II). The curve moved toward the line and had the ability to pass below the line as the blood became more desaturated. This area between the curve and line (Fig. 4.6, aII) was the largest for 100% saturation and decreased, eventually changing sign, as the blood became desaturated. Changes in the total reflectance from different recordings were compensated for by dividing this saturation-sensitive area by the total area under the line. A partial-signature map of relative oxygen saturation was found from the ratio of these saturation-dependent and saturation-independent areas (Fig. 4.6, aII/AII). The term partial signature refers to the use of only the region of the spectrum between the second pair of isosbestic points. The use of the full signature gave a larger range of values for the same change in saturation and tended to average noise to a greater extent. For each type of map, values representing low-to-high saturation were color-coded as blue, green, yellow, and red. Because spectral changes were referenced to isosbestic points, this method should minimize errors contributed by variation in the slope of the spectral baseline from different recording sites.

4.2.1.8. *Relative saturation indices (RSIs)*

An index of the relative oxygen saturation (RSI) was determined from separate regions of the hyperspectral image containing artery, vein, and the selected areas of the ONH and shown in Fig. 4.7 and Table 4.1. The

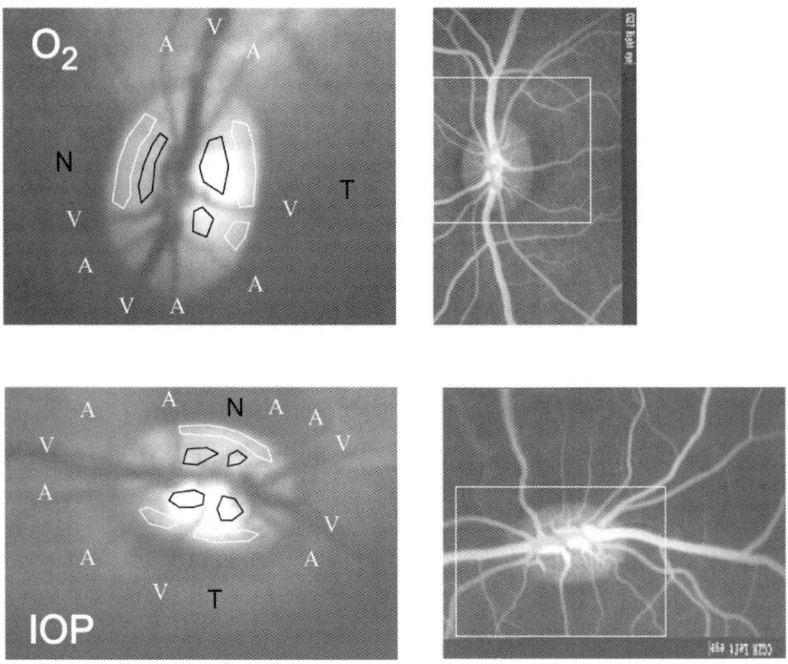

Fig. 4.7. Single-band images (570 nm) of the ONH and vessels from hyperspectral images obtained for oxygen breathing (top) and IOP (bottom) experiments. At right are fluorescein angiograms (venous phase) used to confirm vessel type for each experiment. RSIs in Figs. 4.8 and 4.12 were determined from retinal vessels marked as A (artery) and V (vein), and from the ONH inside areas bounded with white lines (rim) and black lines (cup). Nasal and temporal aspects of the ONH are labeled N and T, respectively. Reprinted with permission from Khoobehi, B., Beach, J., and Kawano, H. Hyperspectral imaging for measurement of oxygen saturation in the optic nerve head. *Invest Ophthalmol Vis Sci* **45**:1464–1472, 2004. ©2004 Association for Research in Vision and Ophthalmology.

Table 4.1. RSIs from the oxygen experiment.

Condition	Artery	Vein	Optic nerve head		
			Temporal	Nasal	Average
Room air*	0.261 ± 0.007	0.159 ± 0.001	0.137 ± 0.007	0.142 ± 0.013	0.139 ± 0.010
Pure oxygen†	0.312 ± 0.002	0.256 ± 0.009	0.151 ± 0.004	0.180 ± 0.007	0.166 ± 0.006
Difference‡	0.051 ± 0.009	0.097 ± 0.010	0.014 ± 0.011	0.038 ± 0.020	0.027 ± 0.016

Unpaired samples of equal variance.
*Average over two-time points during room air breathing.
†Average over five-time points at high O_2.
‡All differences are significant ($P < 0.05$). Data are the means ± S.D.

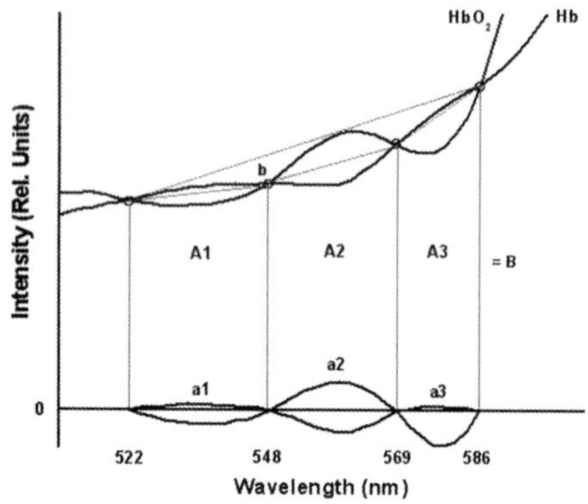

Fig. 4.8. Graphic depiction of the RSI algorithm, showing reference spectral curves for oxyhemoglobin (HbO_2) and deoxyhemoglobin (Hb) and components of the algorithm. Isosbestic points (solid circles) where curves intersect divide the curves into three bands. Short line segments (gray lines) connect isosbestic points. Within each band, areas between the curve and line segment are sensitive to saturation. Polygonal areas under the line segments (A_1, A_2, and A_3) compensate for total light intensity (see text). HbO_2 and Hb curve segments are plotted below after subtracting the polygonal areas, showing the limits of the oxygen-sensitive areas (a_1, a_2, and a_3) for complete saturation and desaturation. The oxygen-sensitive component (OSC) is determined from ratios of the oxygen-sensitive areas, a, and total light areas, A (see text). From the area between the hemoglobin curve and the long gray line segment connecting the first and last isosbestic points (b), and the area under the long gray line segment (B), the blood volume component (BVC) is determined (see text). Reprinted with permission from Khoobehi, B., Beach, J., and Kawano, H. Hyperspectral imaging for measurement of oxygen saturation in the optic nerve head. *Invest Ophthalmol Vis Sci* **45**:1464–1472, 2004. ©2004 Association for Research in Vision and Ophthalmology.

RSIs from retinal vessels and ONH for IOP experiment are shown in Fig. 4.8 and Table 4.2.

4.2.2. *Results*

Figure 4.7 shows the area of the ONH obtained from the 570-nm band in the hyperspectral image for the oxygen concentration images (Fig. 4.7, top-left) and the variable IOP images (Fig. 4.7, bottom-left). Confirmation of vessel type was done by fluorescein angiography (images at right from the venous phase).

4.2.2.1. *Spectral signatures*

Increased oxygen saturation is indicated in Figs. 4.9 and 4.10 when the experimental spectrum changes to match more closely the HbO_2 signature of

Table 4.2. RSIs from the IOP experiment.

Condition	Artery	Vein	Averaged ONH	Temporal cup
Normal IOP*	0.210 ± 0.008	0.139 ± 0.005	0.101 ± 0.001	0.081 ± 0.001
High IOP†	0.030 ± 0.010	0.029 ± 0.010	0.041 ± 0.0002	0.054 ± 0.001
Difference‡	0.180 ± 0.018	0.110 ± 0.015	0.060 ± 0.003	0.027 ± 0.002

Unpaired samples of equal variance.
* Average over three-time points at normal IOP.
†Average over last two-time points at high IOP.
‡All differences are significant ($P < 0.05$). Data are the means ± S.D.
Note: Reprinted with permission from Khoobehi, B., Beach, J., and Kawano, H. Hyperspectral imaging for measurement of oxygen saturation in the optic nerve head. *Invest Ophthalmol Vis Sci* **45**:1464–1472, 2004. ©2004 Association for Research in Vision and Ophthalmology.

Fig. 4.6, with stronger minima at 542 and 577 nm. Desaturation is indicated when the curve more closely resembles the Hb signature having a single spectral minimum.

4.2.2.2. *Oxygen breathing*

The artery (top-left of Fig. 4.10) showed a small increase in the HbO_2 signature with pure O_2, relative to room air. In the vein (top-right), this increase was markedly larger. In the nasal and temporal ONH (bottom-left and -right), pure O_2 increased the HbO_2 signatures, but not to the degree observed in the vessels.

4.2.2.3. *Intraocular pressure*

In the artery (top-left of Fig. 4.10), high IOP sustained for five minutes gradually converted the HbO_2 signature to an Hb signature. In the vein (top-right), the normal IOP curve showed a weak HbO_2 signature. Within one minute after the onset of high IOP, however, the curve was converted to a strong Hb signature. These results suggest that high IOP causes desaturation of the retinal blood supply in both arteries and veins. Increased IOP resulted in only modest increases in total reflectance. High IOP reduced saturation in the ONH microcirculation but to a lesser degree than in the retinal circulation and suggested that saturation was partially restored in some regions.

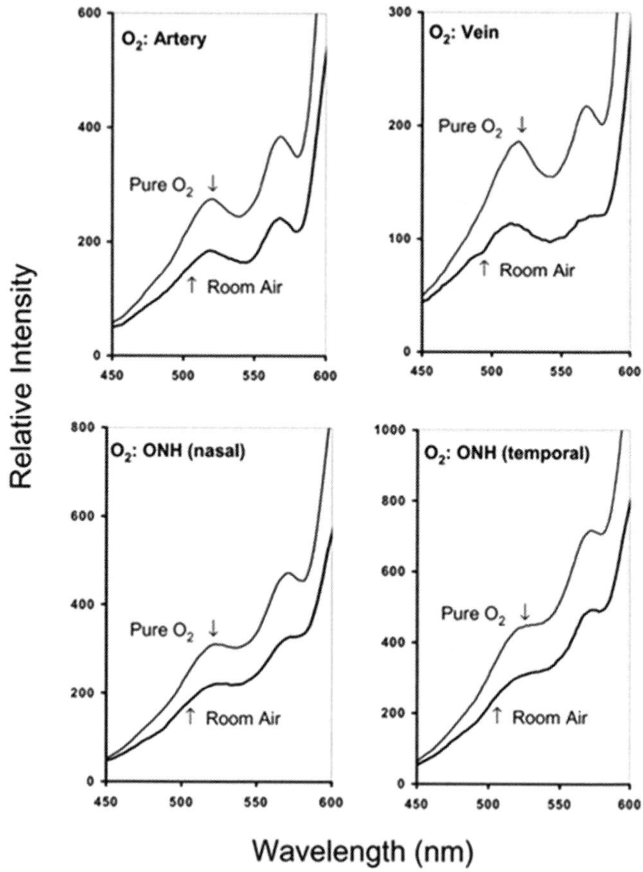

Fig. 4.9. Spectral curves during oxygen breathing experiment. Top row: artery (left) and vein (right). Bottom row: nasal ONH (left) and temporal ONH (right). Bottom curves: room air breathing; top curves: pure oxygen breathing. Reprinted with permission from Khoobehi, B., Beach, J., and Kawano, H. Hyperspectral imaging for measurement of oxygen saturation in the optic nerve head. *Invest Ophthalmol Vis Sci* **45**:1464–1472, 2004. ©2004 Association for Research in Vision and Ophthalmology.

4.2.2.4. *Responses to oxygen breathing*

Figure 4.11 shows spatial changes in the relative saturation of ONH structures during room air breathing (left) and two minutes after switching to pure oxygen (right). The partial-signature maps (Fig. 4.11A) reveal saturation differences; however, structures such as the large vein are more clearly delineated during high saturation in the full-signature maps (Fig. 4.11B, right). Under normal conditions, high saturation areas included outlines of

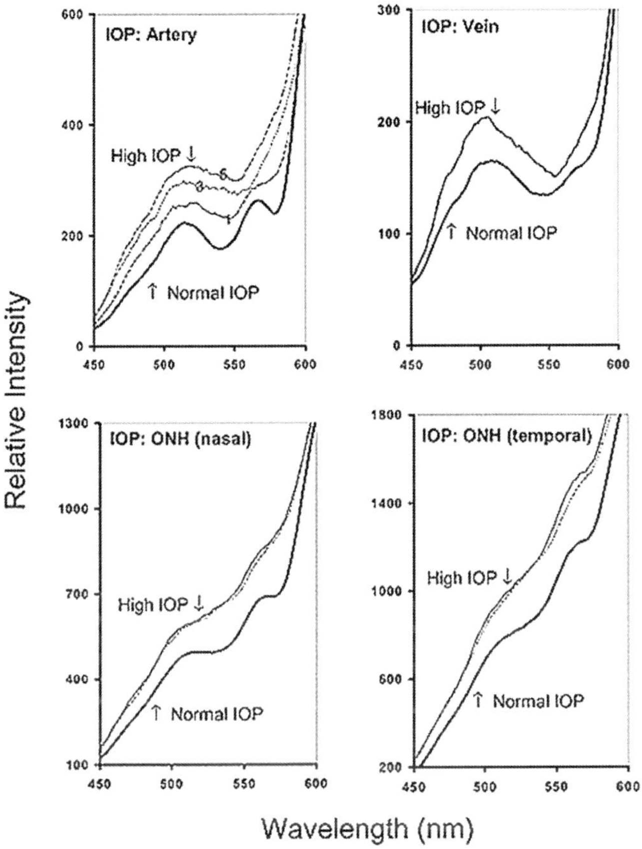

Fig. 4.10. Spectral curves from ONH during the IOP experiment. Top-left: artery at normal IOP and at 1, 3, and 5 minutes after IOP was increased to 60 mmHg; top-right: vein at normal IOP and 1 minute after pressure was increased to 60 mmHg. Bottom-left: nasal ONH; bottom-right: temporal ONH. For the high-IOP curves, the dotted line represents 1 minute and the solid line represents five minutes after IOP was increased to 60 mmHg. Reprinted with permission from Khoobehi, B., Beach, J., and Kawano, H. Hyperspectral imaging for measurement of oxygen saturation in the optic nerve head. *Invest Ophthalmol Vis Sci* **45**:1464–1472, 2004. ©2004 Association for Research in Vision and Ophthalmology. .

arteries out to the ONH boundary. During pure oxygen breathing, saturation increased in the arteries, and new areas of high saturation appeared where veins were located. All structures showed significant increases ($P < 0.05$) in the RSI during pure oxygen breathing (Table 4.1). The increase in the veins was nearly twice (factor of 1.9) that was found in the artery, whereas smaller increases in the ONH (averaged over the cup and rim) were approximately

A

B

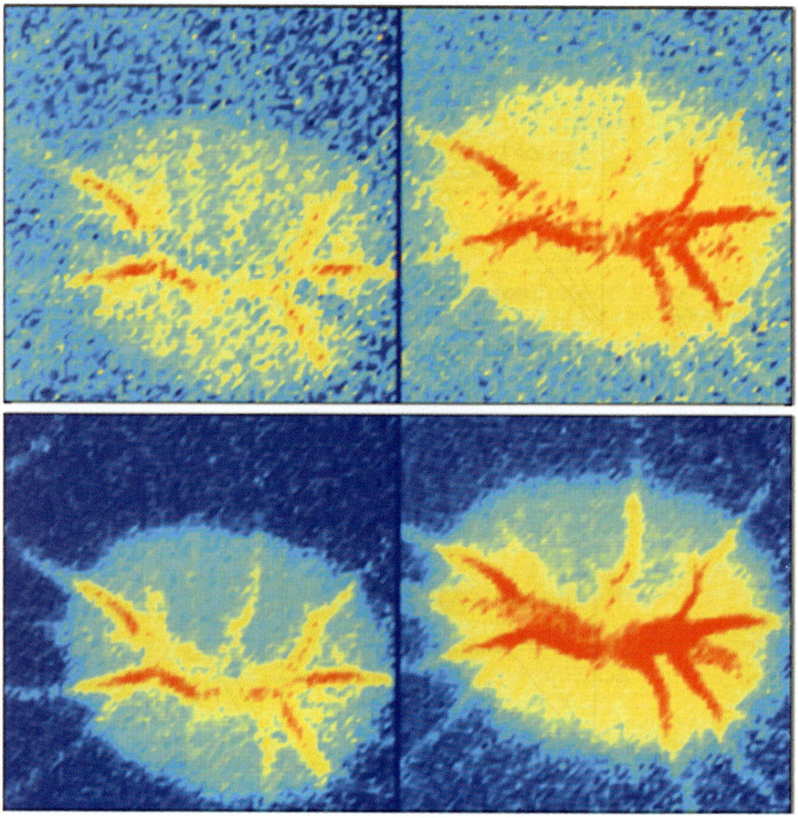

Fig. 4.11. Saturation maps of the oxygen breathing experiment during (left) room air and (right) pure O_2 breathing. (A) Partial-signature maps and (B) full-signature maps. Increasing saturation is indicated by the progression from blue to green to yellow to red. Temporal-to-nasal orientation in each map is top to bottom. Reprinted with permission from Khoobehi, B., Beach, J., and Kawano, H. Hyperspectral imaging for measurement of oxygen saturation in the optic nerve head. *Invest Ophthalmol Vis Sci* **45**:1464–1472, 2004. ©2004 Association for Research in Vision and Ophthalmology.

half (factor of 0.52) that of the artery. A slow decrease in the saturation over time occurred in the vein and ONH RSIs but not in the artery RSIs (Fig. 4.12).

4.2.2.5. *Responses to high IOP*

Changes in saturation at one-minute intervals after switching to high IOP are shown in Fig. 4.13A, bottom row. The high saturation of the arteries

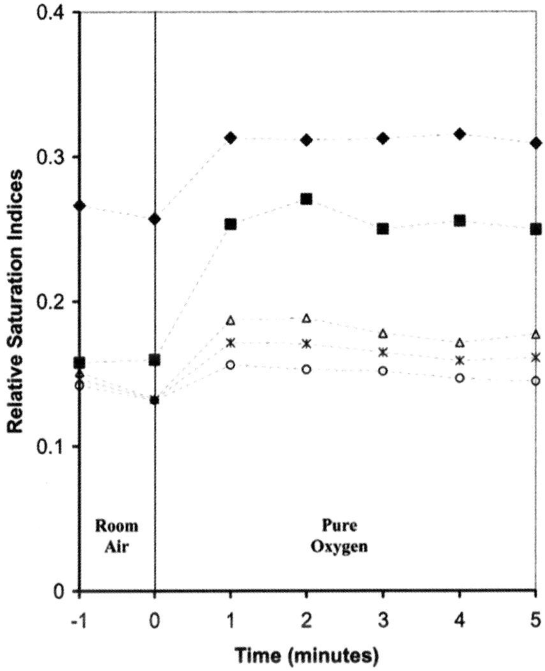

Fig. 4.12. RSIs from retinal vessels and ONH during an oxygen breathing experiment. RSIs were determined from vessel segments inside the ONH and from rim and cup regions, as denoted in Fig. 4.5. Vessel segments (large symbols): artery (filled diamonds), vein (filled rectangles). ONH regions (small symbols): temporal (open circles), nasal (open triangles), average over ONH (x). Breathing pure oxygen began immediately after the second data point (time 0). Reprinted with permission from Khoobehi, B., Beach, J., and Kawano, H. Hyperspectral imaging for measurement of oxygen saturation in the optic nerve head. *Invest Ophthalmol Vis Sci* **45**:1464-1472, 2004. ©2004 Association for Research in Vision and Ophthalmology.

and most of the ONH disappeared after one minute. A gradual return of saturation over the temporal ONH cup was observed from two through four minutes after IOP elevation. High IOP resulted in significant reduction ($P < 0.05$) of the RSI for each structure.

4.2.3. *Discussion*

4.2.3.1. *Pure oxygen breathing experiment*

Switching to pure O_2 strengthened the HbO_2 signature of both types of vessels. Oxygen leakage from the ophthalmic artery could cause the retinal

A

B

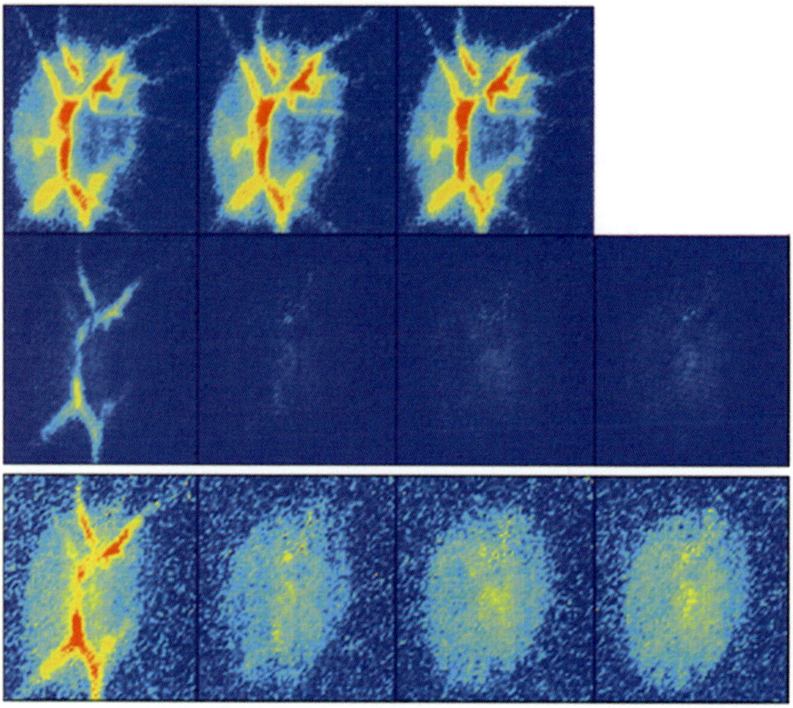

Fig. 4.13. Full-signature saturation maps of IOP experiment. (A) Top row: maps obtained under normal IOP conditions (15 mmHg): bottom row (left to right): 1, 2, 3, and 4 minutes after the onset of high IOP (60 mmHg). (B) Maps with compressed scale showing (left to right) 1, 2, 3, and 4 minutes after the onset of high IOP. Low to high saturation is indicated by the progression from blue to green to yellow to red. For each map, temporal-to-nasal orientation is right to left. Reprinted with permission from Khoobehi, B., Beach, J., and Kawano, H. Hyperspectral imaging for measurement of oxygen saturation in the optic nerve head. *Invest Ophthalmol Vis Sci* **45**:1464–1472, 2004. ©2004 Association for Research in Vision and Ophthalmology.

artery (RA) saturation to be lower than systemic levels. In that case, the response seen in the artery may represent up to an 8% increase in saturation. Because a fixed leakage rate would result in more or less arterial saturation depending on the flow rate, the evaluation of retinal arterial saturation could effectively probe changes in blood flow at the major vessels supplying blood to the inner retina. Pure O_2 strengthened the HbO_2 signature in the ONH, but to a lesser degree than that observed in the vein, as expected if this signature represents the averaged blood saturation in the microcirculation. Under pure O_2 conditions, the ONH and vessel reflectance at the hemoglobin absorption wavelengths was consistently greater than under room air-conditions. Because metabolic changes associated with the progression of retinal disorders presumably alter the oxygen utilization in the tissues, venous saturation maps should be a sensitive probe for disease states. If changes of similar size are present during the state of hypoxia, maps similar to ours should be able to isolate hypoxic areas when the scale is set to operate over the lower venous saturation range.

4.2.3.2. *IOP perturbation experiment*

Raising IOP to 60 mmHg had essentially the opposite effect on blood saturation. At this IOP, the perfusion pressure was very low. Arterial desaturation could have resulted from a slowing or stoppage of flow caused by collapse of the vessel under pressure, during which time oxygen diffused from the vessel. Saturation recovery was seen near the cup of the ONH, which was temporal with respect to the origin of the vessels. Increased reflectance during high IOP can be explained by low blood volume, since high IOP would partially occlude the major surface vessels and vessels feeding the outer ONH microcirculation, causing this area to blanch.

4.2.3.3. *Hyperspectral imaging*

These results demonstrate the ability of HSI to measure relative changes in oxygen saturation of the retinal macro- and microcirculations. However, involuntary eye movement present during fundus examination with bright light and in elderly patients would cause gaps in image data. The ability to scan the image at a faster rate, approximately one second per 100-line scan, would be needed to acquire distortion-free images routinely during fixed gaze in clinical subjects. Acquisition speeds can be significantly improved with multi-tapped frame-transfer CCD cameras and higher speed bus interfaces between the camera and frame buffer.

The present HSI technique enables spectral quantitation to be performed on the ONH over two dimensions, allowing regional changes in saturation to be identified. Significantly, faster recording techniques are needed to achieve a clinically acceptable method for mapping spectral information on the ocular fundus. At that point, HSI should provide a much-needed diagnostic tool for prevention and treatment of retinal disorders.

4.3. Experiment Two

4.3.1. *Methods and Materials*

4.3.1.1. *Animals, anesthesia, blood pressure, and IOP perturbation*

Five normal cynomolgus monkeys, 4–4.5 years of age and 2.5–3 kg body weight were used. The animals were housed in an air-conditioned room (22 ± 1°C and 66 ± 3% humidity) with a 12-hr light–dark diurnal cycle and access to food and water *ad libitum*. Monkeys were anesthetized with intramuscular ketamine (7–10 mg/kg) with xylazine (0.6–1 mg/kg) and intravenous pentobarbital (25–30 mg/kg). Administration of the anesthetics was repeated alternately every 30 min as required to maintain the animal in deep, stage IV anesthesia, as monitored by blood pressure and heart rate. Prior to IOP elevation, a topical anesthetic was given (proparacaine hydrochloride ophthalmic solution, 0.5%; Alcon, Fort Worth, TX, USA). A veterinary blood pressure monitor with a 5-cm pediatric cuff (model 9301Vl; CAS Medical Systems, Branford, CT, USA) was used to record blood pressure every five minutes throughout an imaging session. One eye was dilated and a 27-gauge needle connected to a saline manometer was inserted into the anterior chamber under slit-lamp examination. IOP was controlled by altering the height of the reservoir and measured by means of a tonometer (Tonopen XL; Medtronic, Jacksonville, FL, USA).

4.3.1.2. *Hyperspectral recordings*

HSI was done as previously described.

4.3.1.3. *Spectral determinant of percentage oxygen saturation*

Begin paragraph with sentence: Percent oxygen saturation was found from HSI images using curve fitting methods, as previously described.[73] Before performing the curve fit, the recorded spectrum was transformed by the method of Hammer *et al.*[66] to remove influences of nonhemoglobin light absorption and light scattering. This transformation corrects the recorded curves at three isosbestic wavelengths, 522 nm, 569 nm, and 586 nm, in order to match reference curves of oxygenated and deoxygenated blood.

We modified the transformation to use the three more closely spaced wavelengths at 522 nm, 548 nm, and 569 nm. Saturation was measured at 561 nm, which is a maximum in the difference spectrum. We used this procedure to test the calibration of red cell suspensions. We used the same procedure for *in vivo* spectra at different IOPs, except, here, we determined saturation by least-squares curve fits to oxygenated and deoxygenated reference curves from red-cell suspensions, containing 25 equispaced wavelengths between 522 nm and 569 nm. Curve fits were performed with a Windows software package (MathGrapher 2.0; Springfield Holding b.v., Noordwijk, the Netherlands). Reference spectra of saturated (S_{sat}) and desaturated (S_{desat}) red-cell suspensions were fit to transformed retinal blood spectra (S) using fitting parameters A and B with an additive term (C), as in Eq. (4.9):

$$S = A \times S_{sat} + B \times S_{desat} + C. \tag{4.9}$$

Percentage oxygen saturation was determined by expressing fitting parameters as in Eq. (4.10):

$$\%\mathrm{Sat} = \frac{100 \times A}{(A + B)}, \tag{4.10}$$

where A and B correspond with best-fit coefficients for oxyhemoglobin and deoxyhemoglobin contributions as defined by Eq. (4.9).

4.3.1.4. *Spatial mapping of oxygen saturation: a modification of the previous mapping algorithm incorporating a correction for blood volume*

By following the sign and magnitude of the area constructed between the spectral curve and the three-line segments connecting oxygen-insensitive wavelengths (isosbestic points), an oxygen-sensitive index was obtained. Individual oxygen-sensitive areas were normalized for a total reflected intensity by division by the polygonal areas A_1, A_2, and A_3 under each line segment. The OSC (shown in Fig. 4.2) of the algorithm is given by Eq. (4.11):

$$\mathrm{OSC} = \left(\frac{a_2}{A_2} - \frac{a_1}{A_1} - \frac{a_3}{A_3}\right), \tag{4.11}$$

the value of which increases with saturation. OSC, as defined above, depends on the volume of blood in the recording. Because significantly different blood volume densities exist in vessels and tissue, this difference must be accounted for in order for comparisons from the different structures to be valid. Optic disk blood volumes have previously been found by reflectometry at three wavelengths.[69] We used here the area between the hemoglobin absorption band and a line segment connecting the first and last isosbestic points (see Fig. 4.8) to estimate blood volume. Although this area slightly underestimates the true light absorption from hemoglobin over these wavelengths, it varies directly with the change in blood volume. This quantity is normalized by total light intensity from the area under the line to give the BVC in Eq. (4.12),

$$\mathrm{BVC} = \frac{b}{B}, \tag{4.12}$$

where b is the area between the spectral curve and the line segment under the curve and B is the area under the line segment. The volume-corrected, RSIs independent of hemoglobin concentration is given in Eq. (4.13):

$$\mathrm{RSI}_v = \frac{\mathrm{OSC}}{\mathrm{BVC}}. \tag{4.13}$$

Saturation maps were constructed by applying the algorithm at each image pixel using a MATLAB script. Numerical values of RSI_v, representing the relative saturations of separate structures, were determined from averaged pixels ($n > 1000$) inside the borders of vessels and distinct areas of the ONH. Individual pixel values were assigned to color codes, in the order of blue, cyan, green, yellow, and red, to represent progressively higher saturations. The relationship between RSI_v and percentage saturation was found in order to calibrate the saturation map.

4.3.1.5. *Preparation and calibration of red blood cell suspensions*

Red-cell suspensions were prepared as follows: 50 ml of blood was drawn from the femoral vein of the monkey and separated into equal volumes. The samples were centrifuged at 3000 RPM (13°C) for 20 min. The fluid was carefully aspirated leaving packed red cells, to which was added an equal volume of isotonic saline. The red cells were then resuspended, and this procedure was repeated three times. After rinsing, one sample was exposed to air (oxygenated sample) while to the second sample was added Na-dithionate, until the suspension turned blue. Percentage saturation readings were obtained using a blood gas analyzer (Instrumentation Laboratory, Lexington, MA, USA) with a hemoglobin saturation readout. Suspensions were drawn into 0.2-mm path length, flat capillary tubes (Vitro Dynamics, Rockaway, NJ, USA), sealed on both ends with wax, and hyperspectral images of the samples were recorded. The spectral curve from the red-cell samples were compared with those obtained from hyperspectral recordings of retinal blood during room air and pure oxygen breathings, and during elevated IOP, using the spectral transform methods described above.

4.3.2. *Results*

4.3.2.1. *Calibrated red-cell samples and retinal blood under controlled conditions*

In Fig. 4.14, a vertical line marks the wavelength (561.5 nm) of maximum amplitude change between the curves. Saturations of all samples were determined at this wavelength by the method of Hammer,[66] assuming a value of 100% for the arterial sample during pure oxygen and 0% for the deoxygenated red cells. Red cells reduced with Na-dithionate yielded a hemoglobin saturation reading slightly below zero; this number was replaced by the physically realizable 0%. Table 4.3 gives the saturations of red-cell and retinal-vessel samples that were found by determining the value of the sample curve at the vertical line (Fig. 4.14) as a percentage of the difference between the 100% curve and the 0% curve. For the oxygenated red cells, the value agrees with that of the oximeter to within 1%.

4.3.2.2. *Oxygen saturation of the ONH*

Figure 4.15 contains a representation of typical regions of the ONH in one monkey. Table 4.4 shows values obtained for percentage oxygen saturation within the separate regions of the ONH and overlying retinal vessels at 10 mmHg and 55 mmHg. At 55 mmHg, perfusion pressure was near zero. RA and vein saturation were reduced, respectively, by 35.7% and 6.5% saturation units at the higher pressure. Of the ONH structures, the temporal cup showed the highest saturation at both low and high pressure (77.3 ± 1.0%, 10 mmHg; 60.1 ± 4.0%, 55 mmHg), and the smallest reduction in saturation at high pressure (22.3%) compared with the average reduction

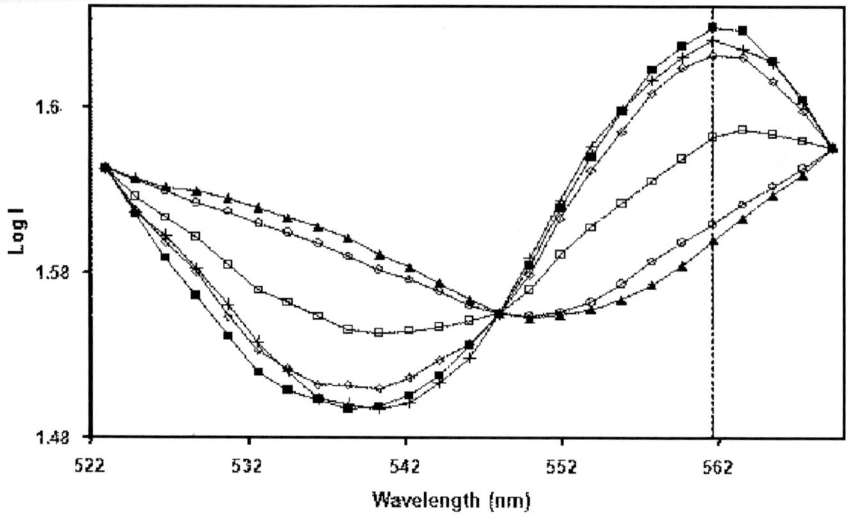

Fig. 4.14. Spectral curves from blood samples: artery (pure O_2) ■; red cells (room air) +; artery (room air) ◇; vein (room air) □; vein (60 mmHg IOP) ^; and red cells (Na-dithionate) ▲. End value wavelengths are isosbestic points (522.9 nm, 569.2 nm). Curves cross at a third isosbestic point (548.0 nm). The curves were aligned as described in Sec. 4.3.1. Comparison of relative positions of the sample curves at the vertical line (561.5 nm) was used to determine saturations in Table 4.3 at a single wavelength. Curve fits to linear combinations of artery (pure O_2) and red cells (dithionate) curves (25 wavelengths) were used to determine vessel and ONH percentage saturation (Table 4.4). Reprinted with permission from Beach, J., Ning, J., and Khoobehi, B. Oxygen saturation in optic nerve head structures by hyperspectral image analysis. *Curr Eye Res* **32**:161–70, 2007. ©2007 Informa Medical and Pharmaceutical Science.

in saturation over the ONH of 39%. At high pressure, the saturation of the temporal cup was 43.1% greater than the average ONH saturation. The saturation findings were not significantly different across monkeys.

In Fig. 4.16, IOP was set to 10, 30, 45, and 55 mmHg. OSC values near the high end of the saturation scale, which represent artery points, all fall well above the line. Although three regression lines had similar slopes, the fourth slope was different. The average regression slope, over all IOPs, was (mean ± SD, $n = 4$) 0.2169 ± 0.054. The goodness-of-fit ranged between 0.2 and 0.74, and the coefficient of variation (SD/mean) was 0.253. These results indicate that OSC does not bear a linear relationship with saturation. After correction for blood volume (right panel), all points fall along straight lines. Over the range of saturation established at the four IOPs, all slopes had similar values. The average regression slope was (mean ± SD, $n = 4$)

Table 4.3. Oxygen saturation in retinal vessels and calibrated red cells.

	% Saturation		
Source	561 nm[a]	Multi wavelength[b]	Co-oximeter[c]
Artery: pure oxygen	100.0	100.0	
Red cells: room air	96.5	92.0	97.2
Artery: room air	88.7	87.5	
Vein: room air	50.4	50.5	
Vein: 60 mmHg	10.3	8.9	
Red cells: Na-dithionate	0.0	0.0	−1.2

[a]Method of Hammer *et al.*[34] using one measurement wavelength.
[b]Least-squares fit to calibrated red cell sample with 25 wavelengths.
[c]From oxygen saturation readout.
Note: Reprinted with permission from Beach, J., Ning, J., and Khoobehi, B. Oxygen saturation in optic nerve head structures by hyperspectral image analysis. *Curr Eye Res* **32**:161–70, 2007.

1.3639 ± 0.0819. The goodness of the fit ranged between 0.74 and 0.98, and the coefficient of variation was 0.06. The linear regression of RSI_v against the percentage saturation, where both quantities are obtained from the same hyperspectral data, defines the calibration of the saturation map.

Figure 4.17 shows color-coded oxygen saturation maps of the ONH and vessels from one monkey at 10 mmHg (left panels) and 55 mmHg (right panels). The color scale is the same for both IOPs. In the upper row at 10 mmHg, the OSC map shows distinctly red-yellow codes for arteries. Veins and ONH tissue are similar (cyan-blue), and the veins are not clearly visible. At 55 mmHg, color codes of all structures correspond with lower saturation. Veins are visible as blue codes. In the middle row, maps of the BVC are shown. At 10 mmHg, volume is highest in arteries and veins (red-yellow codes) and lowest in the ONH (green-cyan codes). At 55 mmHg, all structures show less blood volume. Horizontal patterns of red and yellow in arteries represent volume changes during pulsatile blood flow as the image is scanned vertically. Pulsation is not clearly visible in veins.

Saturation maps of RSI_v after blood volume correction are shown in the bottom row. At 10 mmHg, veins (cyan) are distinguishable from areas

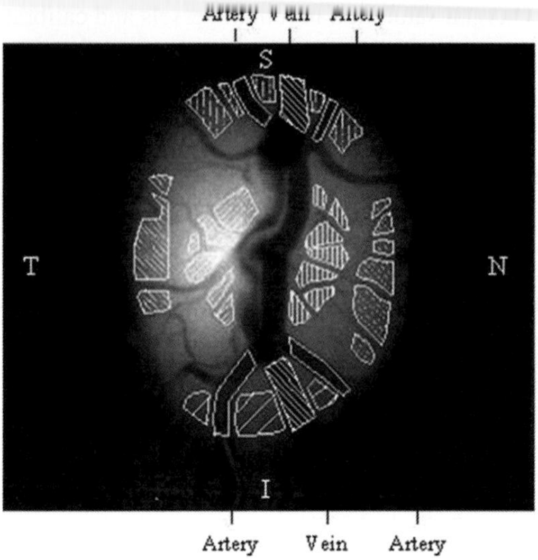

Fig. 4.15. Image of the primate ONH showing analysis regions for vessels (artery and vein segments) and areas of the ONH that include from the rim: superior (S), inferior (I), temporal (T), and nasal (N) areas; and from the cup: temporal and nasal areas, as marked in the figure. Image is oriented with temporal aspect to the left and superior aspect to the top. Vessel types are identified. See Sec. 4.3.1. Reprinted with permission from Beach, J., Ning, J., and Khoobehi, B. Oxygen saturation in optic nerve head structures by hyperspectral image analysis. *Curr Eye Res* **32**:161–70, 2007. ©2007 Informa Medical and Pharmaceutical Science.

of the ONH tissue (combinations of cyan, yellow, and red). Arteries are dark red. Saturation is, thus, lowest in the vein, intermediate in the ONH, and highest in the artery. At 55 mmHg, the order of saturation is the same; however, desaturation of structures is evident: arteries (yellow), ONH tissue (cyan-blue), and veins (deep blue, with more of the vein structure visible). Yellow-red codes in the temporal cup area indicate that this area has a relatively higher saturation at the high pressure. In all maps, stray light reflected from the ONH causes the reflectance from the disk surround to differ from the disk interior, hence, color codes on and off the disk cannot be directly compared.

4.3.3. *Discussion*

The current study is the first, to our knowledge, to report blood oxygen saturation in the ONH structures and map its distribution. At 55 mmHg, the

Table 4.4. Percentage saturation of vessels and optic nerve at low and high IOP.

	10 mmHg		55 mmHg	
Structure	% Sat (N = 56)	% dev. ONH ave.[b]	% Sat (N = 20)	% dev. ONH ave.[b]
Retinal artery	81.8 ± 0.4		46.1 ± 6.2	
Retinal vein	42.6 ± 0.9		36.1 ± 2.5	
ONH average[c]	68.3 ± 0.4		41.9 ± 1.6	
Nasal rim	65.6 ± 0.8	−4.0	36.0 ± 3.2	−14.1
Temporal rim	71.4 ± 0.9	4.6	40.1 ± 3.2	−4.6
Superior rim	61.8 ± 0.6	−9.6	37.5 ± 3.5	−10.6
Inferior rim	64.3 ± 0.5	5.9	33.7 ± 3.1	−19.6
Nasal cup	69.5 ± 0.4	1.8	44.4 ± 3.6	5.8
Temporal cup	77.3 ± 1.0	13.1	60.1 ± 4.0	43.1

[a]Mean ± SE.
[b]Percentage deviation about the average optic nerve head (ONH) saturation at 10 mmHg.
[c]N = 336 at 10 mmHg, 120 at 55 mmHg. Saturations at 55 mmHg differed significantly from those at 10 mmHg in all structures ($p < 0.05$).
Note: Reprinted with permission from Beach, J., Ning, J., and Khoobehi, B. Oxygen saturation in optic nerve head structures by hyperspectral image analysis. *Curr Eye Res* **32**:161–70, 2007. ©2007 Informa Medical and Pharmaceutical Science.

pressure on the disk and vessels is high enough to occlude the blood flow into the disk and retina partially. At this pressure elevation, lowered saturation in the disk tissue and the veins at high IOP could result from decreased blood flow, which results in a greater desaturation of the blood in the presence of a fixed oxygen consumption in tissue. The lowered saturation in arteries was not expected and is not well understood. A possible mechanism, which could reduce the oxygen carried in retinal arteries that are visible on the disk, is oxygen leakage from the central RA. The proximity of the large central artery and vein in the sheath of the optic nerve, behind the eye, may play a role in the effect of high IOP on the retinal arterial SO_2.

Our method evaluates areas under spectral curves with respect to a baseline that does not change with saturation. The advantage of using areas is noise reduction, as significant additive noise appears on spectral curves from

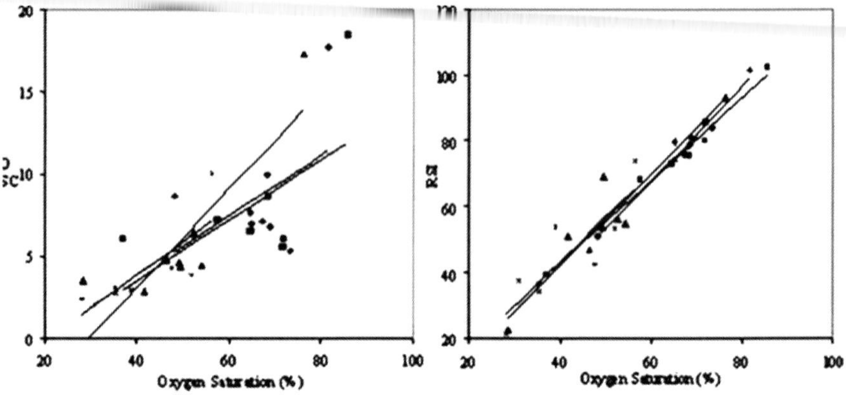

Fig. 4.16. Plots of the OSC of the saturation algorithm (left panel), and the RSI after blood-volume correction (RSIv, right panel), against percentage saturation values found from curve fits. For IOP of 10, 30, 45, and 55 mmHg, goodness of fits to lines were respectively (OSC) 0.204, 0.362, 0.741, and 0.639; (RSIv) 0.966, 0.983, 0.883, and 0.743. See Sec. 4.3.2. Reprinted with permission from Beach, J., Ning, J., and Khoobehi, B. Oxygen saturation in optic nerve head structures by hyperspectral image analysis. *Curr Eye Res* **32**:161–70, 2007. ©2007 Informa Medical and Pharmaceutical Science.

single pixels. The curves shown in Fig. 4.14 were averaged over several hundred pixels and, thus, are virtually noise free. However, the map algorithm operates on a single pixel of the image. Between 8 and 10 spectral points are averaged in each band; hence, there is still noise visible in the saturation maps.

The change in color codes at the edge of the ONH is likely due to different optical properties of nerve and retinal tissue. Significant light is reflected back through vessels from a more robust scattering in the nerve. In the retina, structures behind vessels contain pigments and tend to absorb the light before it is scattered into vessels. Our algorithm does not correct for the different effective optical path lengths of ONH and retina.

4.3.4. *Conclusions*

In animal studies, where the eye can be immobilized, this method should provide new information about oxygen supply and use in the retina and optic disk. Hyperspectral recordings are not yet practical in humans, as periods of 10–30 seconds are required to scan the fundus. Multispectral methods are being developed to reduce the time required to collect imagery from human subjects.

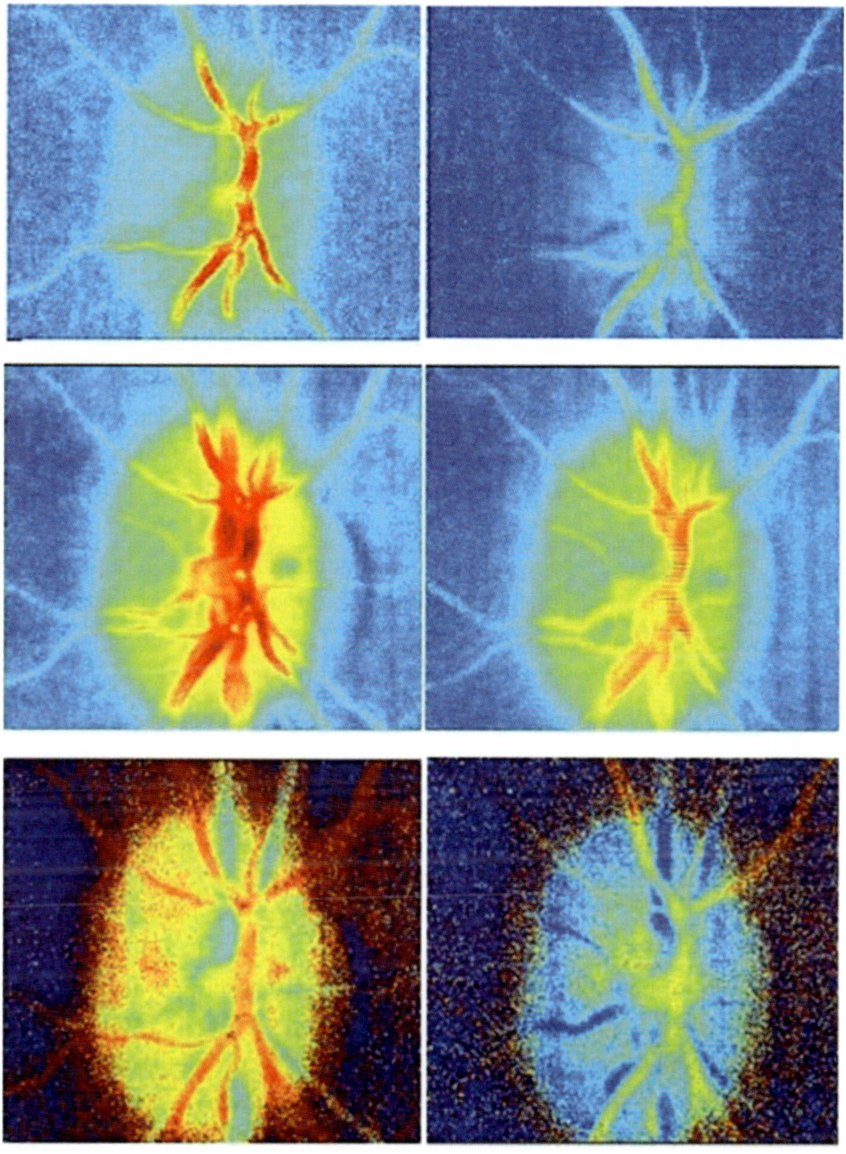

Fig. 4.17. Saturation maps of the ONH and overlying retinal vessels. Top row: OSC of the RSI. Middle row: BVC. Bottom row: RSI map corrected for blood volume. Left panels: 10 mmHg. Right panels: 55 mmHg. See Sec. 4.3.2. Reprinted with permission from Beach, J., Ning, J., and Khoobehi, B. Oxygen saturation in optic nerve head structures by hyperspectral image analysis. *Curr Eye Res* **32**:161–70, 2007. ©2007 Informa Medical and Pharmaceutical Science.

4.4. Experiment Three

4.4.1. *Methods and Materials*

All methods and materials used in Sec. 4.4 have been described in the above experiment except for the following.

4.4.1.1. *Compliance testing*

Compliance testing was performed in each monkey eye three times over a three-week period before starting the HSI studies. Briefly, the position of the ONH, termed mean position of the disc or MPD, was measured at normal IOP (10 mmHg) after 10 and 30 minutes (baseline). Then, IOP was acutely increased to 30 mmHg, and the change in MPD (amount of posterior deformation or movement of the ONH) was determined after 10 and 30 minutes. The compliance test images were obtained by a one-time automatic pre-scan with a 670-diode laser at 4–6 mm depth and 15° scan angle (Heidelberg Retinal Tomograph-II, Heidelberg Engineering, Heidelberg, Germany). The tomograph software (Heidelberg Eye Explorer) computed and automatically set parameters to acquire three-dimensional (3D) topography images from the images obtained in pre-scan. Digitization was performed in frames of 384 × 384 pixels. After the mean topography image was computed automatically from the three topography images, the optic disc margin was defined manually. Finally, the stereometric parameters were computed automatically and compared with the results of previous examinations.

4.4.1.2. *Hyperspectral imaging*

Hyperspectral images were obtained as previously described. During each imaging session, IOP in the dilated eye was set at 10 mmHg (normal), and then raised to 30, 45, and 55 mmHg. Three images were taken at each of two time points: 10 and 30 minutes after IOP elevation. To test reproducibility, each session was repeated five times over a 10-week period. Following the completion of the series of HSI imaging sessions on each monkey, compliance tests were repeated. Following data collection from the first eye, the entire test and image protocol were carried out in the other eye. Thus, 150 images were obtained at each IOP and time point (three images/session × five sessions/eye × two eyes/monkey × five monkeys), and 2400 images obtained in all (150 images/time point-IOP × two time points/IOP × four IOPs/eye × 2 eyes/monkey).

4.4.1.3. *Selection of ONH structures*

Distinct regions within the border of the optic disc were obtained manually from one image (569 nm band) taken from the set of HSI images, where good contrast of the vessels enabled image pixels to be grouped into artery and vein structures, and ONH areas, including the nasal and temporal cup and the rim (nasal, temporal, cup, superior and inferior aspects), as shown in Fig. 4.18.

4.4.1.4. *Statistical methods*

The data were analyzed using a nested analysis of variance (ANOVA).[70] In this ANOVA model, the saturation level of oxygen was the outcome, the site of measurement and IOP were the main effects, and the regions of the retina examined constituted the nested arrangement of treatments. Mean arterial

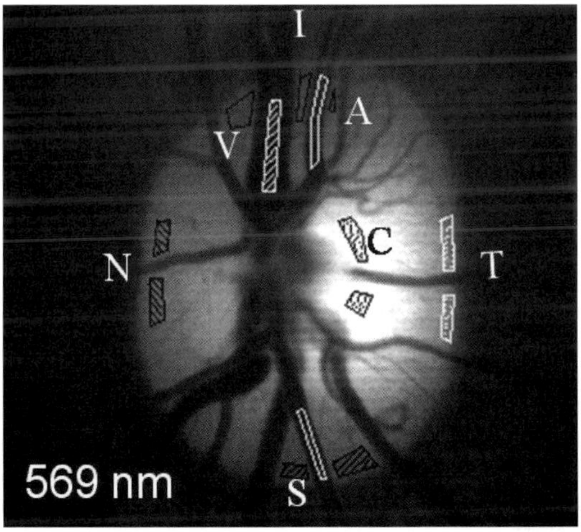

Fig. 4.18. Image of the ONH and overlying retinal vessels obtained at one spectral image (569 nm) of the hyperspectral image set. The areas for calculation of oxygen saturation in each of the seven regions include between 300 and 1000 pixels: A, artery; V, vein; C, cup; T, temporal ONH; N, nasal ONH; S, superior ONH; and I, inferior ONH. Reprinted with permission from Beach, J., Ning, J., and Khoobehi, B. Oxygen saturation in optic nerve head structures by hyperspectral image analysis. *Curr Eye Res* **32**:161–70, 2007. ©2007 Informa Medical and Pharmaceutical Science.

pressure was used as a covariate to rescale the saturation values based on the regression of mean arterial pressure with saturation levels over the applied levels of IOP. Least squares mean values for the site by pressure interaction means were tested for statistically significant differences using protected *t*-tests and an alpha level correction method based on a simulation method.[71]

Analysis of the compliance test images was conducted to determine whether any statistically significant alteration in the MPD (mean position of the disc) occurred due to hypercompliance or increased movement of the ONH during the repeated testing sessions applied to each eye. The statistical design for the evaluation of compliance was a two-way factorial ANOVA with time (0, 10, or 30 minutes within a testing session) and examination date as the main effects and MPD as the outcome. Individual mean values of the time by examination date interaction were separated using the same protected *t*-tests and alpha correction method applied above.[71]

All data management and statistical methods were carried out using programs and procedures of the statistical analysis system (SAS Institute, Cary NC).[72]

4.4.2. *Results*

4.4.2.1. *Compliance testing*

Overall, the change in MPD with increased IOP did not exceed 25 μm for any of the eyes in any of the compliance testing sessions. A comparison of the compliance test results for the first eye of each monkey before and after the HSI sessions showed no damage caused by acute pressure elevations to the ONH.

4.4.2.2. *Blood spectra from ONH structures*

Differences in average spectral amplitudes in Figs. 4.19 and 4.20 were the result of changes in the position of the retina and the pupil alignment caused by elevation of IOP.

4.4.2.3. *Oxygen saturation of ONH structures*

Fifteen measurements (three measurements at 1 min intervals at a given IOP per session × five sessions over a five-week period for each eye of each

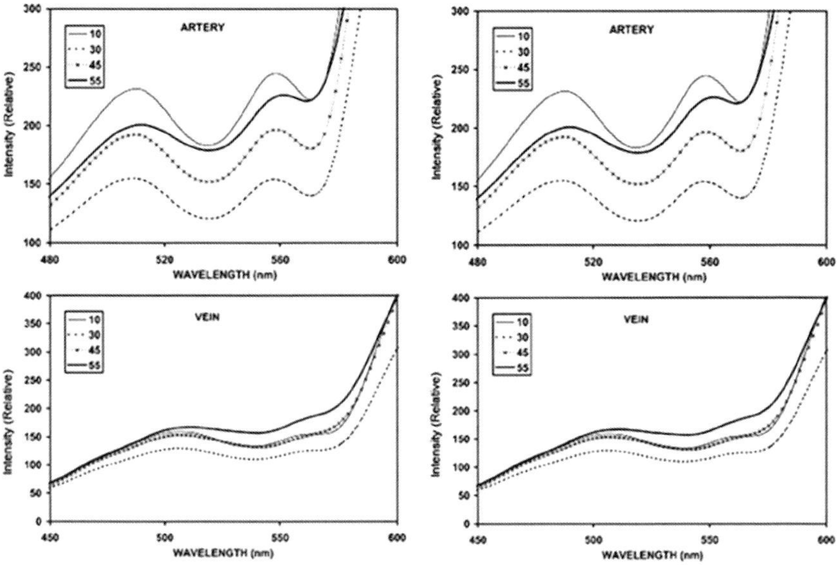

Fig. 4.19. Reflectance spectral curves in artery and vein in monkey 3. Data are from images obtained 30 minutes after IOP was set to 10, 30, 45, and 55 mmHg. Reprinted with permission from Khoobehi, B., Kawano, H., Ning, J., Burgoyne, C.F., Rice, D.A., Khan, F., Thompson, H.W., and Beach, J.M. Oxygen saturation changes in the optic nerve head during acute intraocular pressure elevation in monkeys. In: Manns, F., Soderberg, P.G., and Ho, A., (eds.), *Ophthalmic Technologies XIX*, Proc of SPIE. 7163, 716320. ©2009 SPIE.

monkey) produced standard deviations of mean for the percent O_2 saturation of 1.3–1.6 for each IOP, and a correlation between different measurements of 0.999 ($P < 0.0001$). At the three lowest IOPs in Fig. 4.21, the artery and vein saturations gradually decreased with IOP, the vein dropping at a greater rate. At 55 mmHg, the artery showed a large and significant ($P < 0.0001$) drop in saturation. The rate of decrease for the vein was unchanged, with significantly reduced saturation at 45 and 55 mmHg. In the ONH, saturation of all structures was significantly reduced at 45 mmHg, with the exception of the temporal cup (see Table 4.5) The temporal cup showed a relatively stable saturation versus IOP, similar to that in the retinal vein. In the nasal cup and the rim, saturation was reduced more rapidly with increased IOP, with a range of relative stability observed in the nasal cup between 10 and 45 mmHg that was not present in the rim structures. Table 4.6 shows average values for mean arterial pressure of each monkey over the 10 sessions (five

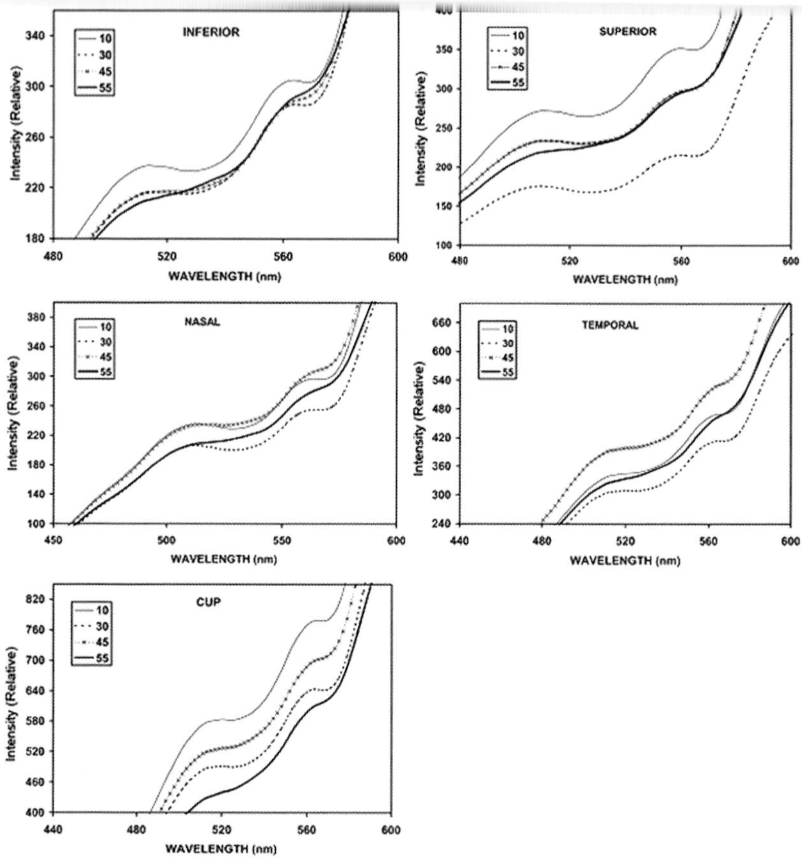

Fig. 4.20. Reflectance spectra from five regions of the ONH in monkey 3. Data are from images obtained 30 minutes after IOP was set to 10, 30, 45, and 55 mmHg. Reprinted with permission from Khoobehi, B., Kawano, H., Ning, J., Burgoyne, C.F., Rice, D.A., Khan, F., Thompson, H.W., and Beach, J.M. Oxygen saturation changes in the optic nerve head during acute intraocular pressure elevation in monkeys. In: Manns, F., Soderberg, P.G., and Ho, A., (eds.), *Ophthalmic Technologies XIX*, Proc of SPIE. 7163, 716320. ©2009 SPIE.

for each eye). At IOP = 55 mmHg, perfusion pressure values approached the single digits.

4.4.2.4. *Oxygen saturation maps*

Figure 4.22 shows the reproducibility of the saturation maps of structures within the optic disc border from three-image sets taken at different times

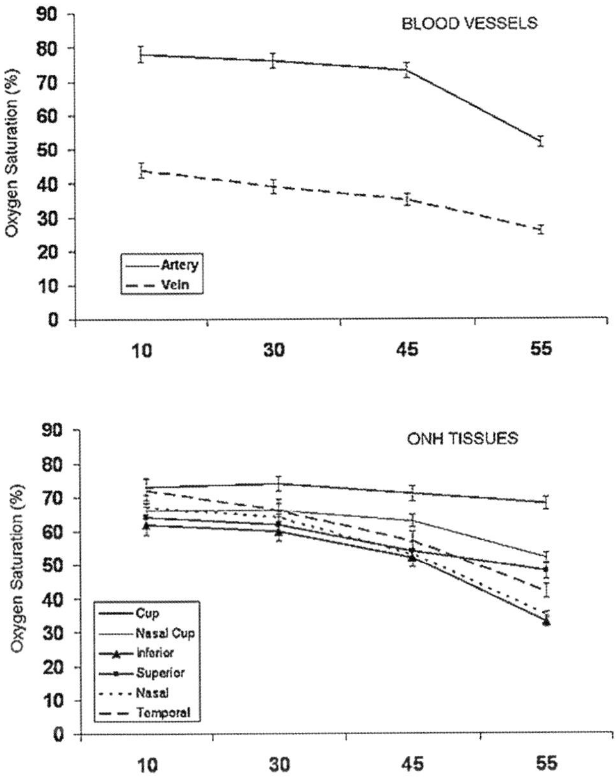

Fig. 4.21. Percentage O_2 saturation (±SE) vs. IOP from retinal vessels overlying the ONH (top) and five regions of the ONH (bottom). $n = 150$ readings for each data point (three readings per eye at each session × five sessions for each eye × two eyes per monkey × five monkeys). Data from Table 4.5. Reprinted with permission from Khoobehi, B., Kawano, H., Ning, J., Burgoyne, C.F., Rice, D.A., Khan, F., Thompson, H.W., and Beach, J.M. Oxygen saturation changes in the optic nerve head during acute intraocular pressure elevation in monkeys. In: Manns, F., Soderberg, P.G., and Ho, A. (eds.) *Ophthalmic Technologies XIX*, Proc of SPIE. 7163, 716320. ©2009 SPIE.

during one-recording session at IOP of 10 mmHg. Color codes represent low to high saturations in the order blue, cyan, green, yellow, and red. High saturation is seen in arteries within the border of the disc and in the ONH tissues surrounding the vessels. Low saturation is visible in veins within the disc border. The presence of choroidal pigments outside the border of the disc alters the relationship between color codes and saturation, hence the color codes in images we present apply only to structures inside the border.[73]

Table 4.5. Mean RSI value for each area at each IOP for all five monkeys.

IOP (mmHg)	Retinal vessels		Optic nerve head					
	Artery	Vein	Nasal	Temporal	Temporal cup	Nasal cup	Superior	Inferior
10	79.026 ± 1.247	43.702 ± 1.247	67.543 ± 1.247	71.764 ± 1.247	73.207 ± 1.247	66.723 ± 1.247	63.548 ± 1.247	62.669 ± 1.247
30	78.424 ± 1.362	38.044 ± 1.362	63.194 ± 1.362	66.625 ± 1.362	73.345 ± 1.362	64.915 ± 1.362	59.842 ± 1.362	59.208 ± 1.362
P value	0.7447	0.0022	0.0187	0.0055	0.9403	0.3276	0.045	0.0611
45	74.171 ± 1.362	34.766 ± 1.362	53.400 ± 1.362	56.769 ± 1.362	70.554 ± 1.362	59.590 ± 1.362	54.144 ± 1.362	52.016 ± 1.362
P value	0.0087	<0.0001	<0.0001	<0.0001	0.1511	0.0001	<0.0001	<0.0001
55	52.455 ± 1.568	25.949 ± 1.568	36.934 ± 1.568	41.305 ± 1.568	63.998 ± 1.568	46.322 ± 1.568	39.581 ± 1.568	33.121 ± 1.568
P value	<0.0001	<0.0001	<0.0001	<0.0001	<0.0001	<0.0001	<0.0001	<0.0001

Values are means ± SE of 150 readings (three readings per eye at each IOP at each session × five sessions for each eye × two eyes per monkey × five monkeys). Readings are from time point 30 minutes after the elevation of IOP. *P* values are derived from comparison with mean RSI at 10 mmHg. Reprinted with permission from Khoobehi, B., Kawano, H., Ning, J., Burgoyne, C.F., Rice, D.A., Khan, F., Thompson, H.W., and Beach, J.M. Oxygen saturation changes in the optic nerve head during acute intraocular pressure elevation in monkeys. In: Manns, F., Soderberg, P.G., and Ho, A. (eds.), *Ophthalmic Technologies XIX*, Proc of SPIE. 7163, 716320. ©2009 SPIE.

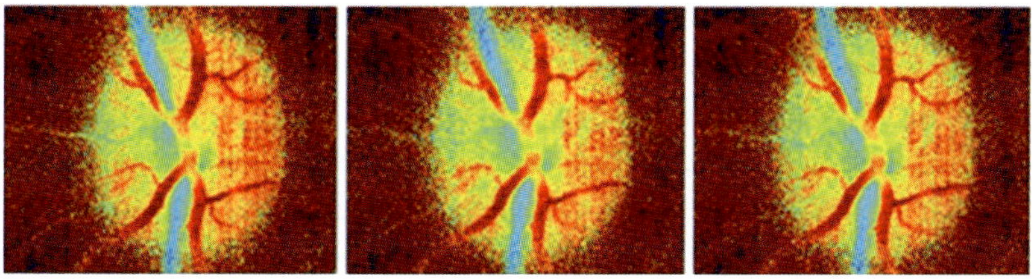

Fig. 4.22. Reproducibility of saturation maps obtained in the same monkey at 10 mmHg from separate HSI recordings on the same day. The position of the ONH has been adjusted to the center of each image. Low-to-high saturation is indicated by the progression through blue–cyan–green–yellow–red. Nasal-to-temporal orientation is left-to-right and inferior-to-superior orientation is top to bottom. Reprinted with permission from Khoobehi, B., Kawano, H., Ning, J., Burgoyne, C.F., Rice, D.A., Khan, F., Thompson, H.W., and Beach, J.M. Oxygen saturation changes in the optic nerve head during acute intraocular pressure elevation in monkeys. In: Manns, F., Soderberg, P.G., and Ho, A. (eds.), *Ophthalmic Technologies XIX*, Proc of SPIE. 7163, 716320. ©2009 SPIE.

The oxygen saturation maps of ONH structures are shown in Fig. 4.23 for increasing IOP. At 30 mmHg, the veins and the ONH areas show slightly decreased saturation, while at 45 mmHg the ONH tissue saturation is substantially reduced. At 55 mmHg, a reduction in the artery is most marked and ONH tissue saturation is further reduced in the rim. The effect of IOP on the oxygen saturations visualized in the saturation maps is in agreement with the percentage saturations in Table 4.5. A linear relationship between numerical values of the saturation map and percentage saturation values obtained by our procedure was established previously.[73] No differences in oxygen saturation levels were seen at 10 and 30 min after IOP level adjustment at any of the IOPs evaluated.

4.4.3. *Discussion*

Percentage oxygen saturation in retinal vessels has previously been reported using vessel oximetry in humans.[74,75] Our values from the monkey retinal vessels at the normal IOP are 10–15 percentage points lower than those that had been previously reported. In the present study, deep anesthesia was needed to maintain a stable eye during HSI collection and IOP elevation. It is likely that under deep anesthesia, systemic saturation could have been lower because of less efficient respiration. As with vessels, the tissue values we report may also be lower than would otherwise be found in animals that were awake.

We previously reported the changes in the oxygen saturation distribution that resulted from experimentally raising IOP from a normal pressure (15 mmHg) to near the perfusion pressure (60 mmHg) for several minutes.[33] Sustained high IOP produced dramatic reductions in blood saturation from the ONH tissue and overlying retinal arteries and veins, overriding the autoregulatory control of blood flow in all of these structures. Autoregulatory responses were observed, however, from the saturation rebound at the cup of the ONH. In the present study, we were interested in determining thresholds for the loss of autoregulation in these different structures resulting from the reduced oxygen saturation that occurred in response to graded increases in IOP.

The findings of this study in vessels were (1) elevation of the IOP to 30–45 mmHg did not result in a significant change in the retinal arterial saturation, whereas further elevation to 55 mmHg caused the saturation of arterial blood to decrease significantly; (2) the same stepwise increases in IOP caused approximately equal reductions in the retinal venous saturation at each step above normal IOP; and (3) between 10 and 45 mmHg where autoregulatory responses were effective, the overall decrease in saturation was larger in the veins than in the arteries, causing the retinal arterio-venous saturation difference (A-V difference) to increase with IOP. The additional oxygen extraction from the vein is consistent with a reduced flow resulting from dropping perfusion pressure. The retinal venous saturation is closely tied to the end-capillary blood saturation of the tissue microcirculation; hence, these responses are consistent with compensatory autoregulation of retinal tissue oxygen saturation at elevated IOP, up to 45 mmHg in our experimental model. A blood flow proportional to the perfusion pressure at high IOP was shown in the retinal and prelaminar ONH in monkeys[76]; in our study, the perfusion pressure dropped to single digits, which would indicate that the blood flow was significantly reduced.

The mechanism by which the retinal arterial saturation is decreased at reduced flow is not known. However, the reports of significant longitudinal gradients in the periarteriolar PO2 in small arteries and arterioles, with essentially no difference between the intra- and extravascular tissue oxygen tensions, have been explained by diffusion of oxygen from these precapillary vessels.[77] More recently, spectral imaging of the saturation distributions in rat-cremaster vessel networks demonstrated oxygen transfer from arteries to veins running in parallel, or at crossover points.[78] Thus, there is an opportunity for oxygen exchange between the central RA and vein, which run together for several millimeters in close proximity, within the distance for oxygen exchange by diffusion, before their entrance from the ONH cup. If oxygen were removed by leakage at a fixed rate, lower volume flow would result in a decrease in the PO2 of the blood supply at the inner retinal arteries. As the dissolved oxygen concentration is reduced, arterial saturation must decrease according to the oxygen dissociation curve. Therefore, the stable relationship we observed between arterial saturation and IOP at 10 and 30 mmHg may reflect autoregulatory control, whereas the nonlinear relationship over the higher range of IOP may, as we speculate, reflect passive flow-dependent mechanisms that could involve oxygen exchange from central RA to retinal vein. There is presently no evidence for this, however.

A relationship between saturation and IOP, which is similar to that found for retinal arteries and veins, was observed in the ONH. However, the blood supply in the ONH is more complex. Our recordings in the ONH rim yielded results similar to those in retinal arteries, showing a stable saturation of the blood supply over the first step in IOP, and reduced saturation at higher IOP. This response is consistent with previous measurements of blood flow changes during acute elevations in IOP in humans.[79] Over the lower range of elevated IOP, the absence of change in saturation at 10 and 30 min after raising the IOP suggests that autoregulation of flow in the ONH rim was present. Reduced oxygen saturation in the higher range of elevated IOP is consistent with reduced blood saturation in the blood supply to this tissue. Since this region receives its blood supply via peripapillary retinal arterioles, which are supplied through the central RA, those saturation responses that were observed in the RA could be also present at the ONH rim. The prelaminar blood supply has been shown to be masked by vessels in the NFL in early arterial phases of fluorescein angiograms in monkeys.[80] Hence, it is likely that our observations of blood saturation in the ONH rim by reflectometry are confined to the NFL, which shares its blood supply with the peripapillary retina.

Within the ONH cup, particularly the temporal aspect, which is central to the spread of the NFL, we observed a steady series of smaller reductions in the blood saturation at each step above normal IOP, similar to the finding in the retinal vein. Since the NFL circulation does not mask the central

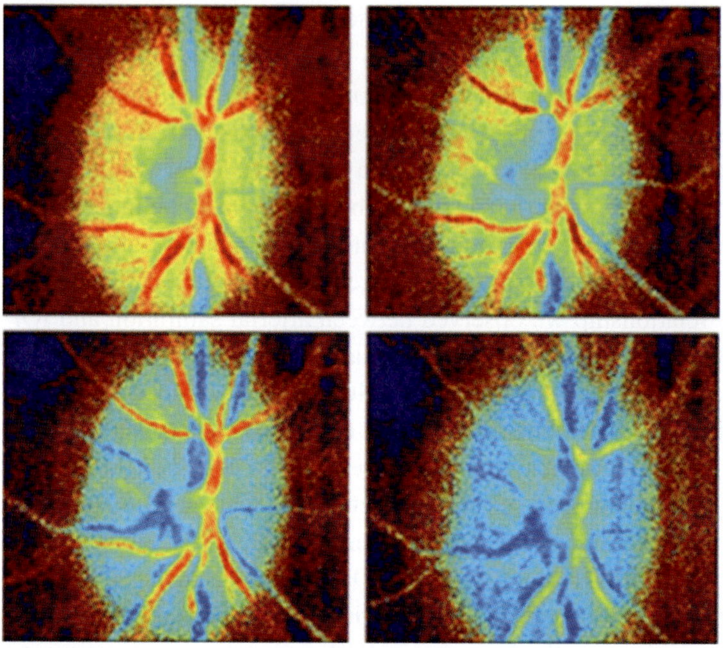

Fig. 4.23. ONH saturation maps for IOP values of 10 mmHg (top-left), 30 mmHg (top-right), 45 mmHg (bottom-left), and 55 mmHg (bottom-right). Low-to-high saturation is indicated by the progression through blue–cyan–green–yellow–red. Nasal-to-temporal orientation is left-to-right and inferior-to-superior orientation is top to bottom. Reprinted with permission from Khoobehi, B., Kawano, H., Ning, J., Burgoyne, C.F., Rice, D.A., Khan, F., Thompson, H.W., and Beach, J.M. Oxygen saturation changes in the optic nerve head during acute intraocular pressure elevation in monkeys. In: Manns, F., Soderberg, P.G., and Ho, A. (eds.), *Ophthalmic Technologies XIX*, Proc of SPIE. 7163, 716320. ©2009 SPIE.

cup, the deeper layers of the ONH, including the prelaminar and laminar regions, would be accessible to optical reflectometry at the central cup. The short posterior ciliary arteries and branches from the pial artery supply the blood flow to these layers of the optic nerve; little flow is derived from the central RA.[3,4] Fluorescein fundus angiography has also demonstrated a blood supply to these regions from the peripapillary choroid.[44] Thus, saturation changes in the temporal cup would not be expected to follow those seen in the rim. Although the relationship between cup saturation and IOP is similar to that of the retinal vein, it may be the result of redistribution of flow from the prelaminar region into the retrolaminar region behind the lamina cribrosa.[81] Our recordings of the oxygen saturation levels of

ONH structures over a range of IOP are consistent with early studies that showed that the NFL could become ischemic during very low perfusion pressure, while the regions of the anterior ONH underlying the NFL could be protected from ischemia.

4.5. Experiment Four

4.5.1. *Methods and Materials*

All methods and materials are the same as in the previous experiment except for theoretical development of oxygen diffusion out of the RA.

We hypothesized that the decrease of O2Sat in RA caused by the increase in IOP can be explained by oxygen diffusion out of the RA alone. Consider a cylindrical RA 0.2 mm in diameter (d), a wall thickness of 0.03 mm (w),[82] surrounded by tissue with oxygen tension (PO2) that is in equilibrium with venous PO2 (see Fig. 4.24). Intaglietta *et al.*[83] have shown that tissue regions between arterioles and venules have essentially uniform tissue PO2. Under normal IOP, some oxygen in the RA blood diffuses through the arterial wall. Two cases were assumed. In the first case, RA and RV were assumed to be separated by arterial wall thickness. The RA and RV are parallel each other for about 5 mm on the optic disk and the tissue underneath the optic disk surface. Thus, oxygen can diffuse from the artery to the tissue through a 0.03 mm layer that is 0.2 mm by 5 mm. For the vessels considered, the wall thickness is several times smaller than the vessel diameter, and the PO2 gradient is approximately constant across the wall. The arterial

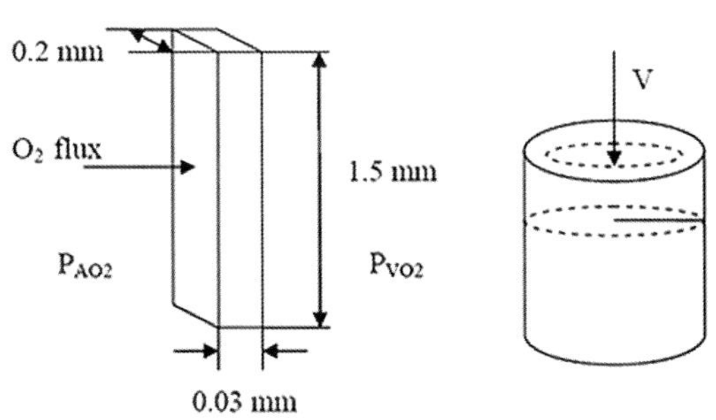

Fig. 4.24. Schematic diagram of the RA diffusion model. Reprinted with permission from Beach, J.M., Ning, J., Khoobehi, B., and Rice, D.A. A simple model of oxygen diffusion out of the retinal artery. In: Manns, F., Soderberg, P.G., and Ho, A. (eds.), *Ophthalmic Technologies XIX*, Proc of SPIE. 7163, 71630S. ©2009 SPIE.

oxygen partial pressure (PAO2) was assumed to be 100 mmHg, venous oxygen partial pressure (PVO2) 50 mmHg, oxygen diffusion coefficient (DO2) 2.4×10^{-3} mm^2/s,[84] oxygen solubility (α) 0.0284 μl/ml/mmHg,[85] and all parameters were at body temperature and 1 atm pressure. A list of glossary and symbols is in the Appendix.

Figure 4.24 is a schematic diagram of the RA diffusion model.

According to Fick's law, the flux of oxygen across the arterial vessel wall for $\Delta P = (PAO2 - PVO2)$ per unit, area can be calculated as:

$$\begin{aligned} F(O_2) &= \frac{\alpha D(\Delta P)}{w} \\ &= \frac{\alpha D(P_{AO2} - P_{VO2})}{w} \\ &= \frac{0.0284 \frac{\mu l}{ml \cdot mmHg} \times 2.4 \times 10^{-3} \frac{mm^2}{s} \times (100 - 50)\,mmHg}{0.03\,mm} \\ &= 0.1136 \times \frac{\mu l \cdot mm}{ml \cdot s} \\ &= 0.1136 \times \frac{\mu l \cdot mm}{10^3 \times mm^2 \cdot mm \cdot s} = 1.14 \times 10^{-4} \frac{\mu l}{mm^2 \cdot s} \end{aligned} \tag{4.14}$$

Total flux of oxygen across the arterial wall

$$\begin{aligned} (Q1(O2)) &= F(O2) \times \text{area} \\ &= 1.14 \times 10^{-4}\,\mu l/mm^2/s \times (0.2\,mm \times \pi \times 5\,mm) \\ &= 3.6 \times 10^{-4}\,\mu l/s. \end{aligned} \tag{4.15}$$

Under normal IOP, we assume that retinal arterial blood flow is well mixed and in a steady state, and red cell velocity (V) is 9.28 mm/s.[4] Blood flow and oxygen consumption are assumed to be distributed homogeneously. The amount of oxygen flowing in blood along the artery (Q(O2)) is the product of oxygen concentration ([O2]) and blood flow (Q), which is the product of blood velocity and the area of the artery (A_{artery}). In the discussion that follows, volumetric flow rate Q is the variable of significance, but we pose it as $V \cdot A$ because both V and A are more readily determined by optical methods. One gram of hemoglobin combines with 1.34 ml of oxygen, and 100 ml of blood contains 15 g of hemoglobin, thus carrying

15 × 1.34 = 20.1 ml oxygen. The amount of oxygen carried by the red cells is expressed as percentage saturation. For a normal 40% hematocrit, 100% saturation corresponds to 20.1 ml O2/100 ml blood. There is only a little oxygen dissolved in the plasma; therefore, the total oxygen in 100 ml of blood is slightly more than 20.1 vol% when saturated:

$$\begin{aligned} Q(O_2) &= [O_2] \times Q = [O_2] \times V \times A_{\text{artery}} \\ &= 20.1\% \times \frac{9.28\,\text{mm}}{\text{s}} \times \pi \left(\frac{0.2}{2}\,\text{mm}\right)^2 \\ &\approx 0.059\frac{\text{mm}^3}{\text{s}} = 0.059\frac{\mu\text{l}}{\text{s}}. \end{aligned} \tag{4.16}$$

The percentage of oxygen diffuses out of the retinal arterial wall, as a surface is total flux of oxygen across the retinal wall divided by the amount of oxygen flowing in blood along the artery. The percentage of oxygen diffuses out of the RA wall surface is: $3.6 \times 10^{-4} \times 100\%/0.059 = 0.6\%$, about 1 part in 120.

Thus, the amount of oxygen that diffuses across the arterial wall on the optic disc is small when IOP is normal (less than 1%). Next, we assumed an extreme case, in which the central RA is in proximity with the central retinal vein for a diffusion distance of about 10 mm before they penetrate the optic disk. We assumed that along this length the central RA is surrounded by tissue with PO2 that is in equilibrium with venous PO2. We assumed that oxygen diffuses across the entire wall to the surrounding tissue. The total wall area (A_{wall}):

$$A_{\text{wall}} = 10\,\text{mm} \times \pi \times d = 6.28\,\text{mm}^2. \tag{4.17}$$

Assuming O2Sat does not change along the artery, the total oxygen leak by diffusion through the arterial wall is:

$$1.14 \times 10^{-4}\,\mu\text{l/mm}^2/\text{s} \times 6.28\,\text{mm}^2 = 7.16 \times 10^{-4}\,\mu\text{l/s}. \tag{4.18}$$

The percentage of oxygen diffusing out of the retinal arterial wall surrounded by tissue (Q2(O2)) is:

$$\frac{Q_2(O_2)}{Q_2(O_2)} \times 100\% = \frac{7.16 \times 10^{-4}}{0.059} \times 100\% \approx 1.21\%, \text{ or one part in 80.} \tag{4.19}$$

4.5.2. *Results*

Mean percentage O2Sat of the RAs was 78.9% at IOP of 10 mmHg, with no significant change when IOP was raised to 30 mmHg. O2Sat decreased to 74.1% at 45 mmHg ($P = 0.01$), and further decreased to 51.5% ($P < 0.0001$) at 55 mmHg IOP. From the modeling, the highest percentage of oxygen that can be lost visa diffusion through the arterial wall is 1.27% when retinal arterial blood velocity is normal. When IOP was raised to 55 mmHg, perfusion pressure (pressure difference between mean arterial blood pressure and IOP) decreased.[73] We have assumed that the retinal arterial blood velocity drops to 1.0 mm/s at 55 mmHg. Substituting this value into the equations above, the percentage of oxygen that can be lost via diffusion through the arterial wall at most is about 38%. To have a drop of 35.7% in retinal arterial O2Sat, the model predicts that retinal arterial blood velocity needs to be less than 0.33 mm/s.

4.5.3. *Discussion*

Figure 4.25 shows that oxygen diffusion from the RA increased as the blood velocity was decreased by the elevation of IOP. There was not much difference in percentage O2Sat when IOP was raised from 10 to 30 mmHg; however, when IOP was raised to 45 mmHg, percentage O2Sat began to decrease and decreased significantly at an IOP of 55 mmHg. This nonlinear relationship between IOP and percentage O2Sat is consistent with the model predictions for oxygen diffusion in Fig. 4.5. This result was not expected; thus, we developed the model to interpret the data.

In the two studies of Khoobehi *et al.*[33] and Beach *et al.*,[73] IOP of 10–15 mmHg was considered normal in anesthetized cynomolgus monkeys. Morita[86] reported a diastolic pressure of 55 ± 11 mmHg and a systolic pressure of 102 ± 18 mmHg (mean arterial blood pressure is about 70.67 mmHg) in 15 normal male cynomolgus monkeys under general anesthesia. The blood flow in the ONH is usually calculated as perfusion pressure divided by the resistance to flow. Practically, perfusion pressure in the optic disc is equal to mean arterial blood pressure in the ONH vessels minus IOP.[5] For the HSI data acquisition in this study, the animals were deeply anesthetized, which decreased the mean blood pressure to the range of 50–60 mmHg. Thus, at an IOP of 55 or 60 mmHg, perfusion pressure in the ONH is low,

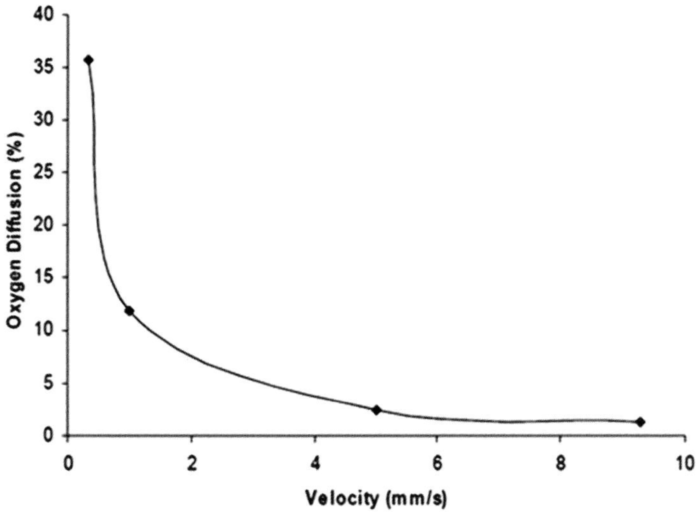

Fig. 4.25. Maximum percentage of oxygen that may diffuse out of the retinal arteries as RA blood velocity decreases from 9.28 mm/s to 0.33 mm/s. Reprinted with permission from Beach, J.M., Ning, J., Khoobehi, B., and Rice, D.A. A simple model of oxygen diffusion out of the retinal artery. In: Manns, F., Soderberg, P.G., and Ho, A. (eds.), *Ophthalmic Technologies XIX*, Proc of SPIE. 7163, 71630S. ©2009 SPIE.

and ONH blood flow could be significantly reduced or completely stopped, which may contribute to the significant diffusion of oxygen out of RA. However, experimental verification is needed.

What other factors may decrease O2Sat along the RA when the IOP is at 55 mmHg? It may be that the decrease in perfusion pressure at high IOP reduces the RA blood flow sufficiently to lower O2Sat. It is unlikely that significant changes in tissue oxygen consumption occur between normal and high IOP. The red blood cell and the arterial wall consume oxygen. A review by Tsai *et al.*[87] concluded that the vascular wall has a high rate of oxygen consumption, positioning an oxygen sink between blood and tissue regulating oxygen release. A fixed rate of oxygen consumption that is maintained during significant reductions in the blood flow could also result in lowering the artery saturation through diffusion.

Our results show that diffusion of oxygen from retinal arteries could cause the observed RA O2Sat decrease during high IOP. Further investigation is needed to evaluate blood flow changes over the range of IOP studied.

4.6. Experiment Five

4.6.1. *Methods and Materials*

All methods and materials are the same as in the previous experiment except for the following.

4.6.1.1. *Modification of the fundus camera*

Retinal images presented in this chapter were taken by a modified Topcon TRC-50EX fundus camera, with a lens and a c-mount through the vertical path of the camera. The modification enables monochromatic fundus illumination and records over 400–720 nm range of wavelengths. The CCD camera was positioned in the polaroid mounting part. A liquid crystal tunable filter (LCTF) 7-nm bandwidth filter was placed in the path of the illumination.[73] An Apogee U47 16-bit digital camera was positioned inside the fundus camera, which can provide low readout noise images. Hyperspectral images were taken through the vertical viewing port by an imaging spectrograph and digital camera (model VNIR 100; Photon Industries Inc., Mississippi Stennis Space Center, USA) across the fundus image (Figs. 4.3 and 4.4).

4.6.1.2. *Sodium fluorescein dye injection*

In order to enhance the blood vessel appearance in angiograms, a sodium fluorescein dye was injected into the monkey's arm. Angiograms were acquired 1–2 minutes after the intravenous injection. When the dye circulated through the retinal arteries, capillaries, and veins, it was progressively eliminated from the vasculature, while staining the ONH and lesions. The biomedical image acquisition software developed by HyperVisual with Photon Industries, Mississippi Stennis Space Center, USA assessed the changes in the relative blood oxygen saturation in the retinal ONH vessels (Figs. 4.5 and 4.6). Then, saturation maps were calculated from the oxygenated and deoxygenated hemoglobin spectral signatures.

4.6.1.3. *Automatic control point detection*

The objective of the adaptive exploratory algorithm is to estimate the initial good-guess of the control points representing the target images' salient features.[88] The following information is available for the adaptive exploratory algorithm. (1) 2D images' resolutions and sizes (width: m pixels; height: n pixels). (2) Corresponding binary $m \times n$ map with 1 standing for white and 0 standing for black. (3) Edge detection algorithm is available for retinal vessel detection with 1 standing for nonedge pixels and 0 standing for edge pixels. It is assumed that both of the reference and the input images do not have huge rotation, scaling, or translation, and, thus, the same features on each image are close to each other.

A global adaptive threshold developed by Otsu[89] is employed to convert the gray level colors to black-and-white. The threshold is a normalized intensity value that lies in the range [0, 1]. Otsu's method chooses the optimal threshold to minimize the intraclass variance of the black and white pixels and to maximize between-class variance in a grayscale image. The algorithm calculates the statistics of the image itself to set the threshold and use the histogram to choose its value at some percentile as a reference value of region strength. Therefore, Otsu's method is nonparametric and nonsupervised. Canny edge detector[90] is employed to extract the ONH vasculature from the binary images. Canny's method detects the edges at the zero-crossings of the second directional derivative of the image. In order to make the localization of magnitude maxima accurate, a filter is defined by optimizing a performance index, which enhances real positive and real negative.

The retinal salient features preserved in the Canny edges are the retinal vessel bifurcations, from which the control points are selected using the adaptive exploratory algorithm (Figs. 4.26, 4.27, 4.28, and 4.29). Initial good-guess of the control points ensures fused image generated at an efficient computational time. Bad control point selection will significantly increase the computation cost, or even cause the image fusion fail. Vessels or some particular abnormalities make images not necessarily matching the retina structures. Even when structure and function correspond, the abnormality still happens sometimes if inconsistence exists between structural and functional changes. Furthermore, angiogram images usually have higher resolution and are rich in information, whereas fundus and the oxygen saturation images have lower resolution and are indeed abstract with some details or even missing some small vessels. Practically, those situations are unavoidable and create difficulties in extracting the control points because the delineation of the vein boundaries may not be precise.

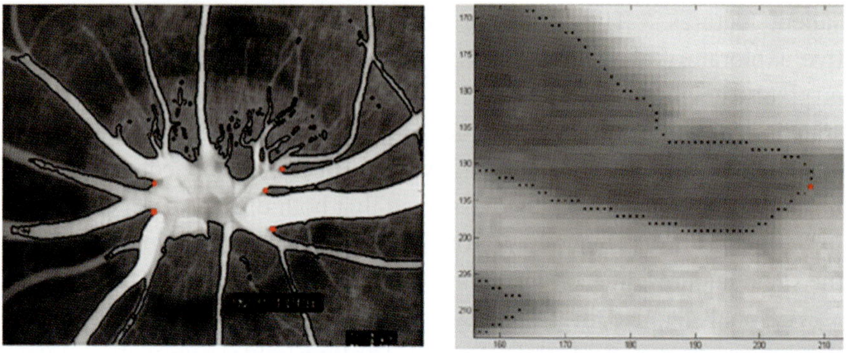

Fig. 4.26. Angiogram image control point selection. Left — five control points selected by using the adaptive exploratory algorithm; Right — enlarged image of the control point selection at the vessel bifurcation point. Reprinted with permission from Cao, H., and Khoobehi, B. Angiogram, fundus, and oxygen saturation optic nerve head image fusion. In: Farkas, D.L., Nicolau, D.V., and Leif, R.C. (eds.), *Imaging, Manipulation, and Analysis of Biomolecules, Cells, and Tissues VII*, Proc of SPIE. 7182, 71821U. ©2009 SPIE.

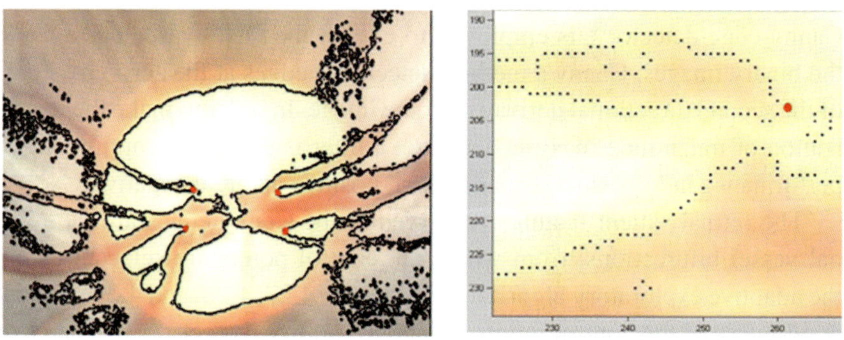

Fig. 4.27. Fundus color image control point selection. Left — 4 control points selected by using adaptive exploratory algorithm; Right — enlarged image of the control point selection at the vessel bifurcation point. Reprinted with permission from Cao, H. and Khoobehi, B. Angiogram, fundus, and oxygen saturation optic nerve head image fusion. In: Farkas, D.L., Nicolau, D.V., and Leif, R.C. (eds.), *Imaging, Manipulation, and Analysis of Biomolecules, Cells, and Tissues VII*, Proc of SPIE. 7182, 71821U. ©2009 SPIE.

For control point detection on the median filtering output, the median filtering is applied toward the black and white ONH oxygen saturation image. Canny edge detection and the adaptive exploratory algorithm are then applied toward the output image after the noise removal for the control point detection.

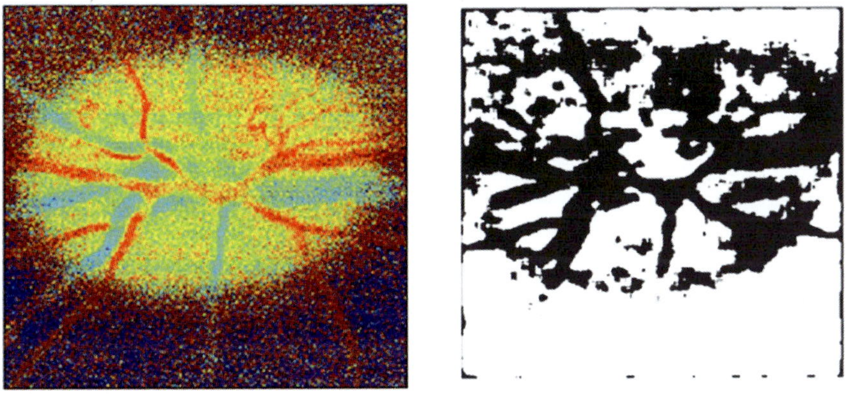

Fig. 4.28. ONH oxygen saturation image (left) and its black and white image with median filter applied (right). Reprinted with permission from Cao, H. and Khoobehi, B. Angiogram, fundus, and oxygen saturation optic nerve head image fusion. In: Farkas, D.L., Nicolau, D.V., and Leif, R.C. (eds.), *Imaging, Manipulation, and Analysis of Biomolecules, Cells, and Tissues VII*, Proc of SPIE. 7182, 71821U. ©2009 SPIE.

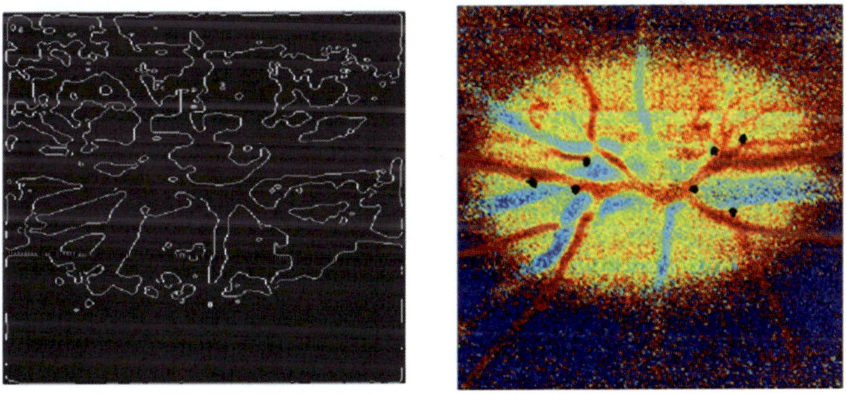

Fig. 4.29. ONH oxygen saturation image's Canny edge (left) and the control point selection by using the adaptive exploratory algorithm (right). Reprinted with permission from Cao, H. and Khoobehi, B. Angiogram, fundus, and oxygen saturation optic nerve head image fusion. In: Farkas, D.L., Nicolau, D.V., and Leif, R.C. (eds.), *Imaging, Manipulation, and Analysis of Biomolecules, Cells, and Tissues VII*, Proc of SPIE. 7182, 71821U. ©2009 SPIE.

4.6.1.4. *Fused image optimization*

An optimization procedure is required to adjust the initial good-guess control points in order to achieve the optimal result. The process can be formulated as a heuristic problem of optimizing an objective function

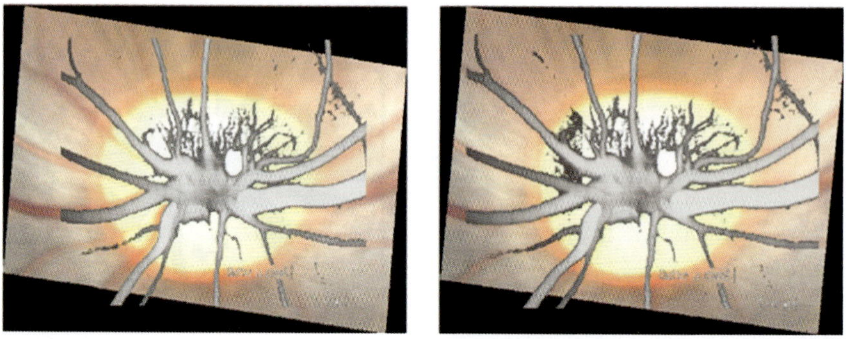

Fig. 4.30. Fused ONH image from angiogram and the fundus color images prior to the optimization (left) and the improved fused image after the iterative optimization (right). Reprinted with permission from Cao, H. and Khoobehi, B. Angiogram, fundus, and oxygen saturation optic nerve head image fusion. In: Farkas, D.L., Nicolau, D.V., and Leif, R.C. (eds.), *Imaging, Manipulation, and Analysis of Biomolecules, Cells, and Tissues VII*, Proc of SPIE. 7182, 71821U. ©2009 SPIE.

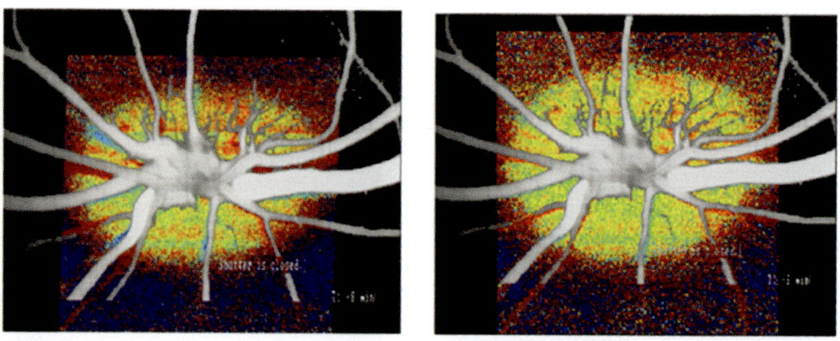

Fig. 4.31. Fused ONH image from angiogram and the oxygen saturation images prior to the optimization (left) and the improved fused image after the iterative optimization (right). Reprinted with permission from Cao, H. and Khoobehi, B. Angiogram, fundus, and oxygen saturation optic nerve head image fusion. In: Farkas, D.L., Nicolau, D.V., and Leif, R.C. (eds.), *Imaging, Manipulation, and Analysis of Biomolecules, Cells, and Tissues VII*, Proc of SPIE. 7182, 71821U. ©2009 SPIE.

that maximizes the Mutual-Pixel-Count between the reference and input images.[91] The algorithm finds the optimal solution by refining the transformation parameters in an ordered way. By maximizing the objective function, one image's vessels are supposed to be well overlaid onto those of the other image (Figs. 4.30 and 4.31).

Mutual-pixel-count measures the ONH vasculature overlapping for corresponding pixels in both images. It is assumed that the retinal vessels are represented by 0 (black pixel) and background is represented by 1 (white pixels) in the binary 2D map. When the vasculature pixel is transformed (u, v) coordinates on the input image correspond to the vasculature pixel's coordinates on the reference image, the MPC is incremented by 1. MPC is assumed be maximized when the image pair is perfectly geometrically aligned by the transformation. The ideal case is that all zero pixels of the input image are mapped onto zero pixels of the reference image. The problem can be mathematically formulated as the maximization of the following objective function:

$$f_{\mathrm{mpc}}(x, y, u, v) = \sum_{I_{\mathrm{ref}}(x,y)=1 \text{ and } I_{\mathrm{input}}(u,v)=1}^{u,v \in ROI} I_{\mathrm{input}}(T_x(u, v), T_y(u, v)), \quad (4.20)$$

where f_{mpc} denotes the value of the mutual-pixel-count. T_x and T_y are the transformations for u and v coordinates of the input image. The ROI (region of interest) is the vasculature region where the MPC is calculated on.

Ideally, f_{mpc} appears to contain a global maximum surrounded by the numerous local maxima. During the iteration, the reference image control point coordinates are fixed. Only input image control point coordinates are subject to adjustment. The coordinate adjustments are iteratively implemented until one of the following convergence criteria is reached:

(1) Predefined maximum number of loops is reached and
(2) The updated f_{mpc} is smaller than ε, i.e.

$$|f^{n+1}_{\mathrm{mpc}}(x, y, u, v) - f^{n}_{\mathrm{mpc}}(x, y, u, v)| < \varepsilon, \quad (4.21)$$

where ε is a very small nonnegative threshold.

4.7. Conclusion

The new algorithm presented in this chapter, which consists of the adaptive exploratory algorithm for the control point detection and heuristic optimization fusion, is reliable and time efficient. The new approach has been applied on three ophthalmologic modalities of nonhuman primate eyes: angiogram, fundus, and oxygen saturation images. It has achieved an excellent result by giving the visualization of fundus or oxygen saturation image with a complete angiogram overlay.

By locking the multi-sensor images in one place, the algorithm allows ophthalmologists to match the same eye over time to get a sense of disease progress and pinpoint surgical tools to increase accuracy and speed of the surgery. Early detection can allow timely treatment to prevent further vision loss, and to prolong effective years of usable vision. Therefore, multi-sensor retinal image registration and fusion, as a novel method in ophthalmology, is able to assist ophthalmologists in performing retinal abnormality analysis and disorder early detection. The new algorithm can be easily expanded to human or animals' 3D eye, brain, or body image feature extraction, registration, and fusion.

References

1. Cioffi, G.A., Orgul, S., Onda, E., Bacon, D.R., and Van Buskirk, E.M. An *in vivo* model of chronic optic nerve ischemia: the dose-dependent effects of endothelin-1 on the optic nerve microvasculature. *Curr Eye Res* **14**:1147–1153, 1995.
2. Cioffi, G.A. and Sullivan, P. The effect of chronic ischemia on the primate optic nerve. *Eur J Ophthalmol* **9**(suppl 1):S34–S36, 1999.
3. Cioffi, G.A., and Wang, L. Optic nerve blood flow in glaucoma. *Semin Ophthalmol* **14**:164–170, 1999.
4. Wang, L., Cioffi, G.A., and Van Buskirk, E.M. The vascular pattern of the optic nerve and its potential relevance in glaucoma. *Curr Opin Ophthalmol* **9**:24–29, 1998.
5. Hayreh, S.S. Factors influencing blood flow in the optic nerve head. *J Glaucoma* **6**:412–425, 1997. [Erratum *J Glaucoma* **7**(1), 71, 1998.]
6. Onda, E., Cioffi, G.A., Bacon, D.R., and Van Buskirk, E.M. Microvasculature of the human optic nerve. *Am J Ophthalmol* **131**:452–453, 1995.
7. Olver, J.M., Spalton, D.J., and McCartney, A.C. Quantitative morphology of human retrolamina optic nerve vasculature. *Invest Ophthalmol Vis Sci* **35**:3858–3866, 1994.
8. Hayreh, S.S. Blood supply of the optic nerve head. *Ophthalmologica* **210**:285–295, 1996.
9. Braun, R.D., and Linsenmeier, R.A. Retinal oxygen tension and the electroretinogram during arterial occlusion in the cat. *Invest Ophthalmol Vis Sci* **36**:523–541, 1995.
10. Sebag, J., Delori, F.C, Feke, G.T., and Weiter, J.J. Effects of optic atrophy on retinal blood flow and oxygen saturation in humans. *Arch Ophthalmol* **107**:222–226, 1989.
11. Stefansson, E., Landers, M.B., and Wolbarsht, M.L. Oxygenation and vasodilatation in relation to diabetic and other proliferative retinopathies. *Ophthalmic Surg* **14**:209–226, 1982.
12. Stefansson, E., Machemer, R., and de Juan, E., Jr, *et al.* Retinal oxygenation and laser treatment in patients with diabetic retinopathy. *Am J Ophthalmol* **113**:36–38, 1992.
13. Tiedeman, J.S., Kirk, S.E., Srinivas, S., and Beach, J.M. Retinal oxygen consumption during hyperglycemia in patients with diabetes without retinopathy. *Ophthalmology* **105**:31–36, 1998.
14. Beach, J.M., Schwenzer, K.J., Srinivas, S., *et al.* Oximetry of retinal vessels by dual-wavelength imaging: calibration and influence of pigmentation. *J Appl Physiol* **86**:748–758, 1999.

15. Crittin, M., Schmidt, H., and Riva, C.E. Hemoglobin oxygen saturation (SO2) in the human ocular fundus measured by reflectance oximetry: preliminary data in retinal veins. *Klin Monatsbl Augenheilkd* **219**:289–291, 2002.
16. Delori, F.C. Noninvasive technique for oximetry of blood in retinal vessels. *Appl Optics* **27**:1113–1125, 1988.
17. Hickam, J.B., Frayser, R., and Ross, J.C. A study of retinal venous blood oxygen saturation in human subjects by photographic means. *Circulation* **27**:375–385, 1963.
18. Jensen, P.J. Non-invasive retinal oximetry in normal human subjects. *Computer Assisted Fundus Image Analysis*. Turin, Italy March 28–30, 2003.
19. Laing, R.A., Cohen, A.J., and Friedman, E. Photographic measurements of retinal blood oxygen saturation: falling saturation rabbit experiments. *Invest Ophthalmol Vis Sci* **14**:606–610, 1975.
20. Schweitzer, D., Hammer, M., Kraft J., *et al. In vivo* measurement of the oxygen saturation of retinal vessels in healthy volunteers. *IEEE Trans Biomed Engr* **46**:1454–1465, 1999.
21. Schweitzer, D., Thamm, E., Hammer, M., and Kraft, J. A new method for the measurement of oxygen saturation at the human ocular fundus. *Int Ophthalmol* **23**:347–353, 2001.
22. Smith, M.H., Denninghoff, K.R., Hillman, L.W., and Chipman, R.A. Oxygen saturation measurements of blood in retinal vessels during blood loss. *J Biomed Opt* **3**:296–303, 1998.
23. Uzumcu, M., Vos, F.M., Vossepoel, A.M., and van der Heijde, G.L. Theoretical analysis of a spectrophotometric technique for measuring oxygen saturation in retinal vessels. *Proceedings of ASCI 2000, 6th Annual Conference of the Advanced School for Computing and Imaging*, Lommel, Belgium, June, 14–16, van Vliet, L.J., Heijnsdijk, W.J., Kielman, T., Knijnenburg, P.M.W. ASCI, Delft 117–121, 2000.
24. Chamot, S.R., Cranstoun, S.D., Petrig, B.L., *et al.* Blood PO2 and blood flow at the optic disc. *J Biomed Opt* **8**:63–69, 2003.
25. Riva, C.E. Noninvasive measurement of oxygen tension in the optic nerve head. *Curr Opin Ophthalmol* **9**:56–60, 1998.
26. Shonat, R.D., Wilson, D.F., Riva, C.E., and Cranstoun, S.D. Effect of acute increases in intraocular pressure on intravascular optic nerve head oxygen tension in cats. *Invest Ophthalmol Vis Sci* **33**:3174–3180, 1992.
27. Shonat, R.D., and Kight, A.C. Oxygen tension imaging in the mouse retina. *Ann Biomed Eng* **31**:1084–1096, 2003.
28. Braun, R.D., Linsenmeier, R.A., and Goldstick, T.K. Oxygen consumption in the inner and outer retina of the cat. *Invest Ophthalmol Vis Sci* **36**:542–554, 1995.
29. Pedersen, D.B., Koch, Jensen P., la Cour, M., *et al.* Carbonic anhydrase inhibition increases retinal oxygen tension and dilates retinal vessels. *Graefes Arch Clin Exp Ophthalmol* **243**:163–168, 2005.
30. Linsenmeier, R.A., and Braun, R.D. Oxygen distribution and consumption in the cat retina during normoxia and hypoxemia. *J Gen Physiol* **99**:177–197, 1992.
31. Berkowitz, B.A. Adult and newborn rat inner retinal oxygenation during carbogen and 100% oxygen breathing: comparison using magnetic resonance imaging delta PO2 mapping. *Invest Ophthalmol Vis Sci* **37**:2089–2098, 1996.

32. Berkowitz, B.A., Kowluru, R.A., and Frank, R.N., *et al.* Subnormal retinal oxygenation response precedes diabetic-like retinopathy. *Invest Ophthalmol Vis Sci* **40**:2100–2105, 1999.
33. Khoobehi, B., Beach, J.M., and Kawano, H. Hyperspectral imaging for measurement of oxygen saturation in the optic nerve head. *Invest Ophthalmol Vis Sci* **45**:1464–1472, 2004.
34. The Diabetic Retinopathy Study Research Group. Preliminary report on effects of photocoagulation therapy. *Am J Ophthalmol* **81**:383–396, 1976.
35. Early Treatment Diabetic Retinopathy Study design and baseline patient characteristics: ETDRS report number 7. *Ophthalmology* **98**:741–756, 1991.
36. Stefansson, E., Landers, M.B. III, and Wolbarsht, M.L. Increased retinal oxygen supply following pan-retinal photocoagulation and vitrectomy and lensectomy. *Trans Am Ophthalmol Soc* **79**:307–334, 1981.
37. Wolbarsht, M.L., and Landers, M.B. III. The rationale of photocoagulation therapy for proliferative diabetic retinopathy: a review and a model. *Ophthalmic Surg* **11**:235–245, 1980.
38. Stefansson, E. Ocular oxygenation and the treatment of diabetic retinopathy. *Surv Ophthalmol* **51**:364–380, 2006.
39. Ahmed, J., Braun, R.D., Dunn, R., Jr., and Linsenmeier, R.A. Oxygen distribution in the macaque retina. *Invest Ophthalmol Vis Sci* **34**:516–521, 1993.
40. Cohen, L.H., and Noell, W.K. Relationships between visual function and metabolism. In: Graymore, C.N. (ed.), *Biochemistry of the Retina*, New York: Academic Press, 36–50, 1965.
41. Haugh, L.M., Linsenmeier, R.A., and Goldstick, T.K. Mathematical models of the spatial distribution of retinal oxygen tension and consumption, including changes upon illumination. *Ann Biomed Eng* **18**:19–36, 1990.
42. Wang, L., Kondo, M., and Bill, A. Glucose metabolism in cat outer retina: effects of light and hyperoxia. *Invest Ophthalmol Vis Sci* **38**:48–55, 1997.
43. Witmer, A.N., Vrensen, G.F., Van Noorden, C.J., and Schlingemann, R.O. Vascular endothelial growth factors and angiogenesis in eye disease. *Prog Retin Eye Res* **22**: 1–29, 2003.
44. Alder, V.A., Cringle, S.J., and Brown, M. The effect of regional retinal photocoagulation on vitreal oxygen tension. *Invest Ophthalmol Vis Sci* **28**:1078–1085, 1987.
45. Stefansson, E., Hatchell, D.L., Fisher, B.L., Sutherland, F.S., and Machemer, R. Panretinal photocoagulation and retinal oxygenation in normal and diabetic cats. *Am J Ophthalmol* **101**:657–664, 1956.
46. Lowry, O.H., Roberts, N.R., and Lewis, C. The quantitative histochemistry of the retina. *J Biol Chem* **220**:879–892, 1956.
47. Lowry, O.H., Roberts, N.R., Schulz, D.W., Clow, J.E., and Clark, J.R. Quantitative histochemistry of retina, II: enzymes of glucose metabolism. *J Biol Chem* **236**:2813–2820, 1961.
48. Pournaras, C.J., Tsacopoulos, M., Strommer, K., Gilodi, N., and Leuenberger, P.M. Scatter photocoagulation restores tissue hypoxia in experimental vasoproliferative microangiopathy in miniature pigs. *Ophthalmology* **97**:1329–1333, 1990.

49. Molnar, I., Poitry, S., Tsacopoulos, M., Gilodi, N., and Leuenberger, P.M. Effect of laser photocoagulation on oxygenation of the retina in miniature pigs. *Invest Ophthalmol Vis Sci* **26**:1410–1414, 1985.
50. Budzynski, E., Smith, J.H., Bryar, P., Birol, G., and Linsenmeier, R.A. Effects of photocoagulation on intraretinal PO2 in cat. *Invest Ophthalmol Vis Sci* **49**:380–389, 2008.
51. Vanderkooi, J.M., Maniara, G., Green, T.J., and Wilson, D.F. An optical method for measurement of dioxygen concentration based upon quenching of phosphorescence. *J Biol Chem* **252**:5476–5482, 1987.
52. Wilson, D.F., Vanderkooi, J.M., Green, T.J., Maniara, G., Defeo, S.P., and Bloomgarden, D.C. A versatile and sensitive method for measuring oxygen. *Adv Exp Med Biol* **215**:71–77, 1987.
53. Wilson, D.F., Rumsey, W.L., Green, T.J., and Vanderkooi, J.M. The oxygen dependence of mitochondrial oxidative phosphorylation measured by a new optical method for measuring oxygen concentration. *J Biol Chem* **263**:2712–2718, 1988.
54. Rumsey, W.L., Vanderkooi, J.M., and Wilson, D.F. Imaging of phosphorescence: a novel method for measuring oxygen distribution in perfused tissue. *Science* **241**:1649–1651, 1988.
55. Vanderkooi, J.M. and Berger, J.W. Excited triplet states used to study biological macromolecules at room temperature. *Biochem Biophys Acta* **976**:1–27, 1989.
56. Wilson, D.F., Pastuszko, A., DiGiacomo, J.E., Pawlowski, M., Schneiderman, R., and Delivoria-Papadopoulos, M. Effect of hyperventilation on oxygenation of the brain cortex of new-born piglets. *J Appl Physiol* **70**:2691–2696, 1991.
57. Shonat, R.D., Wilson, D.F., Riva, C.E., and Pawlowski, M. Oxygen distribution in the retinal and choroidal vessels of the cat as measured by a new phosphorescence imaging method. *Appl Optics* **31**:3711–3718, 1992.
58. Schweitzer, D., Hammer, M., and Scibor, M. (1996) Imaging spectrometry in ophthalmology-principle and applications in microcirculation and in investigation of pigments. *Opthalmic Res* **28**(suppl 2):37–44, 1992.
59. Pittman, R.N. and Duling, B.R. A new method for the measurement of percent oxyhemoglobin. *J Appl Physics* **38**:315–20, 1975.
60. Pittman, R.N. and Duling, B.R. Measurement of percent oxyhemoglobin in the microvasculature. *J Appl Physics* **38**:321–27, 1975.
61. Zheng, L., Golub, A., and Pittman, R.N. Determination of PO2 and its heterogeneity in single capillaries. *Am J Physiol* **271**(1 Pt 2):H365–H372, 1996.
62. Harris, A., Dinn, R.B., Kagemann, L., and Rechtman, E. A review of methods for human retinal oximetry. *Ophthalmic Surg Lasers Imaging* **34**:152–164, 2003.
63. Hardarson, S.H., Basit, S., Jonsdottir, T.E., Eysteinsson, T., Halldorsson, G.H., Karlsson, R.A., Beach, J.M., Benediktsson, J.A., and Stefansson, E. Oxygen saturation in human retinal vessels is higher in dark than in light. *Invest Ophthalmol Vis Sci* **50**:2308–2311, 2009.
64. Vilser, W., Nagel, E., and Lanzl, I. Retinal vessel analysis-new possibilities. *Biomed Tech* (Berl) **47**(Suppl 1, Pt 2):682–685, 2002.
65. Hammer, M., Riemer, T., Vilser, W., Gehlert, S., and Schweitzer, D. A new imaging technique for retinal vessel oxymetry-principles and first clinical results in patients

with retinal arterial occlusion and diabetic retinopathy. *Ophthalmic Technologies* XIX SPIE 7163:71630P:1–12, 2009.

66. Delori, F.C., and Pflibsen, K.P. Spectral reflectance of the human ocular fundus. *Appl Opt* **28**:1061–1077, 1989.
67. Hardarson, S.H., Gottfredsdottir, M.S., Halldorsson, G.H., Karlsson, R.A., Benediktsson, J.A., Eysteinsson, T., Beach, J.M., Alon, H.A., and Stefansson, E. Glaucoma filtration surgery and retinal oxygen saturation. *Invest Ophthalmol Vis Sci* **50**:5247–5250, 2009.
68. Hammer, M., Thamm, E., and Schweitzer, D. A simple algorithm for *in vivo* ocular fundus oximetry compensating for non-haemoglobin absorption and scattering. *Phys Med Biol* **47**:233–238, 2002.
69. Delori, F.C. Reflectometry measurements of optic disc blood volume. In: Lambrou, G.N., and Greve, E.L. *Ocular Blood Flow in Glaucoma*, Kugler and Ghedini Publications, Berkeley, CA, 155–166, 1989.
70. Milliken, G.A. and Johnson, D.E. *Analysis of Messy Data. Volume I: Designed Experiments*, New York: Van Nostrand Reinhold Company, 1984.
71. Edwards, D., and Berry, J.J. The efficiency of simulation-based multiple comparisons. *Biometrics* **43**:913–928, 1987.
72. SAS Institute. *SAS/STAT® 9.1 User's Guide.* Cary, NC: SAS Institute, Inc., 2004.
73. Beach, J., Ning, J., and Khoobehi, B. Oxygen saturation in optic nerve head structures by hyperspectral image analysis. *Curr Eye Res* **32**:161–170, 2007.
74. Schweitzer, D., Hammer, M., Kraft, J., *et al. In vivo* measurement of the oxygen saturation of retinal vessels in healthy volunteers. *IEEE Trans Biomed Eng* **46**:1454–1465, 1999.
75. Berkowitz, B.A. Adult and newborn rat inner retinal oxygenation during carbogen and 100% oxygen breathing. Comparison using magnetic resonance imaging delta PO2 mapping. *Invest Ophthalmol Vis Sci* **37**:2089–2098, 1996.
76. Geijer, C., and Bill, A. Effects of raised intraocular pressure on retinal, prelaminar, laminar, and retrolaminar optic nerve blood flow in monkeys. *Invest Ophthalmol Vis Sci* **18**:1030–1042, 1979.
77. Duling, B.R. and Berne, R.M. Longitudinal gradients in periarteriolar oxygen tension. A possible mechanism for the participation of oxygen in local regulation of blood flow. *Circ Res* **27**:669–678, 1970.
78. Kobayashi, H. and Takizawa, N. Imaging of oxygen transfer among microvessels of rat cremaster muscle. *Circulation* **105**:1713–1719, 2002.
79. Riva, C.E., Hero, M., Titze, P., and Petrig, B. Autoregulation of human optic nerve head blood flow in response to acute changes in ocular perfusion pressure. *Graefes Arch Clin Exp Ophthalmol* **235**:618–626, 1997.
80. Hayreh, S.S., Revie, I.H., and Edwards, J. Vasogenic origin of visual field defects and optic nerve changes in glaucoma. *Br J Ophthalmol* **54**:461–472, 1970.
81. Wise, G.N., Dollery, C.T., and Henkind, P. *The Retinal Circulation.* Harper & Row, New York, 1971.
82. Intaglietta, M., Johnson, P.C., and Winslow, R.M. Microvascular and tissue oxygen distribution. *Cardiovasc Res* **32**:632–643, 1996.
83. Truskey, G.A., Yuan, F., and Katz, D.F. *Transport Phenomena in Biological Systems.* Prentice Hall, Upper Saddle River, N.J., 2004.

84. Weerappuli, D.P. and Popel, A.S. A model of oxygen exchange between an arteriole or venule and the surrounding tissue. *J Biomech Eng* **111**:24–31, 1989.
85. Shoelson, B. Automated quantification of ocular hemodynamics. Ph.D. thesis. Tulane University, 1998.
86. Morita, H. Ventricular wall thickness and blood pressure values in normal cynomolgus monkeys. *J Vet Med Sci* **57**:1045–1048, 1995.
87. Tsai, A.G., Friesenecker, B., Cabrales, P., Hangai-Hoger, N., and Intaglietta, M. The vascular wall as a regulator of tissue oxygenation. *Curr Opin Nephrol Hypertens* **15**:67–71, 2006.
88. Cao, H., Brener, N., Thompson, H., Iyengar, S.S., and Ye, Z. A novel automated retinal image fusion using adaptive exploratory algorithm and mutual-pixel-count maximization. *IEEE 40th Southeastern Symposium on System Theory (SSST 2008)*; New Orleans, LA, USA March 16–18: pp. 122–126, 2008.
89. Otsu, N. A threshold selection method from gray-level histograms. *IEEE Transactions on Systems, Man, and Cybernetics* SMC-9:62–66, 1979.
90. Canny, J. A computational approach to edge detection. *IEEE Trans Pattern Anal Machine Intelligence* **8**:679–698, 1986.
91. Cao, H. Automated fusion of multi-modality biomedical images. *Proc SPIE* **7182**:71821U-8, 2008.

Chapter 5
Automated Localization of Eye and Cornea Using Snake and Target-Tracing Function

Jen-Hong Tan*, Ng, E.Y.K.*, Rajendra Acharya†, U. and Chee, C.‡

Thermography is a noninvasive thermal imaging technique that is widely utilized in the medical and engineering arenas. Infrared (IR) thermography captures IR radiation from the surface of an object and illustrates the corresponding surface temperature as a thermogram. This technique is capable of detecting subtle temperature changes in vascular tissue, and, hence, is important in the detection of ocular diseases.

Localizing the eye and cornea is difficult in an IR thermogram, as the boundary between the iris and sclera is not discernable in a thermal image. This work presents an algorithm that correctly localizes the eye and cornea in an IR thermogram using the snake algorithm and target-tracing function. The target-tracing function is minimized in this algorithm to determine the shape and location of the initial contour, which in latter produces the best localization. With this function, user does not need to determine the region where the snake algorithm should proceed to converge beforehand. The corneal center and radius are estimated from the snake points that have successfully localized the eye feature. In this work, 181 normal and 88

*School of Mechanical and Aerospace Engineering, College of Engineering, Nanyang Technological University, 50, Nanyang Avenue, Singapore 639798.
†Department of Electronics and Computer Engineering, Ngee Ann Polytechnic, Singapore.
‡Medical & Surgical Retinal Centre Department of Ophthalmology, National University Hospital, Singapore 119074.

abnormal ocular IR thermal images are used, and the proposed algorithm localizes the eye and cornea accurately in 84% of all 269 IR thermograms.

The most common causes of blindness around the world, according to an estimation released by the World Health Organization (WHO) in 2002, are cataracts (47.8%), glaucoma (12.3%), age-related macular degeneration (AMD) (8.7%), trachoma (3.6%), corneal opacity (5.1%), and diabetic retinopathy (4.8%) [http://www.owsp.org/problem.htm]. The chances of an individual developing an ocular disease rise with age. Normal vision can become impaired if early and immediate treatment of an eye disease is not received. Among those age-related diseases, cataract is the most common eye abnormality. Cataracts are caused by the clouding of the ocular lens. Iridocyclitis, corneal haze, and dry eye are other commonly seen anterior segment eye abnormalities. The early detection of these ocular impairments and the subsequent proper treatments can prevent vision loss.

5.1. Introduction to Thermography

Thermography is a noninvasive and fast thermal imaging technique. This temperature measuring technology has been applied in thermo-fluid dynamics,[1] polygraph testing,[2] environmental monitoring,[3] and agriculture.[4] Medical field researchers have used it to diagnose breast cancer,[5,6] to manage neuropathic pain,[7,8] impotence,[9] and thyroid gland disease,[10] and to record ocular surface temperature (OST), either as a single image or as a sequence, for the studies of ocular physiology and pathology.[11]

Temperature profiles across the anterior ocular surface were first examined in detail using a wide-field of color-coded thermal imagers in 1989.[12] Previously, invasive investigating techniques, such as a needle probe, were the only options to the study of OST. Topical anesthesia is needed during measurement, and discrepancies in temperature reading can be up to 6°C.[13] These methods can be traumatic and can induce unnecessary ocular blood flow.

Hence, a detailed study of ocular diseases was not possible with the invasive investigating techniques. Nor was it possible with early IR thermometry. However, as the technology of thermography advances, with higher precision and resolution, IR thermal imaging can be used to study ocular physiology, such as the relationship between the physical properties of the anterior eye and OST[14] and the relationship between OST and ocular diseases, with greater ease and accuracy.

The wide field color-coded IR thermal imager revealed physiological correlations between oral temperature and age and OST. Normal individuals with a higher oral temperature showed a higher OST.[15] The temperature difference between the central cornea and limbus averaged 0.45°C.[12,15] In the experiment, corneal surface temperature decreased at a value of 0.001°C per year as test subjects aged,[16] and the intraocular difference in corneal surface temperature was less than 0.6°C for normal subjects.[15]

Subjects with iridocyclitis illustrated a higher corneal surface temperature under the inspection of thermogram. In some cases, the corneal surface temperature was 2.2°C higher, where the average temperature for normal subjects was 35.4 ± 0.1°C.[17] For dry eye[18] and herpes zoster ophthalmicus,[19] the affected eye was observed to be warmer than the unaffected eye.

In primary open-angle glaucoma (POAG), the temperature of the infected eye was lower than the temperature of the healthy eye. The temperature of five anatomical points (the internal canthus, the external canthus, the half-way point between the internal canthus and nasal limbus, the center of the cornea, and the half-way point between the temporal limbus and external canthus) were lower on the thermograms of POAG subjects than on the thermograms of normal individuals.[20] Those temperatures were correlated to the resistivity index, and the effect of retrobulbar hemodynamics on OST was highlighted.[20] In another examination, those eyes that suffered from ischemic central retinal vein occlusion were said to be cooler than those that did not suffer from ischemic central retinal vein occlusion.[21] An investigation in carotid artery stenosis (CAS) revealed a negative correlation between the degree of CAS and the OST.[22]

The application of IR thermal imaging to diagnose ocular diseases is still in the nascent stage, and has not been as successful in this field, as it has been in diagnosing breast cancer and vascular diseases. The main obstacle for the application of IR thermal imaging is the lack of sophisticated analysis and methods to localize the eye and cornea systematically and automatically.

The cornea is commonly manually located on the IR thermogram, and OST or corneal surface temperature is acquired either through a group of separated points or through points within the region identified as the corneal area.[11] The boundary between the iris and sclera is not discernible on the thermogram. Therefore, a fully automated localization of the cornea is difficult.

A study by Efron *et al.* in which a wide field of color-coded IR thermal imagers was first utilized to capture OST, approached the above issue by estimating the geometric center of the cornea and measuring the temperature for every 0.5 mm increments on either side of the ocular anterior surface horizontally. Similar measures were proposed in latter studies, to acquire temperatures of five anatomical locations along the horizontal meridian of the anterior surface.[20,21,23] This problem was also approached using one[24,25] or five 10 × 10 pixels boxes[15,18,22] put along the horizontal meridian, a squared 20 × 20 pixels box,[26] or a circular region [6,27] placed at the center of an eye.

As a step further, Acharya *et al.* developed a semi-automated algorithm to obtain a corneal surface temperature.[28] In this method, the thermogram was converted to gray-scale, and manually cropped to include only the eye. The resultant image matrix was resized to a standard 400 × 200 pixels, and the cornea was located, with a radius of one-fourth of the length of the eye.

This study proposed a fully automated algorithm to localize the eye and cornea, and acquire OST without any human intervention. It used the snake algorithm to delineate the boundary of the eye. Snake and other deformable template matching techniques, such as spline-based shape matching, diffusion snake, and active shape model, are local optimizers that require a good initial position for correct localization. In many of the algorithm applications, users are required to determine the initial position and shape for a snake. In this study, given the eye captured may not fall at the center of the thermogram, proper initialization in the snake algorithm is necessary to prevent incorrect localization.

However, user intervention in initialization is not required in this algorithm, nor are users required to monitor the expansion or convergence of the snake. The above issue is solved by the target-tracing function.[29] The target-tracing function is formulated so that the resultant snake which accurately localizes the boundary of an eye and will give the function the smallest value. A search for minimum on that function is performed in the algorithm over a number of possible converged snakes to deliver the final correct delineation. The corneal center and its radius are obtained based on the final resultant snake points.

The chapter has been organized as follows: in Sec. 5.2, we present the data acquisition process. In Sec. 5.3, we explain the methods and in Sec. 5.4, we present the results of the systems. The discussion on our data analysis is presented in Sec. 5.5. Finally, we conclude the chapter in Sec. 5.6.

5.2. Data Acquisition

Two sets of thermograms were taken for this study. The first set consisted of 181 thermal images taken from 91 normal subjects aged 14–76 years. These thermograms were collected at The Biomedical Centre, Ngee Ann Polytechnic, Singapore. Before data collection, the subjects were asked to relax for 10 minutes. Participants who had any of the following conditions were excluded:

- an eye disease,
- a history of serious eye disease,
- poor general health, and
- a history of ocular or facial surgery.

The second set consisted of images from 44 subjects who each had at least one eye abnormality. The details of the participant eye health are tabulated in Table 5.1. The data collections were conducted at Clinic E, National University Hospital, Singapore. Participants were not allowed to drink hot liquids, rub their eyes, or use teardrops within 20 minutes before the thermal shooting.

Table 5.1. Recruited subjects in this study.

Cases	No. of subjects	Percentage
Cataract	4	8.8
Post-cataract	5	11.1
Dry eye syndrome	5	11.1
Diabetes and diabetic complication	9	20.0
Blepharitis	3	6.6
Episcleritis, conjunctivitis	5	11.1
Others	14	31.1
Total	44	100.0

Ocular thermograms for both sets were obtained in a controlled environment where the temperature was kept at $25 \pm 1°C$ with a mean humidity of 78%. varioTHERM head II (Germany) (http://www.jenoptik-ir.com/) was the instrument used to capture the ocular IR thermal images in this study. The camera was placed 50 cm from the chin rest where subjects rested their chins for image taking. These thermal images were stored in the irbis format, which records temperature values after each thermogram shooting, and exported to jpeg format with a size of 256×256 pixels, by the built-in post-processing software for further processing.

5.3. Methods

The automated localization of the eye and cornea was achieved by using the snake algorithm and the target-tracing function minimized by the genetic algorithm.[30,31] Snake,[32] or active contour, is a series of points (dubbed snake points) moving under an external force field to lock onto nearby edges. We use it to delineate the edges of the eye under the external force field termed gradient vector flow (GVF).[33] However, an accurate localization can be achieved only if the starting contour is closed to the feature of interest and is an appropriate shape. In this algorithm, the problem was overcome by proposing the target-tracing function and solved by using a genetic algorithm to search for the starting contour that is capable of correctly locating the eye. Afterwards, the corneal radius and its center are acquired using the snake that has localized the eye.

5.3.1. *Snake and GVF*

Traditionally, snake is a curve $\mathbf{x}(s) = [x(s) \quad y(s)]$, where $s \in [0, 1]$, defined on an image domain.[32] It moves across the image spatial domain under a combination of external and internal forces to minimize the following energy functional:

$$E = \int_0^1 \frac{1}{2}[\alpha|\mathbf{x}'(s)|^2 + \beta|\mathbf{x}''(s)|^2] + E_{\text{ext}}(\mathbf{x}(s))ds, \tag{5.1}$$

and the solution is given by:

$$\begin{cases} x_t = (A + \gamma I)^{-1}(\gamma x_{t-1} - f_x(x_{t-1}, y_{t-1})) \\ y_t = (A + \gamma I)^{-1}(\gamma y_{t-1} - f_y(x_{t-1}, y_{t-1})) \end{cases}. \tag{5.2}$$

α andβ are parameters that control the length and rigidity of the snake[33]; $\mathbf{x}'(s)$ and $\mathbf{x}''(s)$ are the first and second derivatives with respect to the contour parameter, s. γ is a step-size, and A is a penta-diagonal banded matrix,[32]

$$A = \begin{bmatrix} L & M & N & & & & J & K \\ K & L & M & N & & & & J \\ J & K & L & M & N & & & \\ & J & K & L & M & N & & \\ & & \ddots & \ddots & \ddots & \ddots & \ddots & \\ & & & J & K & L & M & N \\ N & & & & J & K & L & M \\ M & N & & & & J & K & L \end{bmatrix},$$

where $J = \beta$, $K = -\alpha - 4\beta$, $L = 2\alpha + 6\beta$, $M = -\alpha - \beta$, $N = \beta$.

E_{ext} refers to the external energy, and $\nabla E_{\text{ext}} = [f_x(x, y) \quad f_y(x, y)]$. The external energy function opted for this work is the GVF field,[33] $E_{\text{ext}} = \mathbf{v}(x, y) = [u(x, y), v(x, y)]$. The GVF is a dense vector field that minimizes the functional

$$\varepsilon = \iint \mu(u_x^2 + u_y^2 + v_x^2 + v_y^2) + |\nabla f_e|^2 |\mathbf{v} - \nabla f_e|^2 dxdy, \qquad (5.3)$$

where μ is a parameter regularizing the first and second terms in the integrand and $f_e(x, y)$ is the edge map.[30,31]

$$f_e = -\left\{|\nabla[G_{\sigma_1}(x, y) * I(x, y)]| + r_e \cdot G_{\sigma_2}(x, y) * |\nabla_{\text{sob}} I(x, y)|\right\}, \quad (5.4)$$

where r_e is a control parameter and ∇_{sob} is the Sobel gradient operator. G_{σ_1} and G_{σ_2} are Gaussian blur functions with a standard deviation of σ_1 and σ_2, respectively.[30,31] The proposed edge map combines two slightly different types of image derived forces to smoothen the ocular thermogram and to preserve the subtle pixel difference in the lower eyelid region.[30,31]

The solution to Eq. (5.3) is given by:

$$\begin{cases} u_t(x, y, t) = \mu\nabla^2 u(x, y, t) - [u(x, y, t) - f_{e_x}(x, y)] \\ \qquad\qquad \times [f_{e_x}(x, y)^2 + f_{e_y}(x, y)^2] \\ v_t(x, y, t) = \mu\nabla^2 v(x, y, t) - [v(x, y, t) - f_{e_y}(x, y)] \\ \qquad\qquad \times [f_{e_x}(x, y)^2 + f_{e_y}(x, y)^2] \end{cases}. \qquad (5.5)$$

During the GVF snake algorithm run, the snake of interest often gets trapped in the border of the image due to the artifacts introduced during the calculation of Gaussian blur and GVF force field.[30,31] This problem is resolved by expanding the image matrix, using the following formula and performing a number of iterative calculations. Let A be an image matrix, where the size is $m \times n$. The expanded image after each iteration, labeled matrix B (with a size of $[m+2] \times [n+2]$) is obtained by Eq. (5.5).[30,31]

$$A = \begin{bmatrix} a_{1,1} & a_{1,2} & \cdots & a_{1,n} \\ a_{2,1} & a_{2,2} & \cdots & a_{2,n} \\ \vdots & \vdots & \ddots & \vdots \\ a_{m,1} & a_{m,2} & \cdots & a_{m,n} \end{bmatrix},$$

$$B = \begin{bmatrix} b_{1,1} & b_{1,2} & \cdots & b_{1,n+1} & b_{1,n+2} \\ b_{2,1} & b_{2,2} & \cdots & b_{2,n+1} & b_{2,n+2} \\ \vdots & \vdots & \ddots & \vdots & \vdots \\ b_{m+1,1} & b_{m+1,2} & \cdots & b_{m+1,n+1} & b_{m+1,n+2} \\ b_{m+2,1} & b_{m+2,2} & \cdots & b_{m+2,n+1} & b_{m+2,n+2} \end{bmatrix}.$$

Then:

$$\begin{cases} (b_{i,j})_{2\le i\le m+1;2\le j\le n+1} = (a_{i,j})_{1\le i\le m;1\le j\le n} \\[2ex] b_{1,j} = a_{1,j-1} - \mu\left(a_{1,j-1} - \dfrac{a_{1,1}+a_{1,2}+\cdots+a_{1,n}}{n}\right), \\ \qquad\qquad \text{for } j = 2, 3, \ldots, n+1 \\[2ex] b_{m+2,j} = a_{m,j-1} - \mu\left(a_{m,j-1} - \dfrac{a_{m,1}+a_{m,2}+\cdots+a_{m,n}}{n}\right), \\ \qquad\qquad \text{for } j = 2, 3, \ldots, n+1 \\[2ex] b_{i,1} = a_{i-1,1} - \mu\left(a_{i-1,1} - \dfrac{a_{1,1}+a_{2,1}+\cdots+a_{m,1}}{m}\right), \\ \qquad\qquad \text{for } i = 2, 3, \ldots, m+1 \\[2ex] b_{i,n+2} = a_{i-1,n} - \mu\left(a_{i-1,n} - \dfrac{a_{1,n}+a_{2,n}+\cdots+a_{m,n}}{m}\right), \\ \qquad\qquad \text{for } i = 2, 3, \ldots, m+1 \\[2ex] b_{1,1} = \dfrac{b_{1,2}+b_{2,1}}{2}, b_{1,n+2} = \dfrac{b_{1,n+1}+b_{2,n+2}}{2}, b_{m+2,1} \\ \quad = \dfrac{b_{m+1,1}+b_{m+2,2}}{2}, b_{m+2,n+2} = \dfrac{b_{m+1,n+2}+b_{m+2,n+1}}{2}. \end{cases}$$

5.3.2. *Target Tracing Function and Genetic Algorithm*

The initial contour in this work is constructed by two parabolas,[30,31] as illustrated in Fig. 5.1. They are obtained by:

$$\begin{aligned}\mathbf{C}(x_c, y_c, p_1, p_2, w) &= [C_u(s) \quad C_l(s)]^T \\ &= \mathbf{C}(s),\end{aligned} \tag{5.6}$$

where

$$\begin{cases} C_u(s) = [y_u(s) \;\; x_u(s)] = \left[\dfrac{(x_u(s) - x_c)^2}{-4p_1 w} + (y_c + p_1 w) - (h_u - y_c) \;\; x_u(s)\right], \\ \qquad\qquad\qquad\qquad\qquad\qquad\qquad\qquad \text{for } 0 < s \le 0.5 \\ C_l(s) = [y_l(s) \;\; x_l(s)] = \left[\dfrac{(x_l(s) - x_c)^2}{4p_2 w} + (y_c - p_2 w) - (h_l - y_c) \;\; x_l(s)\right], \\ \qquad\qquad\qquad\qquad\qquad\qquad\qquad\qquad \text{for } 0.5 < s \le 1 \end{cases}$$

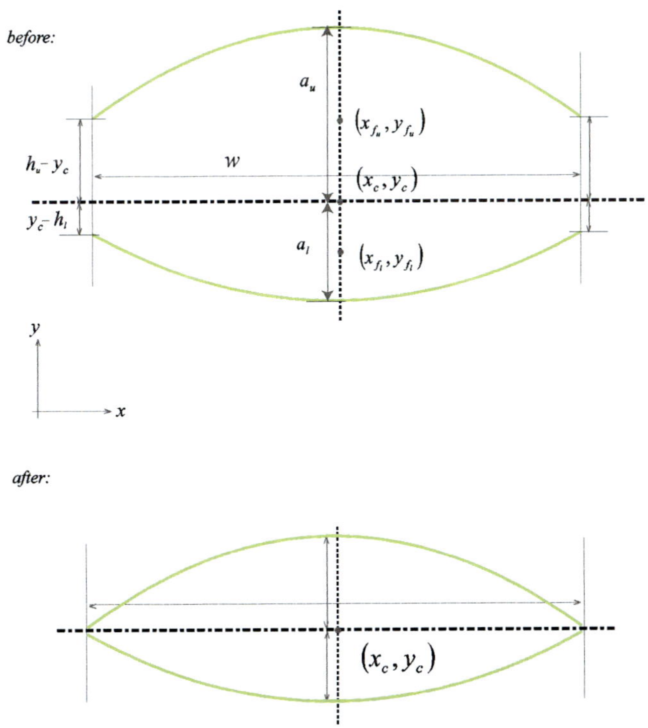

Fig. 5.1. Setting up of initial contour.

and

$$p_1 = \frac{a_u}{w}, \quad p_2 = \frac{a_l}{w}, \quad h_u = \frac{(w/2)^2}{-4p_1 w} + (y_c + p_1 w),$$

$$h_l = \frac{(w/2)^2}{4p_2 w} + (y_c - p_2 w).$$

$C(s)$ is a series of snake points that consists of two parabolas, and is $\mathrm{x}(s) = [x(s)y(s)]$ in discrete form. These points move under Eq. (5.2) and stop shifting after some number of iterations when Eq. (5.1) reaches its minimum. For each iteration, an equal-distance redistribution of points is performed after the calculation of Eq. (5.2) to reduce the convergence time. Let ${}^v\mathrm{C}(s) = {}^v\mathrm{C}(x_c, y_c, p_1, p_2, w)$ denotes a snake that deforms from initial contour $\mathrm{C}(s) = \mathrm{C}(x_c, y_c, p_1, p_2, w)$ and converges to minimum in its own total energy, the target-tracing function is defined as[30,31]:

$$\begin{aligned} y_{ft} &= F_{tt}(x_c, y_c, p_1, p_2, w) \\ &= \int \left[\left\| \frac{d^2}{ds^2} I_{sc}({}^v\mathrm{C}(s)) \right\| \times c_1 + I_{sc}^2({}^v\mathbf{C}(s)) + |{}^v\mathbf{C}'(s)| \times c_2 \right] ds \\ &\quad + c_3 \times \frac{\int I_{sc}^2({}^v\mathbf{C}(s))ds}{\int \|{}^v\mathbf{C}'(s)\| ds}, \end{aligned}$$

in which

$$I_{sc} = \frac{f_\epsilon(x, y) - \min f_\epsilon}{\max f_\epsilon - \min f_\epsilon},$$

$$\frac{d^2}{ds^2} I_{sc}({}^v\mathrm{C}(s)) \approx I_{sc}({}^v\mathrm{C}(s_{n+1})) - 2 \cdot I_{sc}({}^v\mathrm{C}(s)) - I_{sc}({}^v\mathrm{C}(s_{n-1})) \qquad (5.7)$$

where c_1, c_2, and c_3 are control parameters. The ${}^v\mathrm{C}(s)$ that gives the minimum in target-tracing function corresponds to the snake that has converged and accurately localized the eye of interest. The search for the minimum on the target-tracing function is performed over the variables x_c, y_c, p_1, p_c, and w, and is solved by a genetic algorithm.

A genetic algorithm is a stochastic search technique renowned for its capability of solving tough problems, e.g. objective functions that do not have the desired properties for function optimization. A genetic algorithm starts with a population that has a fixed number of individuals generated at random. For a single generation, each individual is denoted by a chromosome in that population, representing one of the potential solutions to the problem. The fitness function is used to evaluate the fitness of each individual and assign a fitness value to the individual. Individuals with a greater fitness value are more likely to be reproduced in the future. These individuals evolve generation by generation, according to "survival of the fittest," through reproduction, crossover, and mutation. These procedures run iteratively until the best fit individual, or the converged snake in this work, gives the minimum in the fitness function (in this work target-tracing function is the fitness function).[30,31]

Figure 5.2 illustrates a 2D manifold spreading across the spatial domain of an image. In Fig. 5.2, the x- and y-axes are the same as the x and y directions defined in the snake equation; the values of the z-axis are defined as the values of I_{sc}. The dark solid line that falls on the elliptical-like lower region is a snake that has accurately localized the eye. From Fig. 5.2, it can be observed that, as a snake converges under the combination of internal and external forces, it is sliding along the manifold to reach the minimum of its total potential energy, $I_{sc}^2(^v\mathrm{C}(s))$.

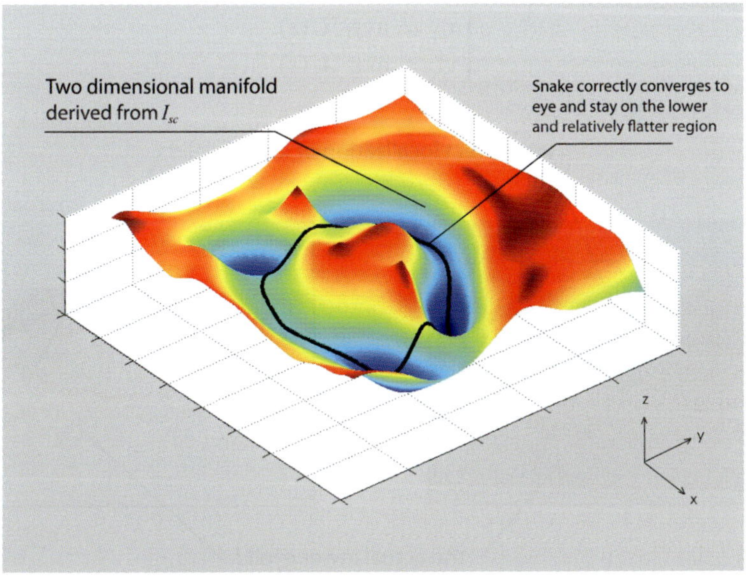

Fig. 5.2. A converged snake lying on a 2D manifold.

For most cases, the edges of an eye in an ocular thermogram generate an elliptical dark ring in I_{sc}. This dark ring is a low and relatively level region on the 2D manifold, as illustrated in Fig. 5.2. The term $I_{sc}^2(^v\mathrm{C}(s))$ represents the total potential energy of all snake points, and the term $\left|\frac{d^2}{ds^2}I_{sc}(^v\mathrm{C}(s_n))\right|$ determines the flatness of the region of the manifold where a snake stays, with respect to x-y plane. Hence, for a snake that has converged and stayed on the eye edges, it gives a minimum value for the term $\left|\frac{d^2}{ds^2}I_{sc}(^v\mathrm{C}(s_n))\right|$ and $I_{sc}^2(^v\mathrm{C}(s))$.

However, in actual practice, the problem as illustrated in Fig. 5.3 appears occasionally. Hence, the term $\frac{\int I_{sc}^2(^v\mathrm{C}(s))ds}{\int \|^v\mathrm{C}'(s)\|ds}$ is added to promote the selection of the snake falling on the preferred region. The term $\|^v\mathrm{C}'(s)\|$ is input to prevent the snake from growing larger than the edges of eye.

5.3.3. *Locating Cornea*

The cornea is assumed to be circular and since snake points $^v\mathrm{C}(s) = [^xC(s) \quad ^yC(s)]$ are equi-distant, the location of its center can be calculated through

$$\begin{cases} x_{cc} = \text{avg}\ ^xC(s) \\ y_{cc} = \text{avg}\ ^yC(s) \end{cases}. \tag{5.8}$$

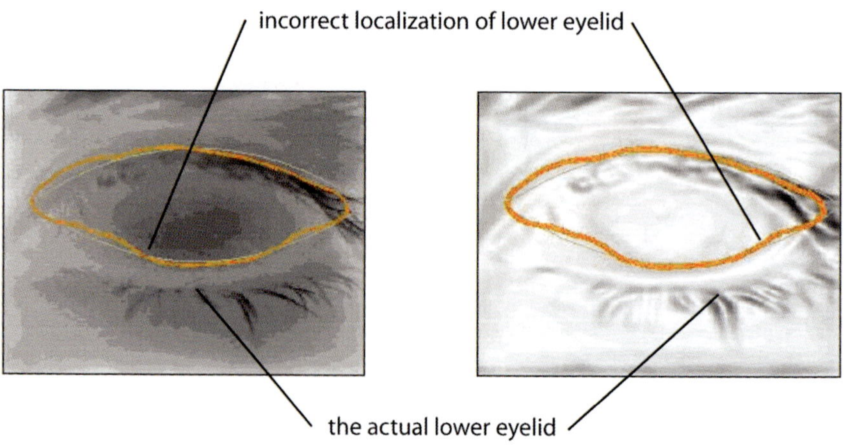

Fig. 5.3. Incorrect localization of lower eyelid by snake (occasionally).

Let the radius of cornea be r_c

$$r_c = rt \times dX, \tag{5.9}$$

and define

$$dX = \max{}^{x}C(s) - \min{}^{x}C(s) \tag{5.10}$$

and

$$dY = \max{}^{y}C(s) - \min{}^{y}C(s). \tag{5.11}$$

Consider two cases in which one of the snakes correctly localizes the eye and another snake converges only to the cornea (not visible in thermogram), as illustrated in Fig. 5.4. For a snake to correctly converge to the eye edges, the corresponding dX and dY are almost equal to the width and the height of the eye. If the corneal radius is roughly 0.25 times the width of an eye, the dX/dY for the eye would be two (assume the cornea touches the upper and lower eyelid). The rt, then, has to be 0.25 for Eq. (5.9) to be true.

In the second case, the dX/dY is one, and rt is 0.5 for the equation to be true. From these cases, we propose a formula to determine the corneal radius adaptively, based on simple interpolation.

$$\frac{2-1}{0.25-0.5} = \frac{\frac{dX}{dY}-1}{rt-0.5}.$$

Hence,

$$rt = \left(\frac{dX}{dY} - 1\right)(-0.25) + 0.5. \tag{5.12}$$

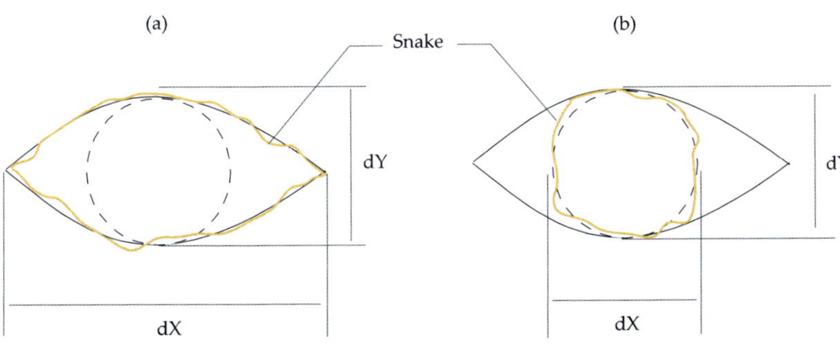

Fig. 5.4. (a) An eye correctly localized by snake and (b) an eye incorrectly localized by snake.

Table 5.2. Summary of results on the proposed algorithm.

Cases	Normal	Abnormal	Percentage
No. of images the algorithm correctly localizes eye and cornea	152	75	84
No. of images the algorithm incorrectly localizes eye and cornea	29	13	16
Total	181	88	100

5.4. Results

The results of the proposed algorithm are tabulated in Table 5.2. Two hundred and sixty-nine thermograms were collected from 135 subjects. The localization was said to be "correct" if the cornea located by the algorithm fell approximately at the center of the eye and its diameter value was close to the height of the eye. This algorithm showed a success rate of 84%; Fig. 5.5 illustrates some of these results.

Table 5.3 shows a tabulation of the number of incorrect localizations of the eye and cornea with respect to the age group. The rate of failure was 2.6 times greater in the normal subjects who were more than 35 years old compared to the younger peers.

5.5. Discussion

In this work, human intervention on starting contours was replaced by the minimization of the target-tracing function. The genetic algorithm searches for the minimum of the target-tracing function. In most cases, it took less than one minute for the genetic algorithm to get the minimum of the target-tracing function on a PC equipped with an Intel core dual processor.

The population size for each generation in a genetic algorithm is set to 15 individuals. The criteria to stop it from further continuation are: (i) no improvement in the fitness function (target-tracing function) (Eq. (5.7)) for three consecutive generations, (ii) no improvement in the fitness function for 100 seconds, and (iii) the number of generations has reached 18.

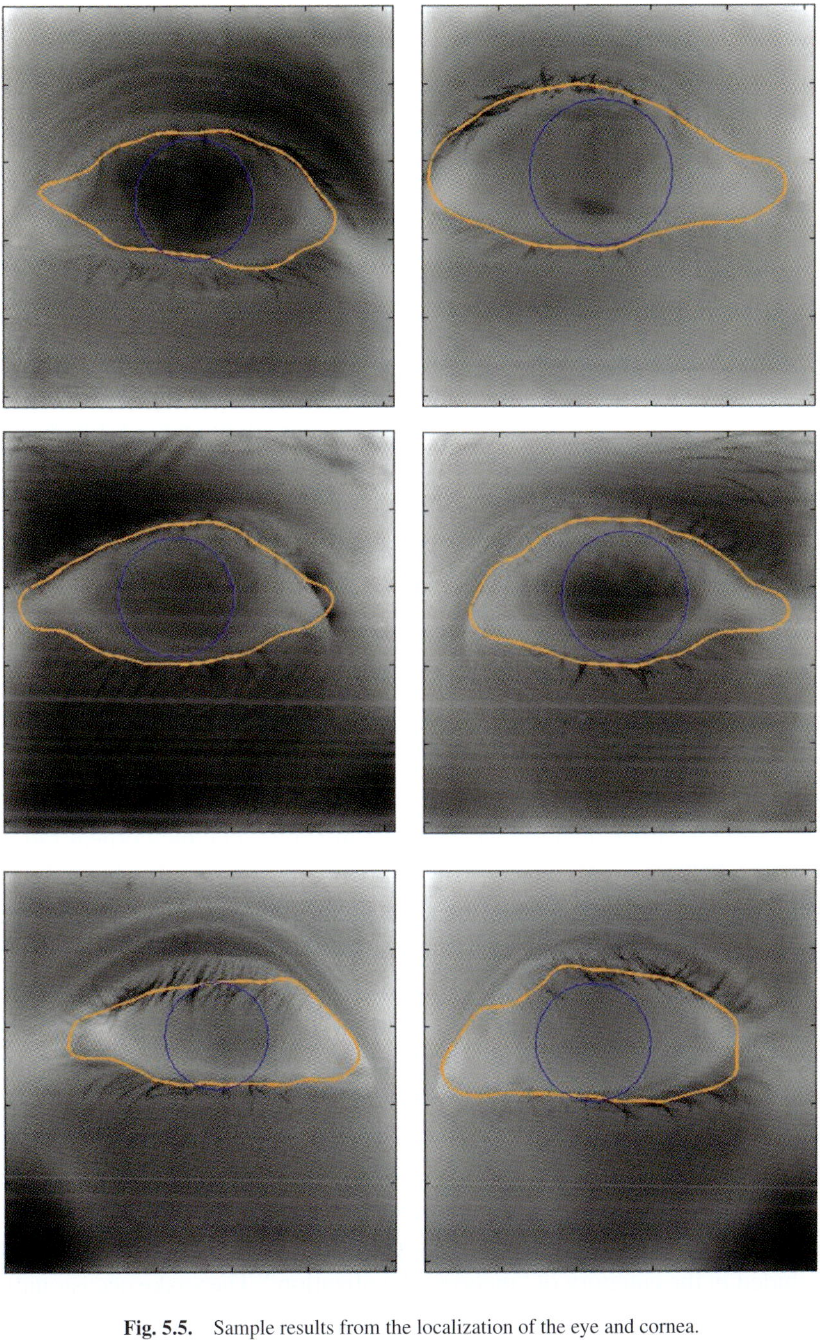

Fig. 5.5. Sample results from the localization of the eye and cornea.

Table 5.3. Number of incorrect localization by age groups.

Groups	No. of incorrect localization	Percentage
Aged 35 and below	8	28
Aged above 35	21	72
Total	29	100

The movement of snake toward minima in this study is highly dependent on σ_1, the standard deviation of the Gaussian blur function. Larger values in that term result in an over-smoothened image, giving a GVF force field that will shrink a snake into the cornea-sclera region instead of sticking to the eye edges. On the contrary, a smaller value in the term gives a rougher image, and the resulting GVF force field often provides unnecessary friction to the snake movement.

Equations (5.9)–(5.12) were proposed for the determination of the corneal center and radius. They were to evaluate the snake points returned from the target-tracing function and adaptively decide the corneal center and the corresponding radius. However, in practice, these equations require extra attentions in two extreme cases.

First, in cases where the corneal diameter is calculated through Eqs. (5.9) and (5.12) is larger than the dY obtained in Eq. (5.11), the corneal radius is set to $dY/2$. Second, the corneal radius is often too small when dX/dY is larger than 2.3. In this case, the radius value is again approximately set to $dY/2$.

From Table 5.3, it is apparent that the failure rate of this algorithm was higher in normal subjects older than 35 years. Observations revealed that the facial temperature profile of subjects older than 35 years is more complex than that of the young. This extra complexity hampers the target-tracing function from correctly selecting the snake that most closely delineates the eye edges.

For subjects who have an epicanthic fold (or "double eyelid"), the algorithm may not localize the eye and cornea accurately. This type of error is included in the category of "incorrect localization." The snake occasionally stays on the epicanthic fold instead of on the actual eyelid. The corneal radius in this case will be larger, and the area of OST acquisition will cover part of the eyelashes.

Literature covers a number of automated methods developed for ocular detection. Asteriadis *et al.* utilized edge-related geometrical information of an eye and its surrounding area to locate the eye.[34] The variance projection function (VPF) was developed and applied to the similar application with encouraging results.[35] Furthermore, the computational complexity in their method is relatively low.

Feng *et al.*[36] detected an eye window based on the extraction of multi-cues from a gray level, and the precise iris and eye corner were located by VPF and eye variance filter. This method has been tested on 930 face images by the MIT AI laboratory, and the accuracy was 92.5%. Jee *et al.* detected eye pairs using edge detection, binary information, and support vector machines, and achieved an accuracy of more than 92%.[37] That system is fast and reliable, and can be used for face detection.

Research has also been performed in the tracking of eye movements in a sequence of images. For example, the time-adaptive self-organizing map (TASOM)-based active contour models (ACMs) tracks eye movements.[38] This method detected the boundaries of the human ocular sclera and tracked its movements. Eye features, such as the iris center or eye corners, were located using information regarding the iris edge. It showed a good performance in general and a better performance than that of the gradient vector field snake-based method.

The above-mentioned algorithm works well for optical facial images, but may not work for IR thermal images. In many cases, the lower eyelid is not clearly discernible in thermogram images. The eyelash, rather than the eyelid, defines the boundary for the upper eyelid. Moreover, the boundary defining iris and cornea are almost absent in nearly all thermograms. Therefore, we propose this automated algorithm, by snake and target-tracing function, to locate the eye and cornea in an IR thermogram.

5.6. Conclusion

IR thermography is a possible alternative approach for the diagnosis of ocular diseases. In this study, we have proposed a method to identify a cornea automatically using a snake and target-tracing function. This algorithm can identify a cornea with an accuracy of roughly 84%. This accuracy can be further improved by incorporating factors, such as the complexity of facial surface temperature due to aging and epicanthic fold, into target-tracing function.

References

1. Meola, C. and Carlomagno, G.M. Recent advances in the use of infrared thermography. *Meas Sci Technol* **15**:R27–R58, 2004.
2. Pavlidis, I. and Levine, J. Thermal image analysis for polygraph testing. *IEEE Eng Med Biol Mag* **21**:56–64, 2002.
3. Cehlin, M., Moshfegh, B. and Sandberg, M. Visualization and measuring of air temperatures based on infrared thermography. *Proc 7th Int Conf Air Distribution in Rooms ROOMVENT*, Reading, UK, 2000.
4. Wisniewski, M., Lindow, S., and Ashworth, E. Observations of ice nucleation and propagation in plants using infrared video thermography. *Plant Physiol* **113**:327–334, 1997.
5. Ng, E.Y.K. and Chen, Y. Segmentation of breast thermogram: improved boundary detection with modified snake algorithm. *J Mech Med Biol* **6**:123–136, 2006.
6. Ng, E.Y.K., Tan, J.H., Ooi, E.H., Chee, C., and Acharya, U.R. Variations of ocular surface temperature with different age groups. *Image Modeling of Human Eye Artech House*, 2008.
7. Hooshmand, H., Hashmi, M., and Phillips, E.M. Infrared thermal imaging as a tool in pain management an 11 year study: I. *Thermology International* **11**:53–65, 2001.
8. Hooshmand, H., Hashmi, M., and Phillips, E.M. Infrared thermal imaging as a tool in pain management an 11 year study: II clinical applications. *Thermology International* **11**:119–129, 2007.
9. Merla, A., Romani, G.L., Tangherlini, A., Romualdo, S.D., Proietti, M., Rosato, E., Aversa, A., and Salsano, F. Penile cutaneous temperature in systemic sclerosis: a thermal imaging study. *Int J Immunopathol Pharmacol* **20**:139–144, 2007.
10. Helmy, A., Holdmann, M., and Rizkalla, M. Application of thermography for non-invasive diagnosis of thyroid gland disease. *IEEE Trans Biomed Eng* **55**:1168–1175, 2008.
11. Tan, J.H., Ng, E.Y.K., Rajendra, A.U., and Chee, C. Infrared thermography on ocular surface temperature: a review. *Infrared Physics & Technology*, 2009.
12. Efron, N., Young, G., and Brennan, N. Ocular surface temperature. *Curr Eye Res* **8**:901–906, 1989.
13. Rosenbluth, R.F. and Fatt, I. Temperature measurements in the eye. *Exp Eye Res* **25**:325–341, 1977.
14. Purslow, C. and Wolffsohn, J.S. The relation between physical properties of the anterior eye and ocular surface temperature. *Optom Vis Sci* **84**:197–201, 2007.
15. Morgan, P.B., Soh, M.P., Efron, N., and Tullo, A.B. Potential applications of ocular thermography. *Optom Vis Sci* **70**:568–576, 1993.

16. Morgan, P.B., Soh, M.P., and Efron, N. Corneal surface temperature decrease with age. *Cont Lens Anterior Eye* **22**:11–13, 1999.
17. Stefanie, P.B. Thermotopography Shows 'Enormous Promise' for Diagnosis and Treatment of Eye Diseases.
18. Morgan, P.B., Tullo, A.B., and Efron, N. Infrared thermography of the tear film in dry eye. *Eye* **9**:615–618, 1995.
19. Tullo, A.B., Cardona, G., Morgan, P.B., and Efron, N. Ocular and facial thermography in herpes zoster ophthalmicus and post-herpetic neuralgia. *Invest Ophthalmol Vis Sci* **37**:S49, 1996.
20. Galassi, F., Giambene, B., Corvi, A., and Falaschi, G. Evaluation of ocular surface temperature and retrobulbar haemodynamics by infrared thermography and colour Doppler imaging in patients with glaucoma. *Br J Ophthalmol* **91**:878–881, 2007.
21. Sodi, A.A., Giambene, B.A.D., Falaschi, G.B., Caputo, R.C., Innocenti, B.B., Corvi, A.B., and Menchini, U.A. Ocular surface temperature in central retinal vein occlusion: preliminary data. *Eur J Ophthalmol* **17**:755–759, 2007.
22. Morgan, P.B., Smyth, J.V., Tullo, A.B., and Efron, N. Ocular temperature in carotid artery stenosis. *Optom Vis Sci* **72**:850–854, 1999.
23. Galassi, F., Giambene, B., Corvi, A., Falaschi, G., and Menchini, U. Retrobulbar hemodynamics and corneal surface temperature in glaucoma surgery, *Int Ophthalmol* **28**:399–405, 2008.
24. Murphy, P.J., Morgan, P.B., Patel, S., and Marshall, J. Corneal surface temperature change as the mode of stimulation of the non-contact corneal aesthesiometer. *Cornea* **18**:333–342, 1999.
25. Betney, S., Morgan, P.B., Doyle, S.J., and Efron, N. Corneal temperature changes during photorefractive keratectomy. *Cornea* **16**:158–161, 1997.
26. Mori, A., Oguchi, Y., Okusawa, Y., Ono, M., Fujishima, H., and Tsubota, K. Use of high-speed, high-resolution thermography to evaluate the tear film layer. *Am J Ophthalmol* **124**:729–735, 1997.
27. Chiang, H.K., Chen, C.Y., Cheng, H.Y., Chen, K.-H., and Chang, D.O. Development of infrared thermal imager for dry eye diagnosis. *Proceedings of SPIE — The International Society for Optical Engineering*, San Diego, CA, USA, 2006.
28. Acharya, U.R., Ng, E.Y.K., Gerk, C.Y., and Tan, J.H. Analysis of normal human eye with different age groups using infrared images. *J Med Syst* **33**:207–213, 2008.
29. Tan, J.H., Ng, E.Y.K., and Rajendra, A.U. Automated detection of eye and cornea on infrared thermogram using snake and target tracing function coupled with genetic algorithm. *Quantitative Infrared Thermography International Journal*, 2009.
30. Tan, J.H., Ng, E.Y.K., and Rajendra, A.U. Automated detection of eye and cornea on infrared thermogram using snake and target tracing function coupled with genetic algorithm. *Quantitative Infrared Thermography International Journal* **6**:21–36, 2009.
31. Tan, J.H., Ng, E.Y.K., and Rajendra, A.U. Detection of eye and cornea on IR thermogram using genetic snake algorithm. *9th International Conference on Quantitative Infrared Thermography*, Krakow, Poland, pp. 143–150, 2008.
32. Kass, M., Witkin, A., and Terzopoulos, D. Snakes: active contour models. *Int J Comput Vis* 321–331, 1988.
33. Xu, C. and Prince, J.L. Snakes, shapes, and gradient vector flow. *IEEE Trans Image Process* **7**:359–369, 1998.

34. Asteriadis, S., Nikolaidis, N., Hajdu, A., and Pitas, I. A novel eye-detection algorithm utilizing edge-related geometrical information. *14th European Signal Processing Conference (EUSIPCO06)*, Florence, Italy, 2006.
35. Feng, G.C. and Yuen, P.C. Variance projection function and its application to eye detection for human face recognition. *Pattern Recognit Lett* **19**:899–906, 1998.
36. Feng, G.C. and Yuen, P.C. Multi-cues eye detection on gray intensity image. *Pattern Recognit* **34**:1033–1046, 2001.
37. Jee, H., Lee, K., and Pan, S. Eye and face detection using SVM. *Proceedings of Intelligent Sensors, Sensor Networks and Information Processing Conference*, pp. 577–580, 2004.
38. Khosravi, M.H. and Safabakhsh, R. Human eye sclera detection and tracking using a modified time-adaptive self-organizing map. *Pattern Recognit* **41**:2571–2593, 2008.

Chapter 6

Automatic Diagnosis of Glaucoma Using Digital Fundus Images

Rajendra Acharya, U.*, Oliver Faust*, Zhu Kuanyi*, Tan Mei Xiu Irene*, Boo Maggie*, Sumeet Dua†, Tan Jen Hong‡ and Ng, E.Y.K.‡

Glaucoma is a progressive optic neuropathy that is caused by an increase of intraocular pressure (IOP) in eye. It mainly affects the optic disc by enlarging the cup size. If undiagnosed and not treated at an early stage, it can lead to blindness. Glaucoma is diagnosed through optical coherence tomography (OCT) and Heidelberg retinal tomography (HRT) and both methods are expensive. In this chapter, we present an improved method to diagnose glaucoma based on digital fundus images. This method makes use of digital image-processing techniques, such as preprocessing, image segmentation, and morphological operations, to detect both optic disc and blood vessels tree. Furthermore, these techniques are used to extract features such as cup-to-disc (c/d) ratio, blood vessels area, and the ratio that relates the blood vessels area in both inferior and superior sides to the blood vessel area in the nasal-temporal side. We validated these features with a Gaussian mixture model (GMM) classification system. This system was used to classify normal and glaucoma images. It identifies glaucoma with a sensitivity of 77% and a specificity of 88%.

*Department of ECE, Ngee Ann Polytechnic, Singapore.
†Department of Computer Science, Louisiana Tech University, Ruston, LA 71272, USA.
‡Department of Mechanical and Aerospace Engineering Nanyang Technological University, Singapore.

6.1. Introduction to Glaucoma

Glaucoma is an optic neuropathy in which structural changes in the optic nerve head[1] and a loss of retinal ganglion cells, in some perceivable patterns, occur. Worldwide, glaucoma affects about 0.5% of all people who are 50 years and younger. The percentage increases to 10% for the age group of 80 years old and older. These facts make glaucoma the second leading cause of blindness after diabetic retinopathy.[2] Often, glaucoma is detected only after vision loss occurs. The increment in IOP in an eyeball, at times, inflicts damage on the optic nerve, which carries image information from light receptors in the retina to the brain. To date, this disease is incurable, the best we can achieve with current technology and understanding is to avoid the total loss of vision by early detecting the disease and as far as possible preventing the disease. There are two forms of glaucoma: primary open-angle glaucoma and angle-closure glaucoma. Each type is briefly explained in the following sections.

6.1.1. *Glaucoma Types*

6.1.1.1. *Primary open-angle glaucoma*

Most glaucoma cases are open-angle glaucoma. This type occurs when the aqueous humor in eye drains too slowly over time, leading to a fluid build-up and gradual increase in IOP. Open-angle glaucoma is characterized by the fact that most entrances to the so-called drainage canals are clear, but unfortunately the insides of the canals become clogged, thus creating a problem that is similar to a clogged pipe below the drain in a sink.[3] There is a chance that this clogging causes gradual vision loss. The damage to optic nerve is so slow and painless that many patients will be unaware of the development of this ocular disease. However, if diagnosed and treated, glaucoma usually responds well to medication.[3]

6.1.1.2. *Angle-closure glaucoma*

Angle-closure glaucoma is also known as narrow-angle glaucoma or acute glaucoma. It is both less common and different from open-angle glaucoma. This ocular disease can cause the loss of vision within a day of its onset. Angle-closure glaucoma has two causes either drainage canals are blocked or they are covered. These canals are blocked similarly to a sink being blocked by some objects covering the drain.[4] In the case of angle-closure glaucoma, the iris is often not as wide and open as it should be. In such cases, the pupil dilates too much or too quickly (as when entering a dark room), and its outer edge compresses over the drainage canals. Patients diagnosed with this disease usually have very narrow drainage canals, which may be present at birth. Angle-closure glaucoma is usually treated with surgery. The surgeon cuts away a small portion of the outer iris edge; this helps to unblock the drainage canals.[5]

6.1.2. *Diagnosis of Glaucoma*

Many glaucoma tests are time consuming and require trained personal as well as special equipment. Glaucoma can be diagnosed through ophthalmoscopy, tonometry, and perimetry; a regular glaucoma checkup includes tonometry and ophthalmoscopy.[4] New techniques to diagnose this ocular disease accurately at early stages are urgently needed.

Recent advances in computer-based systems improved the glaucoma-screening process. Imaging systems, such as HRT, scanning laser polarimetry, OCT, and fundus cameras have been extensively used for ocular diagnosis.[6] OCT, HCT, and confocal laser scanning tomography can indicate retinal nerve fiber damage before this damage has a negative effect on the visual field. However, many hospitals cannot afford such equipment, because of the high costs involved in purchasing, maintaining, and using such equipment. Hence, digital fundus cameras are often used as a cheaper alternative tool to diagnose glaucoma. These digital fundus cameras produce digital fundus images, which are subjected to image processing. This image processing yields features such as optic disk and blood vessels. These features can be used to diagnose the disease.[6–9]

Artificial intelligence (AI) has been incorporated into the above procedures in some investigations for the automated diagnosis of glaucoma. A fuzzy set concept was used to handle the uncertainty that is unavoidable in medical diagnosis,[10] and a neuro-fuzzy method was developed to identify both the presence and the absence of glaucoma with a classification efficiency of 75.8%.[11] A feed forward artificial neural network (ANN) was utilized to discriminate the images of the optic nerve heads of normal and glaucoma subjects.[12] This method was able to provide an accuracy of 88.9% and a sensitivity (correct abnormal) of 84.4%. Other machine-learning algorithms, such as linear and quadratic discriminant analyses, support vector machine, and the mixture of Gaussian, multilayer perceptron, and Parzen windows, were also employed in the diagnosis of glaucoma.[13]

We compared the classifier performances by comparing their receiver-operating characteristic (ROC). To be specific, the parameter we used for the comparison was the area under the ROC curves and sensitivities at chosen specificities. They showed that both forward-selection and backward-elimination methodologies improved the identification rate. Therefore, we conclude that the proposed method has the potential to reduce testing time by diminishing the number of visual-field location measurements.

In this chapter, we present an automated system, which detects glaucoma based on three distinct features. The first feature is the c/d ratio, which quantifies changes in the cup area. Usually, a subject with glaucoma has a higher cup area, and, consequently, the optic nerve head shifts toward the nasal. This shift can be measured by taking the distance between the optic disc center and the nerve head under discussion. The second feature is defined as the ratio of this shift and the optic disc diameter. The last feature is defined as the ratio of the total blood vessel area in both inferior and superior sides of the optic disc in relation to the total blood vessel area in nasal and temporal areas; this ratio is the ISNT ratio. Figure 6.1 shows the functional block diagram of the system under discussion.

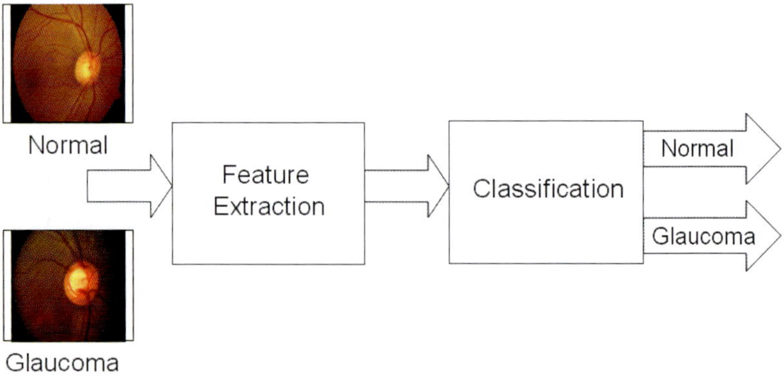

Fig. 6.1. The proposed glaucoma detection system (reprinted from Ref. [14], with kind permission from *Journal of Medical Systems*).

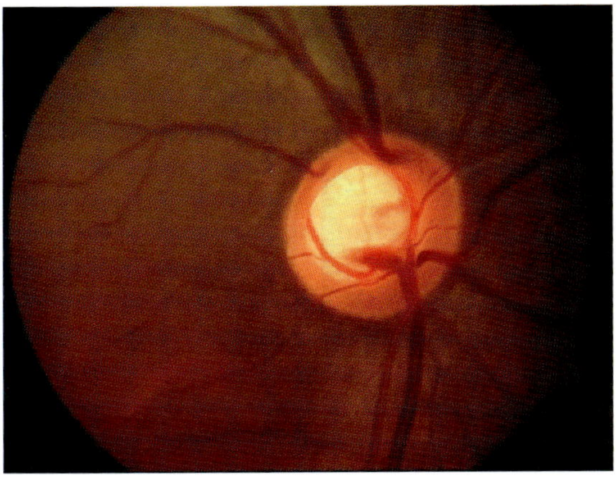

Fig. 6.2. A typical fundus image (reprinted from Ref. [14], with kind permission from *Journal of Medical Systems*).

6.2. Materials and Methods

The fundus images (60 fundus, 30 normal, and 30 glaucoma images) were taken from patients at the Kasturba Medical College, Manipal, India. The subjects were between 20 and 70 years old. The images were graded by physicians from the ophthalmology department of the same hospital. The fundus images were stored in as 720×560 pixel bitmap images. Figure 6.2 shows a typical fundus image, Fig. 6.3(a) shows a normal fundus image, and Fig. 6.3(b) shows a glaucomatous fundus image. Blood vessels in the glaucomatous fundus image are often observed to be swollen.

6.2.1. *c/d Ratio*

The optic cup is enlarged and swollen in a subject suffering from glaucoma. An increase in the c/d ratio is one of the most significant symptoms of glaucoma. Hence, the ratio of the optic cup area to the area of the optic disc was taken as a feature. Figure 6.4 shows the block diagram to compute the c/d ratio.

After analyzing the RGB components of the image, we discovered that it is easier to differentiate the optic disc with the red image component. However, both optic disc and cup could not be easily distinguished in this

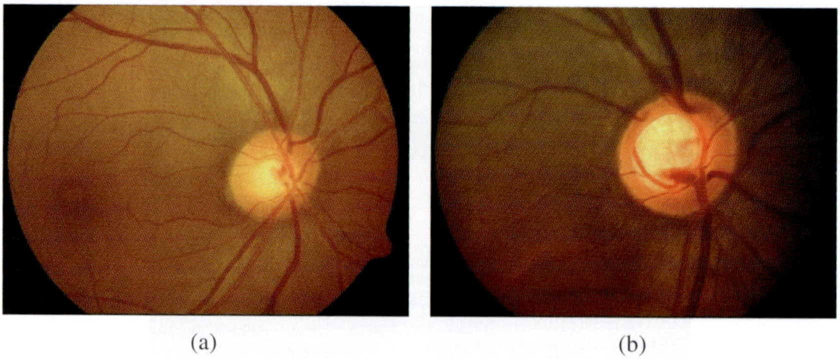

Fig. 6.3. (a) Fundus image of normal eye and (b) fundus image of glaucomatous eye.

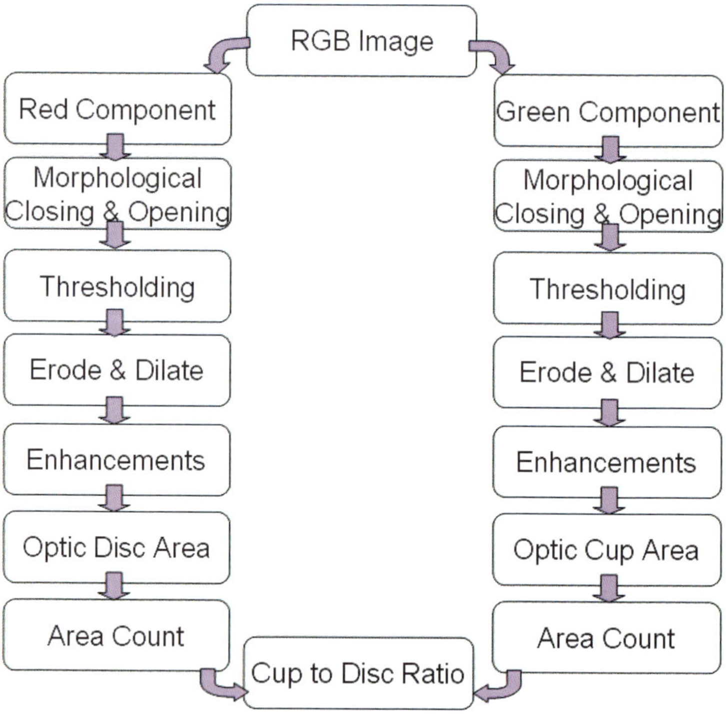

Fig. 6.4. Block diagram of the computation of c/d ratio (reprinted from Ref. [14], with kind permission from *Journal of Medical Systems*).

color component, because there was no well-defined border between these two. By trail and error, we decided that the green component yields the best results for distinguishing the optic cup from the remainder of the image. The contrast of the resultant images on these two channels was enhanced to facilitate feature extraction.

For accurate measurements of both optic disc and cup areas, the blood vessels were removed from the images. This removal was done through a series of morphological operations, namely, dilation, erosion, as well as opening and closing operations.[15] Morphological erosion is used to wear away foreground pixels in specified regions,[7] whereas morphological dilation operations are used to enlarge specific regions of foreground pixels. These techniques remove unwanted bright spots or boundaries in a fundus image.

First, a disc-shaped structuring element of size 15 was created, and a close and open morphological operations were performed both on the red and on the green component images. The structure of image is modified using morphological imaging. Dilation and erosion are widely used on the image. The image grows due to dilation and shrinks because of erosion.[7] The close operation filled the gaps and smoothened the outer edges, whereas the open operation removed small bright spots present.

Then, the boundary of the optic disc and cup was shown using thresholding, in which the eight-bit red and green images were converted to binary images. The thresholding value used was 0.805. Finally, the images were eroded and dilated to further smoothen outer boundaries, as illustrated in Fig. 6.5(a) and (b). The white pixels in the corresponding regions were counted to get the ratio of the area of the optic cup and disc.

6.2.2. *Measuring the Area of Blood Vessels*

A subject with glaucoma will have swollen blood vessels. Hence, the area of blood vessels in a fundus image is an important and useful parameter in the diagnosis of glaucoma. Figure 6.6 shows the flowchart to evaluate the area of blood vessels.

The location of the blood vessel was determined using the green component, as the blood vessels within the optic disc are easier to see than the other components. Image intensity value was complemented, and,

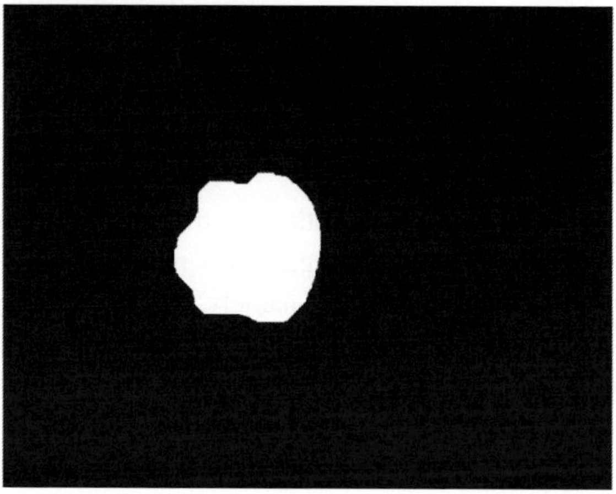

(a)

(b)

Fig. 6.5. (a) Optic disc area and (b) optic cup area.

hence, in the output image, the initial dark areas become bright and vice versa.

A disc-shaped structuring element of size 10 was created and followed by various morphological operation techniques to remove unwanted bright spots, fill the gaps, and smoothen the outer edges. By thresholding with threshold value of 0.1, a binary image was obtained. The blood vessel area

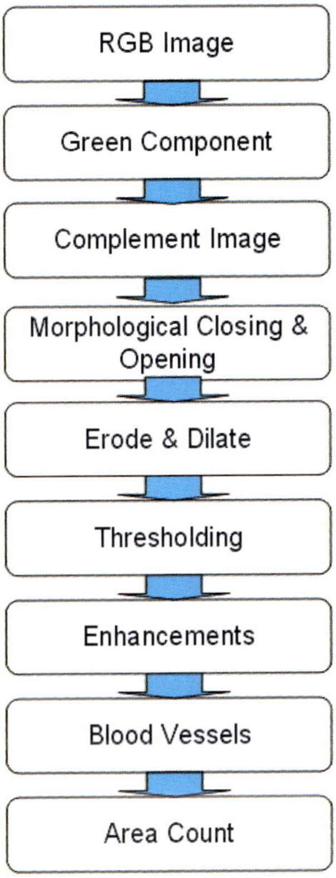

Fig. 6.6. Block diagram to evaluate the area of blood vessels (reprinted from Ref. [14], with kind permission from *Journal of Medical Systems*).

was estimated by summing up the number of white pixels in a specified region, as illustrated in Fig. 6.7.

6.2.3. *Measuring the ISNT Ratio*

Both superior and inferior regions of the optic disc host most of the blood vessels. It has been estimated that about 27% of the optic disc area is covered with blood vessels.[16] A shift in the optic nerve head is likely to cause a small increase in the blood vessel area in both nasal and temporal regions of the optic disc. This shift is likely to cause a decrease in both inferior and superior regions. The ratio between the blood vessel area in both inferior and superior regions and the blood vessel area in

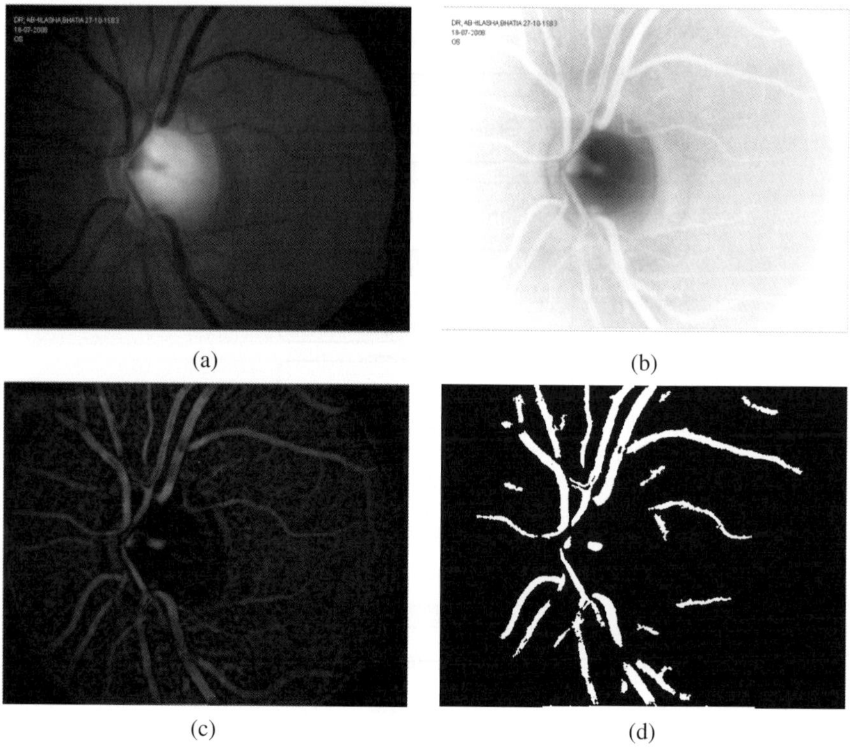

Fig. 6.7. (a) Green component, (b) complemented image, (c) subtracted image, and (d) detected blood vessels for normal fundus image.

both nasal and temporal regions can be taken as another feature. A larger ISNT ratio is a significant feature indicating the presence of glaucoma. Figure 6.8 shows a fundus image where the four quadrants are labeled. The mapping is as follows: upper quadrant maps to superior (S) region, lower quadrant maps to inferior (I) region, the quadrant nearest to the nose maps to nasal (N), and the opposite quadrant maps to the temporal (T) region. The flowchart of the computation of ISNT ratio is illustrated in Fig. 6.9.

For this experiment, blood vessels were extracted, as discussed in Sec. 6.2.2. Then, a mask of size 720×560 was created to identify the inferior, superior, nasal, and temporal regions. Figure 6.10 shows the masks used to identify blood vessels in each of the regions in the optic disc. Next, the mask was overlapped onto the segmented image consisting only of blood

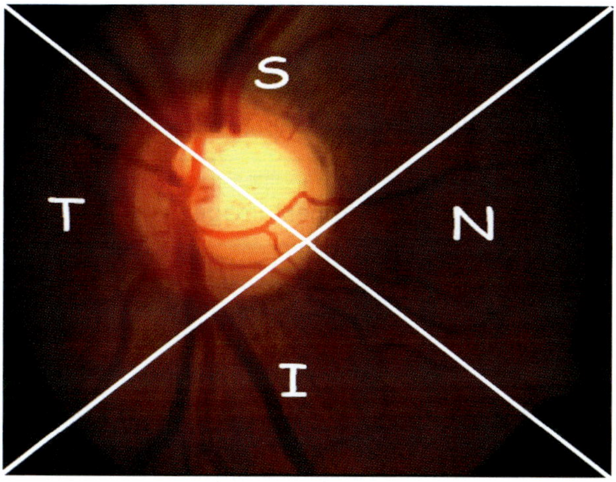

Fig. 6.8. The four quadrants of a fundus image. Clockwise: S denotes the superior quadrant, N denotes the nasal quadrant, I denotes the inferior quadrant, and T denotes the temporal quadrant (reprinted from Ref. [14], with kind permission from *Journal of Medical Systems*).

vessels in order to display the blood vessels. Figure 6.11 illustrates the blood vessels near ISNT.

6.2.4. *Classifier*

The previous section described the feature extractions step of the proposed method. Now, we discuss the classifier GMM, which was used for classification. This classification method can be categorized into one of two types: parametric and nonparametric classifiers. Parametric classifiers use data statistics to implement a so-called best discriminated function. In general, the classification error in both supervised and unsupervised learning methods can be minimized with the parametric approach. GMM uses a set of multidimensional features to estimate a continuous probability density function. A GMM probability density is composed from the sum of N multidimensional Gaussian components. GMM is described by mixture component weights w_i, means μ_i, and covariances $\sum_i$. For a single observation, x, the probability density of a GMM is described by λ:

$$p(x|\lambda) = \sum_{i=1}^{N} w_i g\left(x|\mu_i, \sum_i\right). \tag{6.1}$$

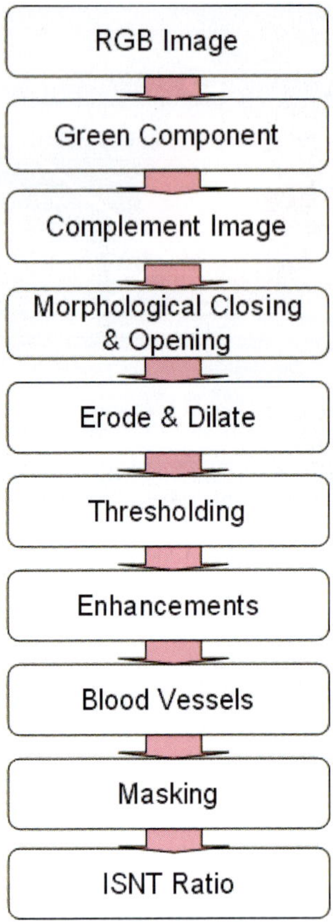

Fig. 6.9. Block diagram of ISNT ratio.

The probability density of a single Gaussian component of D dimensions is given as:

$$g\left(x \middle| \mu_i, \sum_i\right) = \frac{1}{\sqrt{(2\pi)^D \left|\sum_i\right|}} \exp\left(-\frac{1}{2}(x-\mu_i)' \sum_i^{-1} (x-\mu_i)\right). \quad (6.2)$$

The symbol ($'$) represents either vector or matrix transpose. To determine the GMM parameters requires a solution of the maximum likelihood (ML) parameter estimation criterion. The joint likelihood of T independent and identically distributed feature vector observations,

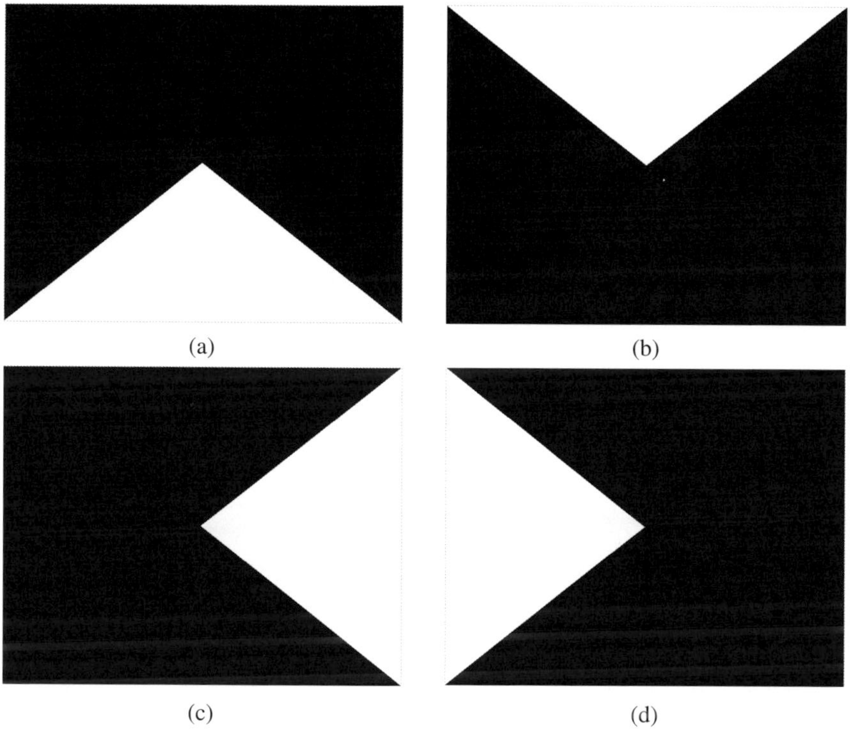

Fig. 6.10. (a) Inferior mask, (b) superior mask, (c) nasal mask, and (d) temporal mask (reprinted from Ref. [14], with kind permission from *Springer Science + Business Medical*).

$X = \{x_1, x_2, x_3, \ldots, x_T\}$, may be specified according to Eq. (6.5).

$$p(X|\lambda) = \prod_{i=1}^{T} p(x_i|\lambda). \tag{6.3}$$

In its logarithmic form, Eq. (6.3) follows as:

$$L(\lambda) = \log p(X|\lambda) = \sum_{i=1} \log p(x_i|\lambda). \tag{6.4}$$

In terms of the mixture component densities, the log-likelihood function to be maximized is defined as:

$$L(\lambda) = \sum_{i=1}^{T} \log \left(\sum_{i=1}^{N} w_i g \left(x_t \middle| \mu_i, \sum_i \right) \right). \tag{6.5}$$

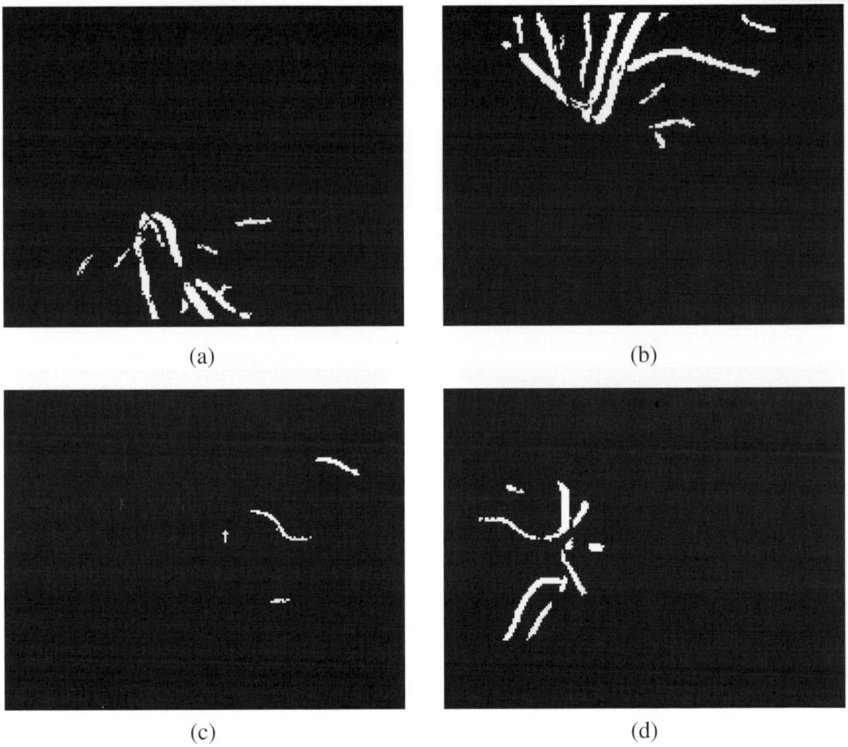

Fig. 6.11. (a) Inferior region, (b) superior region, (c) nasal region, and (d) temporal region.

In this case, ML means that the selected model parameter maximizes the likelihood of the observations. The expectation-maximization (E-M) algorithm is a general method for maximizing the log-likelihood from given observations.[17] The E-M algorithm requires input parameters in the following form:

$$\hat{\lambda} = \left\{ \{\hat{w}_1, \ldots, \hat{w}_N\}, \{\hat{\mu}_1, \ldots, \hat{\mu}_N\}, \left\{ \hat{\sum_i}, \ldots, \hat{\sum_N} \right\} \right\}. \tag{6.6}$$

Based on these parameters, the algorithm will determine new estimates:

$$\lambda_i = \left\{ \{w_i, \ldots, w_N\}, \{\mu_1, \ldots, \mu_N\}, \left\{ \sum_1, \ldots, \sum_N \right\} \right\}, \tag{6.7}$$

such that the following holds $p(X|\lambda) \geq p(X|\hat{\lambda})$.

One application of GMM is to form smooth approximations for arbitrarily shaped densities. GMM provides a useful tool to model the characteristics of multi-model distributed data. Another useful characteristic is the fact that GMM employs a diagonal covariance matrix that is less complex compared to full covariance matrixes that are usually required.[18] This significantly reduces the computational complexity of the algorithm. GMM has been used in many areas such as pattern recognition and classification. In general, this method poses a great success in the areas of identification and verification.

6.3. Results

We start the result discussion by listing both mean and standard deviation of the computed features. Table 6.1 shows this list. The c/d ratio and the area of blood vessels are larger for glaucoma, due to the increase in pressure. This ratio is 0.343 ± 0.245 for a normal subject and 0.503 ± 0.221 for a glaucoma subject. The number of blood vessels for a normal subject is 29254.3 ± 10775.5. This number increases in glaucoma subjects (35746 ± 11443.2). The ISNT ratio is also greater for subjects suffering from glaucoma (1.037 ± 0.021) than the normal subjects (1.024 ± 0.02). We conducted a Student t-test on these two groups (normal subjects and glaucoma subjects) for different features, and the acquired p value was less than 0.03. The low p value indicates that these results are statistically significant.

Table 6.2 illustrates how many samples were used for training and testing the classifier. Furthermore, this table lists also the classification results. In this investigation, 42 images were used for training and the rest (18 images) were used for testing. During classification, only one normal sample was

Table 6.1. Values of three features for normal and glaucoma cases.

Features	Normal	Glaucoma	p value
c/d ratio	0.343 ± 0.245	0.503 ± 0.221	0.01
Blood vessels	29254.3 ± 10775.5	35746 ± 11443.2	0.03
ISNT ratio	1.024 ± 0.02	1.037 ± 0.021	0.02

Table 6.2. Classification results.

Type of image	Number of data sets used for training	Number of data sets used for testing	Correctly classified test data	Percentage correctly classified
Normal	21	9	8	88.89
Glaucoma	21	9	7	77.78
Average				83.33

Table 6.3. Sensitivity, specificity, and positive predictive values for the GMM classifier.

Classifier	TN	TP	FP	FN	Sensitivity	Specificity	Positive predictive accuracy
GMM	8	7	1	2	77.78%	88.89%	87.5%

classified as abnormal and two glaucoma images were classified as normal. The average classification rate was 83.33%.

Table 6.3 shows the sensitivity, specificity, and positive predictive accuracy for the two classes. In the table, we denote true positive (TP) for the number of glaucoma images classified as correctly as glaucoma, true negative (TN) for the number of normal images correctly identified as normal, false negative (FN) for the number of glaucoma samples misclassified as normal, and false positive (FP) for the number of normal image misclassified as glaucoma. Sensitivity is the probability of an abnormal subject being correctly classified as abnormal; specificity is the probability of a normal subject being correctly identified as normal by classifier. The proposed system detects glaucoma with a sensitivity of 77.78% and a specificity of 88.89%. Furthermore, the positive predictive value is 87.50%.

A graphical user interface (GUI) was developed in this work to enable a user to access the algorithm in a user-friendly manner, as illustrated in Fig. 6.12. It comprised input image data, output feature extraction data, a review of the patients' last visit, patient data selection buttons and display,

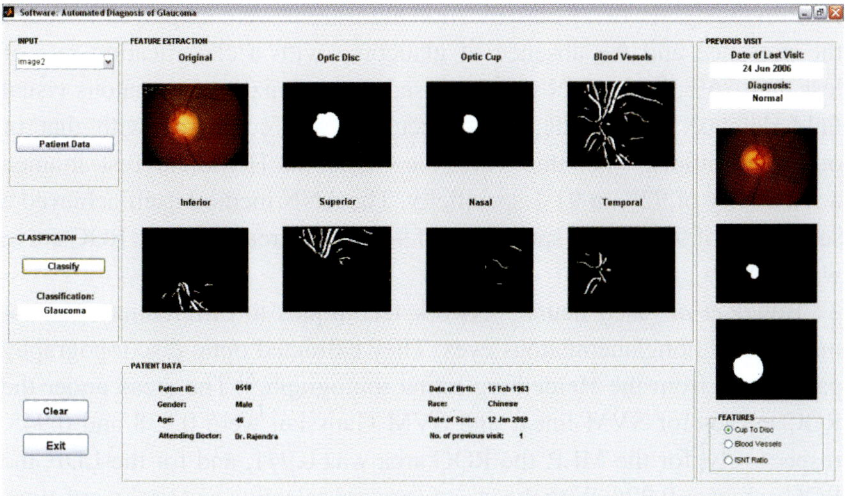

Fig. 6.12. The GUI.

and a classification textbox. Using the *Patient Data* button, the patient image file (Image 2) was loaded. Then, the original image, optic disc, optic cup, blood vessels, inferior blood vessels, superior blood vessels, nasal blood vessels, and temporal blood vessels were displayed. The patient details were also be displayed in the *Patient Data* section of the display. Patient ID, gender, age, attending physician, date of birth, race, date of scan, and previous visit were automatically displayed in the *Patient Data* section.

In addition, the right-hand corner of the display showed an earlier visit image. There is a provision provided to display optic disc and blood vessel images and the ISNT ratio.

6.4. Discussion

We have extracted three features to detect glaucoma automatically. Our features are clinically significant and can identify the disease with an accuracy of 83%. It is important to diagnose glaucoma in an early stage in order to minimize damage to the optic nerve. Only if this damage is minimal, the disease can be effectively treated and the progression of the disease can be prevented.

Previously, six fuzzy classification algorithms were employed to detect the presence and the absence of glaucoma with a classification rate of less than 76%.[10] An ANN was proposed to recognize glaucomatous visual field defects, and its diagnostic accuracy was compared with that of other algorithms.[19] For this work, the Glaucoma Hemifield Test attained a sensitivity of 92% at 91% specificity. The ANN method itself achieved a sensitivity of 93% and a specificity of 94%. The area under the ROC curve was 0.984.

Bowd *et al.* used neural network techniques to differentiate glaucomatous and nonglaucomatous eyes. They extracted optic disc topography parameters from the Heidelberg retina tomograph.[20] The areas under the ROC curves for SVM linear and SVM Gaussian were 0.938 and 0.945, respectively, for the MLP, the ROC area was 0.941, and for the LDF, the ROC area was 0.906. With the use of forward selection and backward elimination optimization techniques, the areas under the ROC curves for SVM Gaussian and the current LDF were increased to approximately 0.96. Hence, the neural network analyses show an increasing diagnostic accuracy of tests for glaucoma.

Recently, Nayak *et al.* have used a novel method for glaucoma detection using the c/d ratio, the ratio of the distance between optic disc center, the optic nerve head to diameter of the optic disc, and the ratio of blood vessels area in inferior-superior quadrants to the area of blood vessel in the nasal-temporal quadrants.[14] The resulting feature vector was fed to a neural network for classification. Their proposed system classified the glaucoma automatically with a sensitivity and specificity of 100% and 80%, respectively.

Our results show a sensitivity of 77.7% and a specificity of 88.8% for the proposed system. We predict that the accuracy of the proposed classification system can be improved by using more parameters, such as textures. In addition, by increasing the number of training and testing images, the result can be further improved. The environmental lighting condition plays an important role in the determination of the classifier performance. Uniform lighting condition set while acquiring a fundus image can also yield better results. This method can serve as an adjunct tool to aid a physician in cross-checking his or her diagnosis.

6.5. Conclusion

In this chapter, a simple novel method for the automatic diagnosis of glaucomatous abnormal eye through fundus images was developed using image-processing techniques and GMM. This proposed system can identify the unknown image with an accuracy of more than 83%. A GUI was developed to show the results and progression of the features. This system can also be used by physicians who would like to cross-check their glaucoma diagnosis.

References

1. Hitchings, R.A. and Spaeth, G.L. The optic disc in glaucoma, ii: Correlation of appearance of the optic disc with the visual field. *Br J Ophthalmol* **61**:107–113, 1977.
2. Bulletin of World Health Organization, http://www.who.int/bulletin/volumes/82/11/resnikoff1104abstract/en/.
3. G.R. Foundation, Primary open angle glaucoma, http://www.glaucoma.org/learn/types.php.
4. Nayak, J., Bhat, P.S., Acharya, U.R., Lim, C.M., and Kagathi, M. Automated identification of different stages of diabetic retinopathy using digital fundus images. *J Med Syst* **32**:107–115, 2008.
5. Heiting, G., Haddrill, M., and Slonim, C. Narrow-angle glaucoma, http://www.allaboutvision.com/conditions/narrow-angle-glaucoma.htm.
6. Acharya, U.R., Chua, K.C., Ng, E.Y.K., Wei, W., and Chee, C. Application of higher order spectra for the identification of diabetes retinopathy stages. *J Med Syst* (in press), 2008.
7. Gonzalez, R.C. and Wintz, P. *Digital Image Processing*. Second edition, Addison-Wesley, Reading, MA, 1987.
8. Nayak, J., Bhat, P.S., Acharya, U.R., Lim, C.M., and Kagathi, M. Automated identification of different stages of diabetic retinopathy using digital fundus images. *J Med Syst* **32**:107–115, 2008.
9. Wong, L.Y., Acharya, U.R., Venkatesh, Y.V., Chee, C., Lim, C.M., and Ng, E.Y.K. Identification of different stages of diabetic retinopathy using retinal optical images. *Inf Sci* **178**:106–121, 2008.
10. Losch, B. Application of fuzzy sets to the diagnosis of glaucoma. *Proceedings of the 18th Annual International Conference of the IEEE Engineering in Medicine and Biology* **4**:1550–1552, 1996.
11. Ulieru, M., Cuzzani, O., Rubin, S.H., and Ceruti, M.G. Application of soft computing methods to the diagnosis and prediction of glaucoma. *Proceedings on IEEE International Conference on Systems, Man and Cybernetics* **5**:3641–3645, 2000.

12. Parfitt, C.M., Mikelberg, F.S., and Swindale, N.V. The detection of glaucoma using an artificial neural network. *Proceedings of 17th Annual Conference IEEE Engineering in Medicine and Biology* **1**:847–848, 1995.
13. Chan, K., Lee, T.-W., Sample, P.A., Goldbaum, M.H., Weinreb, R.N., and Sejnowski, T.J. Comparison of machine learning and traditional classifiers in glaucoma diagnosis. *IEEE Trans Biomed Eng* **49**(9):963–974, 2002.
14. Nayak, J., Acharya, U.R., Bhat, P.S., Shetty, A., and Lim, T.C. Automated diagnosis of glaucoma using digital fundus images. *J Med Syst* **33**:337–346, 2009.
15. Walter, T., Klein, J.C., Massin, P., and Erginay, A. A contribution of image processing to the diagnosis of diabetic retinopathy-detection of exudates in color fundus images of the human retina. *IEEE Trans Med Imaging* **21**:1236–1243, 2002.
16. Greaney, M.J., Hoffman, D.C., Garway-Heath, D.F. *et al.* Comparison of optic nerve imaging methods to distinguish normal eyes from those with glaucoma. *Invest Ophthalmol Vis Sci* **43**:140–145, 2002.
17. Reynolds, D.A., Quatieri, T., and Dunn, R. Speaker verification using adapted Gaussian mixture models. *Digital Signal Process* **10**:19–41, 2000.
18. Nelwamondo, F.V. and Marwala, T. Faults detection using Gaussian mixture models, mel-frequency cepstral coefficients and kurtosis. *Systems, Man and Cybernetics* SMC apos. *IEEE International Conference* **1**:290–295, 2006.
19. Bizios, D., Heijl, A., and Bengtsson, B. Trained artificial neural network for glaucoma diagnosis using visual field data: a comparison with conventional algorithms. *J Glaucoma* **16**:20–28, 2007.
20. Bowd, C., Chan, K., Zangwill, L.M., Goldbaum, M.H., Lee, T.W., Sejnowski, T.J., and Weinreb, R.N. Comparing neural networks and linear discriminant functions for glaucoma detection using confocal scanning laser ophthalmoscopy of the optic disc. *Invest Ophthalmol Vis Sci* **43**:3444–3454, 2002.

Chapter 7
Temperature Distribution Inside the Human Eye with Tumor Growth

Ooi, E.H. and Ng, E.Y.K.

7.1. Introduction to Temperature Distribution

In this study, we create a three-dimensional (3D) model of the human eye and investigate the thermal effects of an eye tumor on ocular temperature distribution. The steady state heat diffusion equation governs temperature inside a healthy ocular region, while the Pennes bioheat equation describes the heat flow inside the eye tumor. The eye model is numerically examined using the boundary element method, and the presence of an eye tumor is found to produce a warmer ocular temperature distribution. A slight thermal asymmetry is observed on the corneal surface. A parametric optimization based on the ANOVA and Taguchi methods is carried out to determine the importance and precedence of the various factors that may affect the ocular surface temperature and to determine the optimal setting of factors that maximize the signal from the eye tumor while isolating the other effects that may be classified as noise.

The temperature of the human body depends on its physiological condition. Around 400 B.C.E., the Greek physician Hippocrates posited that diseases are likely to be detected in the parts of the human body where an excess of heat or cold is found. Modern medicine still relies on temperature

School of Mechanical and Aerospace Engineering, Nanyang Technological University, 50 Nanyang Avenue, Singapore 639798.

monitoring for human health, as illustrated in the common practice of measuring body temperature for fever to determine if a patient is healthy or not. The application of thermal diagnostics extends beyond general practitioners. In breast cancer studies, medical researchers have made successful diagnoses based on the abnormal thermal patterns of the infected breast using IR thermography.[1,2] The application of such thermal diagnostics systems has also been demonstrated in other diseases, such as vascular disorders,[3,4] rheumatism,[5–7] and the diseases that cause fever, such as SARS[6] and bird flu.

The use of ocular temperature to monitor ocular physiology is well documented.[8,9] Ocular abnormalities such as dry eye syndrome and acute inflammation have been found to produce ocular surface temperatures that are different from the temperature of a normal human eye.[8–11] In more severe ocular diseases, such as cataracts, glaucoma, diabetic retinopathy, and eye tumors, no conclusive evidence has suggested that ocular surface temperature may be used as a means for detection. Nonetheless, extensive research and preliminary experimental investigations currently being carried out on patients with glaucoma have revealed promising results.[12,13]

In this study, a mathematical investigation is carried out to examine the thermal effects of eye diseases on the temperature distribution inside the human eye. We seek to determine the changes in ocular temperature distribution. In particular, we seek to measure changes in the corneal surface temperature and to determine the likelihood that these temperature changes will be detected using the current method of ocular thermometry.

Various ocular diseases may infect the human eye. However, in the present study, we shall limit our focus to eye tumors. One of the main reasons eye tumors are selected for the investigations carried out in this study is the availability of a well-established method for simulating the thermal behavior of tumors in biological tissues.[14,15] According to Gautherie *et al.*[16] tumor tissues have a higher rate of metabolic activity that leads to the production of a greater amount of metabolic heat. Additionally, the blood perfusion rate inside tumor tissues has been found to be, in general, greater than the blood perfusion rate of surrounding healthy tissues.[17] These characteristics of tumor tissue enable the thermal effects of tumors to be simulated without much difficulty.

Despite the growing number of mathematical models that investigate the temperature distribution inside the human eye during various ocular circumstances, to the best of our knowledge, no mathematical studies have looked into the effects of an eye tumor on human eye temperature. On the other hand, investigations on the thermal effects of tumors that grow on other parts of the human body, such as the breast and skin, have been carried out by various researchers. Ng and Sudharsan[14,15] developed a finite element model of the human breast to simulate the changes in temperature caused by the growth of a breast tumor. Similar studies have been reported by Hu *et al.*[18] using the finite volume method. Models of the human skin with a tumor growing inside have been developed by Deng and Liu[19,20] using various numerical approaches, such as the dual reciprocity method and the Monte-Carlo method.

In the models of the human breast and skin developed by Ng and Sudharsan,[14,15] Hu *et al.*[18] and Deng and Liu,[19,20] each tumor was modeled as a homogeneous region with distinguishable thermal properties, characterized by a larger metabolic heat generation and blood perfusion rate when compared to the healthy surrounding tissue. Using these modeling approaches, the temperatures of the infected tissues were found to be 0.5–2°C greater than the temperatures of healthy tissues.

In the present study, the same approach used for simulating the thermal behavior of tumors in human breast and skin cells is adopted to investigate the thermal effects of eye tumors on ocular temperature distribution. Simulations are carried out using a 3D model of the human eye. The boundary element method is used to obtain a numerical solution. The presence of terms associated with the metabolic heat generation and blood perfusion rate of the tumor in the governing bioheat equation gives rise to domain integrals in the integral representation derived. These integrals are managed using the dual reciprocity method.[21]

A parametric optimization based on the analysis of variance (ANOVA)[22] and the Taguchi method[23] is carried out to identify the precedence and importance of the various factors that may affect the ocular surface temperature and to determine the optimal setting of factors that maximize the signal from the eye tumor while isolating the other effects that may be classified as noise.

7.2. Classification of Eye Tumors

Various types of eye tumors such as ocular melanoma (melanoma of the iris, ciliary body, and choroid), retinoblastoma, and secondary intraocular tumors (tumors that have spread to the eye from another part of the body) may grow inside the human eye. The focus of this study is limited to choroidal melanoma. Choroidal melanoma grows on the highly vascularized choroid, where the large supply of blood flow makes it ideal for tumor cells to grow and metastasize. When a tumor grows on the choroid, it roughly takes the shape of a dome or a mushroom.[24,25] Dome-shaped choroidal melanomas are usually associated with the early stages of tumor growth, while the mushroom-shaped melanomas often take shape during the later stages.[26]

According to the American Joint Committee on Cancer (AJCC) and the International Union against Cancer (UICC), choroidal melanoma can be classified according to its size (T), its degree of lymph nodes (N), and the presence of metastasis (M). This system of classification is also known as TNM classification. In 2006, the sixth edition of the TNM classifications (TNM6) was published by AJCC and UICC. According to TNM6, four categories of choroidal melanoma can be distinguished based on the T1, T2, T3, and T4 categories. These categories generally describe the largest basal diameter (LBD) and thickness (t) of the tumor. LBD defines the diameter of the tumor at the base, where it is attached to the choroidal surface, while thickness refers to the apical height of the tumor, i.e. the distance from the base to the tip of the tumor.[27]

Table 7.1 summarizes the range of LBD and t of categories T1, T2, T3, and T4 choroidal melanoma.[29] Throughout this study, unless otherwise stated, the phrase "eye tumor" is used to refer to choroidal melanoma.

7.3. Mathematical Model

7.3.1. *The Human Eye*

Figure 7.1, which shows the model of the human eye in the O_{xy} coordinate system, was developed based on the dimensions that have been identified in literature.[29,30] The model consists of six regions: the cornea, the anterior

Table 7.1. Classification of choroidal melanoma based on the T category\label.

Category	Description	
	LBD	t
T1	≤ 10 mm	≤ 2.5 mm
T2	≤ 15 mm	≤ 5.0 mm
T3	>15 mm	>5.0 mm
T4	Tumor grows beyond the eyeball	

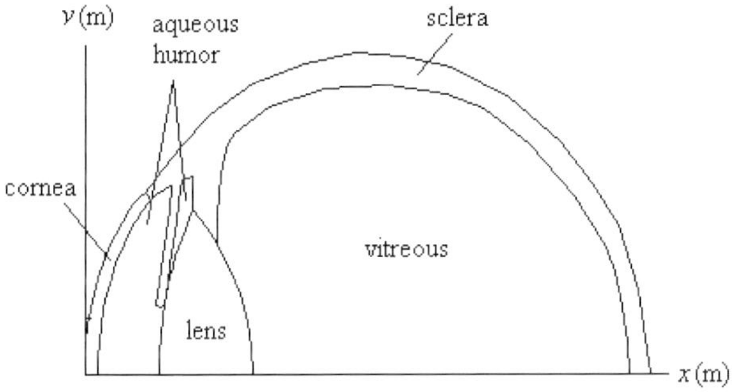

Fig. 7.1. The human eye in the mid-sagittal plane.

chamber, the lens, the posterior chamber, the vitreous, and the sclera. The iris, the retina, and the choroid have been modeled with the sclera as one homogeneous region.[31–33]

To obtain a 3D model, the geometry illustrated in Fig. 7.1 is rotated by 360° around the x-axis (pupillary axis). The 3D model of the human eye generated from the aforementioned step, with a view of its interior structure is shown in Fig. 7.2. The regions occupied by the cornea, the anterior chamber, the lens, the posterior chamber, the vitreous, and the sclera are denoted by R_1, R_2, R_3, R_4, R_5, and R_6, respectively. The exterior boundaries of the model are defined by the surfaces of the cornea, C_1, and sclera, C_2.

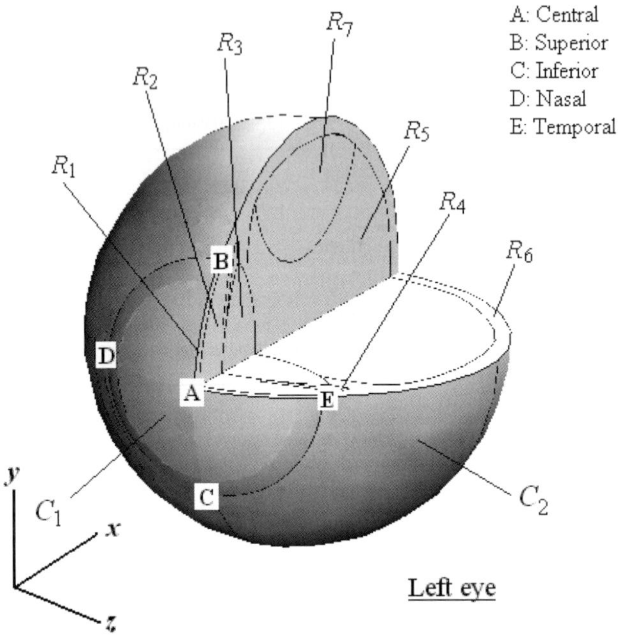

Fig. 7.2. 3D model of the human eye with a view of its interior.

7.3.2. *The Eye Tumor*

The region occupied by the eye tumor is constructed using the same approach as described by Yoriyaz *et al.*[34] A sphere of a particular diameter with its center placed at a specific point on the interface between the vitreous and the sclera is first generated. The region of the sphere that is cut by the interfaces between the sclera and the vitreous is then taken as the region occupied by the tumor, which we denote as R_7. A sketch of the steps involved in the development of the eye tumor model is illustrated in Fig. 7.3. An example of an eye tumor growing at the upper region inside the human eye, developed using the aforementioned steps, is shown in Fig. 7.2.

The model of the eye tumor generated using the steps illustrated in Fig. 7.3 is categorized based on the diameter of the sphere that is used for generating the model. Suppose that the model of the tumor is generated using a 13-mm diameter sphere. Based on the classification tabulated in

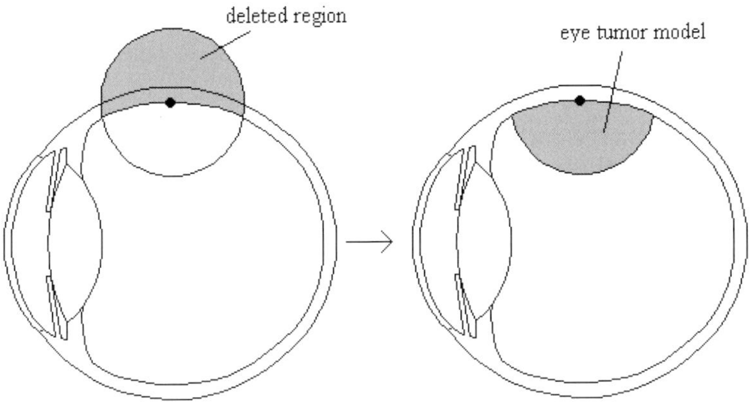

Fig. 7.3. Steps involved in constructing the model of the eye tumor.

Table 7.1, we may then categorize the tumor model as a category T2 eye tumor.

7.3.3. *Governing Equations*

The equation governing the steady state temperature distribution inside the regions of the normal (healthy) human eye in the model shown in Fig. 7.2, is given as:

$$\nabla(\kappa_i \nabla T_i(x, y, z)) = 0 \quad \text{for } i = 1, 2, 3, 4, 5 \text{ and } 6, \tag{7.1}$$

where κ_i and T_i, respectively, denote the thermal conductivity and temperature of region R_i. In Eq. (7.1), each region of the human eye is assumed homogeneous and thermally isotropic. Furthermore, metabolic heat generation, blood perfusion rate, and natural convection inside the anterior chamber are all assumed to have negligible effects on the heat flow inside the human eye.[32,33]

For simplicity, the region occupied by the eye tumor is also assumed homogeneous and thermally isotropic. To account for the metabolic heat generation and the blood perfusion rate inside the eye tumor, heat flow inside region R_7 is taken to be given by the Pennes bioheat equation, which may be mathematically expressed as:[35]

$$\nabla(\kappa_i \nabla T_i(x, y, z)) = \rho_{\text{bl}} c_{\text{bl}} \varpi_{\text{bl}} (T_i(x, y, z) - T_{\text{bl}}) - Q_m, \quad \text{for } i = 7, \tag{7.2}$$

where ρ_{bl} and c_{bl} are the density and specific heat of blood, respectively, T_{bl} is the arterial blood temperature, which we assume to be the same as the body core temperature, ϖ_{bl} is the blood perfusion rate inside the eye tumor, and Q_m denotes the volumetric heat generated from the metabolic activity inside the eye tumor.

7.3.4. *Boundary Conditions*

A few assumptions have to be made prior to the specification of the boundary conditions on the surfaces of the cornea, C_1, and sclera, C_2. First, the presence of an eye tumor inside the human eye is assumed to have no significant effects on the heat transfer between the corneal surface and the environment. Second, blood flow inside the choroid is assumed not to be affected by the growth of a tumor inside the human eye.

Heat loss from the corneal surface to the environment via convection, radiation, and tear evaporation may be mathematically described using

$$-\kappa_1 \frac{\partial T_1}{\partial n} = h_{amb}(T_1 - T_{amb}) + \varepsilon\sigma(T_1^4 - T_{amb}^4) + E_{vap}, \quad \text{on } C_1, \tag{7.3}$$

where the first, second, and third terms on the right-hand side correspond to the heat loss due to convection, radiation, and tear evaporation. Respectively, h_{amb} is the ambient convection coefficient, T_{amb} is ambient temperature, ε is emissivity on the corneal surface, σ is Stefan-Boltzmann constant, E_{vap} is the amount of heat loss due to tears evaporation, and $\partial T_1/\partial n$ is the rate of change of T_1 in the outward unit vector, normal to the external corneal surface, C_1.

On the surface of the sclera C_2, heat flow from blood into the eye may be described using a single heat transfer coefficient,[33] so that:

$$-\kappa_6 \frac{\partial T_6}{\partial n} = h_{bl}(T_6 - T_{bl}), \quad \text{on } C_2, \tag{7.4}$$

where h_{bl} is the heat transfer coefficient between the artificial surrounding and the human eye (also known as blood convection coefficient), T_{bl} is the temperature of the artificial surrounding, which simulates the temperature of blood, and $\partial T_6/\partial n$ is the rate of change of T_6 in the outward unit vector, normal to the external scleroid surface, C_2.

Table 7.2. Thermal properties for each of the eye component and eye tumor.

Parameter	Value	Reference
Thermal conductivity ($Wm^{-1}K^{-1}$)		
Cornea, R_1	0.58	[36]
Sclera, R_2	1.00	[31]
Aqueous humor, R_3	0.58	[36]
Lens, R_4	0.40	[38]
Vitreous, R_5	0.60	[38]
*Tumor, R_6	0.35–0.67	
Ambient convection coefficient ($Wm^{-2}K^{-1}$)	10	
Blood convection coefficient ($Wm^{-2}K^{-1}$)	65	[38]
Ambient temperature (K)	298	
Blood temperature (K)	310	
Tears evaporation rate (Wm^{-2})	40	[36]
Corneal surface emissivity	0.975	[11]
Stefan-Boltzmann constant ($Wm^{-2}K^{-4}$)	5.67×10^{-8}	[39]
*Tumor blood perfusion rate ($m^3s^{-1}m^{-3}$)	0.0014–0.0072	
*Tumor metabolic heat generation (Wm^{-3})	15,000–80,000	

*See Sec. 7.4.

The variation of temperature and heat flux at the interfaces between two contiguous ocular regions, I_{ij} may be described using the continuity condition such that

$$T_i = T_j, \quad \text{and} \quad \kappa_i \frac{\partial T_i}{\partial n} = \kappa_j \frac{\partial T_j}{\partial n} \quad \text{on } I_{ij}, \tag{7.5}$$

where T_i and T_j are temperatures at the interface between regions R_i and R_j, respectively.

The values of the parameters used in Eqs. (7.3) and (7.4) are similar to those in Ooi *et al.*[33] and are summarized in Table 7.2.

7.4. Material Properties

The thermal properties of each component of the human eye shown in Fig. 7.2 are easily obtained from the literature and are listed in Table 7.2. While the thermal properties of each ocular region can be obtained from literature, the same may not be said of the eye tumor. On the other hand, thermal properties of other types of tumors have been measured and reported. To enable simulations in the present study, values of the unknown thermal properties of the eye tumor, specifically, the thermal conductivity, blood perfusion rate, and metabolic heat generation, are chosen to be given over a range of values that are derived based on the experimental data measured from the different types of tumors that grow inside the human body.

Based on the measurements compiled by Jain,[40] the thermal conductivities of various tumors (breast, colon, liver, lung, and pancreatic) that grow on different parts of the human body are found to be in the range of 0.35–0.67 $\mathrm{Wm^{-1}K^{-1}}$. No significant differences in the values of thermal conductivity between the same types of normal and metastatic tumors are observed. Values of the tumor blood perfusion rate that were measured by various researchers using different measuring techniques were compiled and published by Fieldman *et al.*[41] Based on the data measured on breast tumors, lymphomas, anaplastic carcinoma, and differentiated tumors, the tumor blood perfusion rate was found to be in the range of 0.0014–0.0072 $\mathrm{m^3s^{-1}m^{-3}}$.

Unlike the thermal conductivity and blood perfusion rate, metabolic heat generation remains an elusive term in the bioheat equation, where experimental data remains scarce.[40] One of the more established studies that quantified the values of tumor metabolic heat generation is, perhaps, the work carried out by Gautherie[16] on breast tumors. According to Gautherie,[16] the metabolic heat generation of breast tumors remains constant during the phase of exponential growth. Experimental data collected from patients diagnosed with breast tumors revealed that the volumetric metabolic heat generated by the tumor is related to the tumor doubling time by

$$Q_m = \frac{\Lambda}{D_T}, \tag{7.6}$$

where Λ is a constant that has a value of $3.27 \times 10^6\ \mathrm{W \cdot day \cdot m^{-3}}$ and D_T is the tumor doubling time in days, which is defined as the time needed for the tumor to double its volume.[42]

Let us assume that Eq. (7.6) is valid for all types of tumors. Since the doubling time is a unique property of tumors, the only variable in Eq. (7.6) that may change to account for different types of tumors is the constant Λ. Instead of using the range of metabolic heat generation of breast tumor, it may be more appropriate to allow the value of Λ to vary within a certain range. In this study, we shall assume Λ to have a value that ranges from 1×10^6 to 5×10^6 W $\cdot$ day $\cdot$ m^{-3}. For an average eye tumor doubling time of 63 days,[43] the corresponding range of volumetric metabolic heat calculated using Eq. (7.6) is found to be approximately 15,000–80,000 Wm^{-3}.

By establishing the range of thermal conductivity, blood perfusion rate, and metabolic heat generation of various tumors of the human body, we may now assume that the thermal properties of an eye tumor lie within the corresponding range of values given in Table 7.2. The values given in Table 7.2 have been obtained for different types of tumors that grow on various parts of the human body, and this assumption is likely to remain valid, as we do not expect the thermal properties of the tumors to be significantly different from one another.

7.5. Numerical Scheme

7.5.1. *Integro-Differential Equations*

The boundary element method begins with the derivation of the integral representations of the governing equations. Following the steps outlined by Ang,[44] the integral representations of Eqs. (7.1) and (7.2) are respectively given as

$$\begin{aligned}\lambda(\xi_i, \eta_i, \zeta_i)T(\xi_i, \eta_i, \zeta_i) = {} & \iint_{\Gamma_i} T_i(x, y, z)\frac{\partial}{\partial n} \\ & \times [\Phi_{3D}(x, y, z; \xi_i, \eta_i, \zeta_i)]dA(x, y, z) \\ & - \iint_{\Gamma_i} \Phi_{3D}(x, y, z; \xi_i, \eta_i, \zeta_i)\frac{\partial}{\partial n} \\ & \times [T_i(x, y, z)]dA(x, y, z) \\ & \qquad \text{for } (\xi_i, \eta_i, \zeta_i) \in R_i \cup \Gamma_i \text{ and} \\ & \qquad \text{for } i = 1, 2, 3, 4, 5 \text{ and } 6, \end{aligned} \tag{7.7}$$

and

$$\begin{aligned}\lambda(\xi_i, \eta_i, \zeta_i)T(\xi_i, \eta_i, \zeta_i) = &\iint_{\Gamma_i} T_i(x, y, z)\frac{\partial}{\partial n}[\Phi_{3D}(x, y, z; \xi_i, \eta_i, \zeta_i)] \\ &\times dA(x, y, z) - \iint_{\Gamma_i} \Phi_{3D}(x, y, z; \xi_i, \eta_i, \zeta_i)\frac{\partial}{\partial n} \\ &\times [T_i(x, y, z)]dA(x, y, z) \\ &+ \iiint_{R_i} \Phi_{3D}(x, y, z; \xi_i, \eta_i, \zeta_i) \\ &\times [\alpha T_i(x, y, z) - \beta]dV(x, y, z) \\ &\qquad \text{for } (\xi_i, \eta_i, \zeta_i) \in R_i \cup \Gamma_i \text{ and} \\ &\qquad \text{for } i = 7, \end{aligned} \tag{7.8}$$

where $\alpha = (\rho_{\text{bl}}c_{\text{bl}}\varpi_{\text{bl}})/\kappa_i$, $\beta = (\rho_{\text{bl}}c_{\text{bl}}\varpi_{\text{bl}}T_{\text{bl}} + Q_m)/\kappa_i$, Γ_i denotes the curve boundary of region R_i, $dA(x, y, z)$ denotes the area of an infinitesimal portion of surface Γ_i, $dV(x, y, z)$ denotes the volume of an infinitesimal portion of region R_i, $\lambda(\xi_i, \eta_i, \zeta_i) = 1$, if point (ξ_i, η_i, ζ_i) lies at the interior of region R_i, $\lambda(\xi_i, \eta_i, \zeta_i) = 1$, if point (ξ_i, η_i, ζ_i) lies on the smooth part of surface Γ_i, and if $\Phi_{3D}(x, y, z; \xi_i, \eta_i, \zeta_i)$ and $\partial\Phi_{3D}(x, y, z; \xi_i, \eta_i, \zeta_i)/\partial n$ are the fundamental solution of the 3D Laplace equation and its normal derivative, respectively, which are given as:

$$\Phi_{3D}(x, y, z; \xi_i, \eta_i, \zeta_i) = -\frac{1}{4\pi}[(x-\xi_i)^2+(y-\eta_i)^2+(z-\zeta_i)^2]^{-\frac{1}{2}}, \tag{7.9}$$

and

$$\frac{\partial}{\partial n}[\Phi_{3D}(x, y, z; \xi_i, \eta_i, \zeta_i)] = \frac{1}{4\pi}\frac{n_x(x-\xi_i) + n_y(y-\eta_i) + n_z(z-\zeta_i)}{[(x-\xi_i)^2 + (y-\eta_i)^2 + (z-\zeta_i)^2]^{\frac{3}{2}}}, \tag{7.10}$$

where n_x, n_y, and n_z are the components of the outward unit normal vector on Γ_i in the direction of x, y, and z, respectively.

The boundary element method is implemented by dividing the boundary Γ_i into small triangular elements. The domain integral appearing in Eq. (7.8) is converted into an equivalent boundary integral using the dual reciprocity method.[21] For a detailed derivation, one may refer to Ooi *et al.*[45]

7.6. Results

7.6.1. *Numerical Model*

Three categories of eye tumor, T1, T2, and T3, are examined in the present study. T4 is not considered, since the tumor has grown beyond the region defined by the model of the human eye (see Table 7.1). Based on the range of values given in Table 7.1, values of 10 and 15 mm are selected to be the diameter of the sphere used for generating the model of the eye tumor for categories T1 and T2, respectively. A value of 17 mm is assumed for category T3 since the largest value of LBD is not explicitly defined.

For each category of eye tumor examined, investigations are carried out for two locations of tumor growth. In the first location, the tumor is assumed to grow in the positive *y* region of the human eye. Specifically, the center of the sphere used to generate the eye tumor is placed at coordinates (0.013, 0.0105, 0). This example is defined as Case 1. In the second location, which we define as Case 2, the tumor is assumed to grow at the back of the human eye so that it is symmetrically aligned around the pupillary axis. In this case, the center of the sphere is placed at coordinates (0.0244, 0, 0). Locations of the eye tumors inside the human eye in Cases 1 and 2, as viewed in the mid-sagittal plane, are illustrated in Fig. 7.4.

In carrying out the numerical scheme outlined in Sec. 7.5, the boundaries of each region of the human eye are discretized into small triangular elements. In Case 1, 2,212; 2,362; and 2,306 elements are generated in the human eye model with categories T1, T2, and T3 eye tumors, respectively,

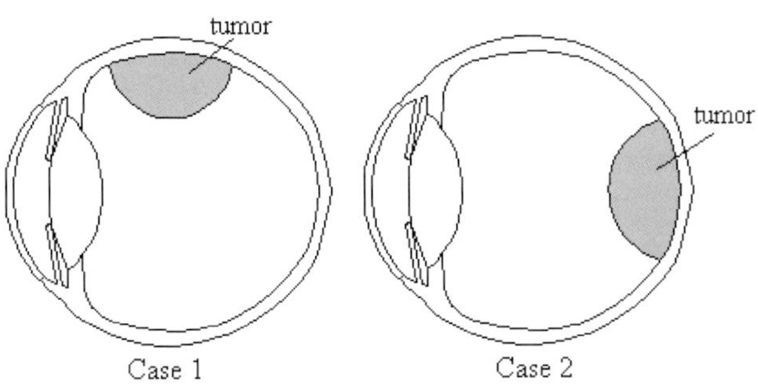

Fig. 7.4. Locations of eye tumor in Cases 1 and 2.

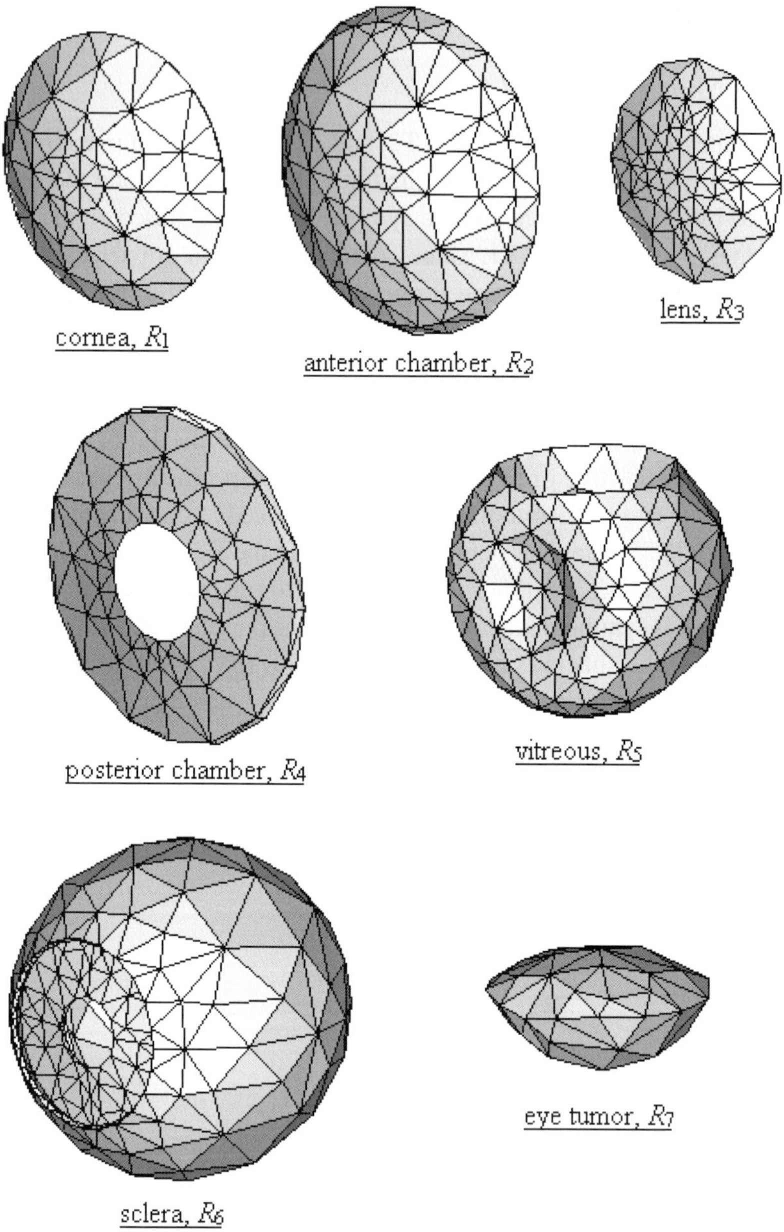

Fig. 7.5. The discretized structure of the cornea, anterior chamber, lens, posterior chamber, vitreous, sclera, and eye tumor (figures are not to scale).

while in Case 2, 2,388; 2,316; and 2,332 elements are generated, respectively. Boundary discretization is implemented using the built-in mesh generator of the partial differential equation calculator, COMSOL Multiphysics. To implement the dual reciprocity method, 100 interior points are selected inside the region occupied by the eye tumor for Cases 1 and 2 and for each category of eye tumor investigated. The meshed structures of each region of the human eye and a category T2 eye tumor in Case 1 are illustrated in Fig. 7.5.

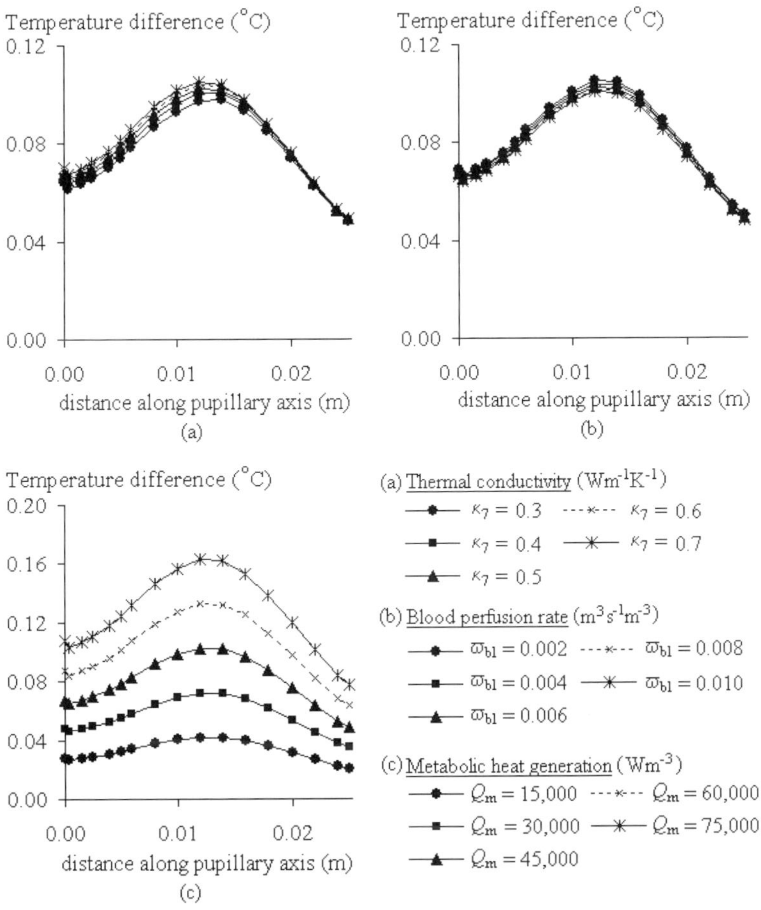

Fig. 7.6. Temperature variations along the pupillary axis of an eye with a category T1 eye tumor in Case 1.

7.6.2. *Case 1*

The effects of eye tumors in categories T1, T2, and T3 on the temperature distribution along the pupillary axis are plotted in Figs. 7.6, 7.7, and 7.8, respectively. Values of the temperature difference shown in Figs. 7.6, 7.7, and 7.8 represent the increases in temperature from the case in which no tumor grows inside the eye.

An increase in the temperature along the pupillary axis is apparent when a tumor grows inside the human eye. The "bell-shaped" curves seen in

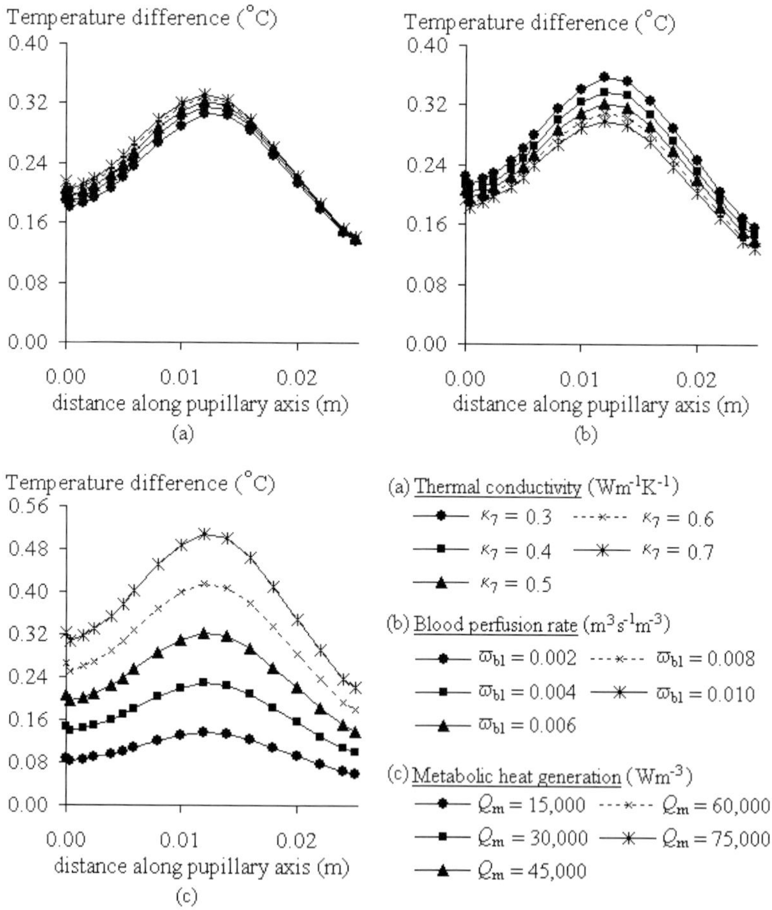

Fig. 7.7. Temperature variations along the pupillary axis of an eye with a category T2 eye tumor in Case 1.

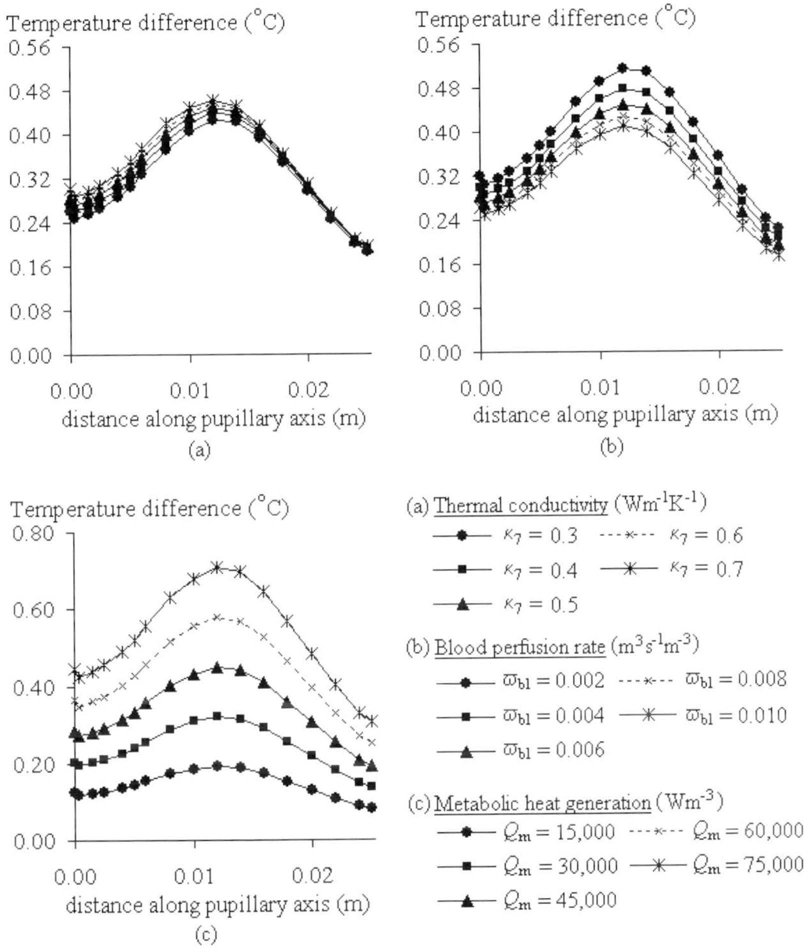

Fig. 7.8. Temperature variations along the pupillary axis of an eye with a category T3 eye tumor in Case 1.

Figs. 7.6, 7.7, and 7.8 imply that the thermal effects of the eye tumor are greater at the vicinity of the tumor region. Similar thermal characteristics have been reported by other researchers who carried out investigations on the human breast and skin.[18–20] The temperature plots in Figs. 7.6, 7.7, and 7.8 show that a category T3 eye tumor produces the largest increase in temperature, and categories T2 and T1 follow. This implies that a larger tumor produces greater warming effects than a smaller tumor, which may be caused by the larger amount of metabolic heat generated by the larger-sized tumor.

In each category of eye tumor examined, changes in metabolic heat generation produce the largest variation in the temperature along the pupillary axis. The effects of the thermal conductivity and blood perfusion rate are roughly the same and in both cases, they are found to be smaller than the effects of metabolic heat generation.

Increases in eye tumor thermal conductivity and metabolic heat generation are found to increase the temperature along the pupillary axis. The opposite is observed for the blood perfusion rate, where a decrease in temperature is predicted. These are true for all categories of eye tumor investigated. A larger thermal conductivity generally encourages more heat transfer between the eye tumor and the surrounding ocular regions.

Similarly, a larger metabolic heat generation suggests that more heat is produced by the eye tumor, so that it causes the temperature of the surrounding ocular regions to increase. The opposite effects of the blood perfusion rate may be explained using Eq. (7.2). The first term on the right-hand side of Eq. (7.2) corresponds to the heat transfer due to blood perfusion, which serves as a medium for transferring excessive heat from the human eye. Therefore, a stronger blood flow will lead to heat loss, thus reducing the temperature inside the human eye.

In the eye with a category T1 eye tumor (Fig. 7.6), variations in the tumor thermal conductivity and blood perfusion rate appear to have minimal effects on the temperature distribution inside the human eye when compared to the metabolic heat generation. The largest increase in temperature is found to be approximately 0.16°C, which occurs when the value of the tumor metabolic heat generation is the highest (75,000 Wm^{-3}). Similar results are observed in eyes with category T2 and T3 eye tumors (see Figs. 7.7 and 7.8). In these cases, however, the effects of the blood perfusion rate are much stronger than the category T1 eye tumor, which is a further indication of the importance of size in governing the temperature increase inside the human eye. The largest increase in temperature caused by the category T2 and T3 eye tumors are found to be approximately 0.5 and 0.7°C, respectively. These numbers are within the range of temperature rise predicted by the models of the human breast and skin.[18–20]

Table 7.3 summarizes the increase in the temperature at the corneal surface caused by the growth of a tumor inside the human eye. The location of the central, superior, inferior, nasal, and temporal points are defined by

Table 7.3. Increase in temperature on the corneal surface in Case 1.

	Temperature increase (°C)			
		Tumor category		
Location	Normal	T1	T2	T3
Central (A)	34.274	0.068	0.206	0.285
Superior (B)	35.477	0.101	0.298	0.404
Inferior (C)	35.472	0.042	0.128	0.181
Nasal (D)	35.474	0.058	0.174	0.243
Temporal (E)	35.473	0.060	0.182	0.253
Average		0.066	0.198	0.273

points A, B, C, D, and E, as shown in Fig. 7.2. The coordinates of these points are given as (0, 0, 0), (0.0025, 0.0064, 0), (0.0025, −0.0064, 0), (0.0025, 0, −0.0064) and (0.0025, 0, 0.0064), respectively. Points B, C, D, and E generally describe the periphery of the cornea. Values in the second column of Table 7.3 represent temperatures at points A, B, C, D, and E of a normal human eye. Results presented in Table 7.3 are obtained for values of eye tumor thermal conductivity, blood perfusion rate, and metabolic heat generation at the control.

Out of the five points analyzed, the increase in temperature at the superior point is found to be the largest for all three categories of eye tumor examined. This increased temperature may be because the superior point is closest to the eye tumor. Consequently, the warming effects from the metabolic heat generated by the eye tumor are greater here than the other points on the corneal surface. At the inferior, the point on the corneal surface farthest from the eye tumor, the increase in temperature is the smallest, further supporting the argument given above. As expected, a larger tumor produces a larger increase in the corneal surface temperature.

In the second column of Table 7.3, points at the periphery of the corneal surface of a normal eye are found to have temperatures that are not significantly different from one another. The temperature at the center, meanwhile, is approximately 0.2°C lower than at the periphery. This implies a symmetrical thermal profile on the corneal surface of a normal human eye. When a tumor grows inside the human eye, the temperature at the superior is approximately 50% and 60% higher than the central and the inferior, respectively. This heat indicates a thermal asymmetry along the vertical geometrical center of the cornea. An average increase of 0.066, 0.198, and 0.273°C are found on the corneal surface of eyes with categories T1, T2, and T3 eye tumor, respectively.

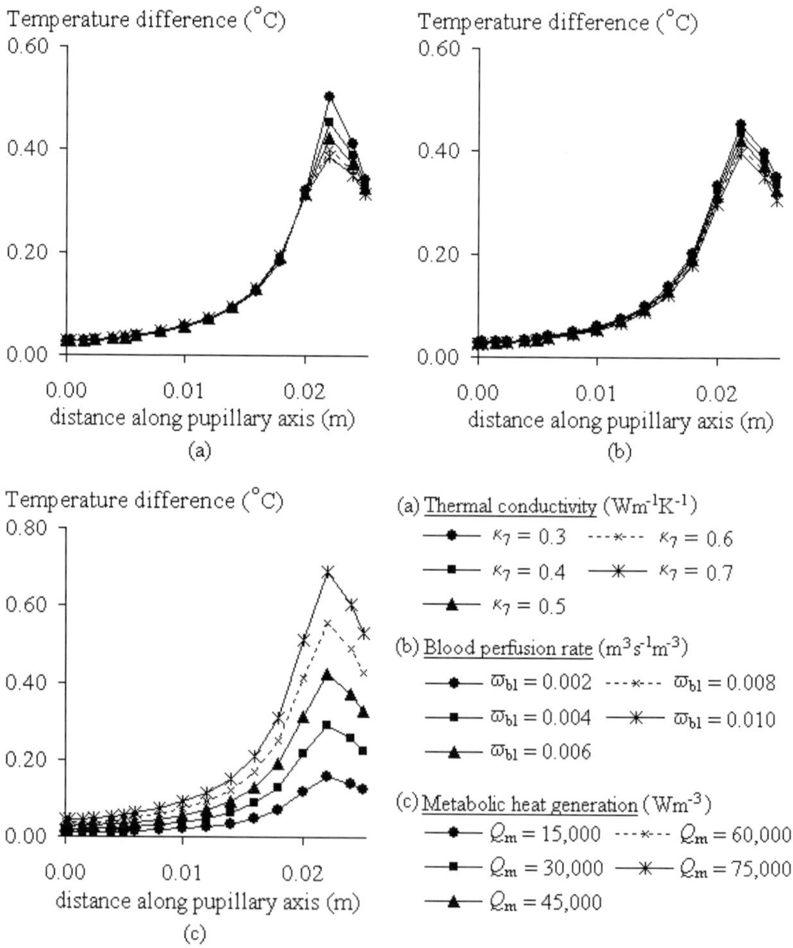

Fig. 7.9. Temperature variations along the pupillary axis of an eye with a category T1 eye tumor in Case 2.

7.6.3. *Case 2*

The thermal effects of categories T1, T2, and T3 eye tumors on the temperature increases along the pupillary axis in Case 2 are plotted in Figs. 7.9, 7.10, and 7.11, respectively. Unlike the "bell-shaped" curves seen in Figs. 7.6, 7.7, and 7.8, in Case 1, the profiles observed in Figs. 7.9, 7.10, and 7.11 show a warmer temperature distribution at the posterior region of the eye. This warmer region is a consequence of the eye tumor being oriented along the pupillary axis. Variations of the temperature at the anterior regions of

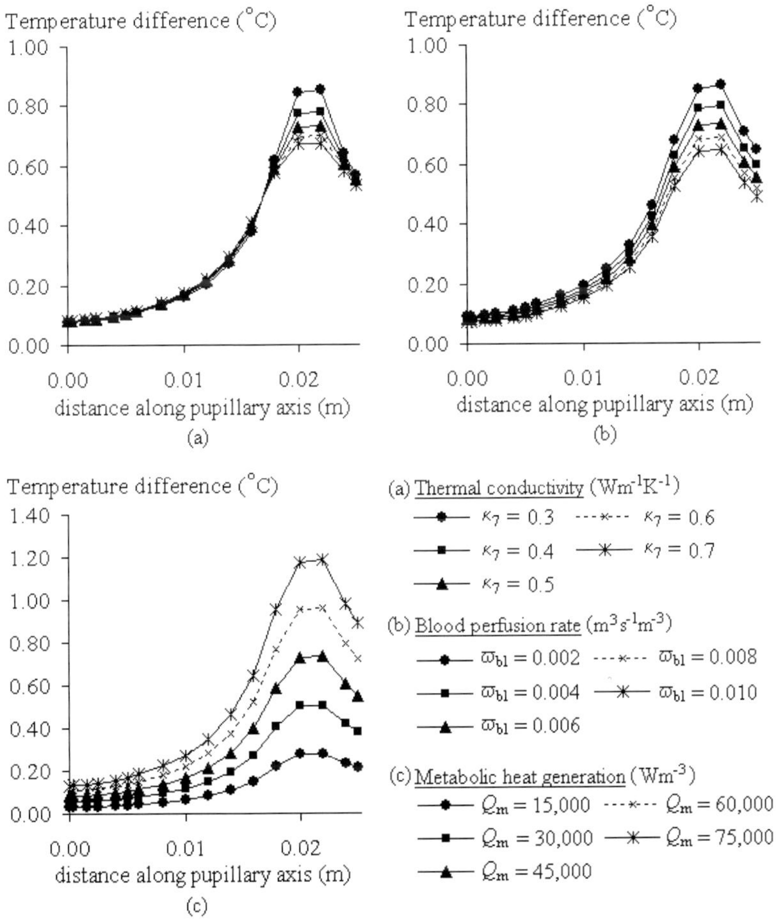

Fig. 7.10. Temperature variations along the pupillary axis of an eye with a category T2 eye tumor in Case 2.

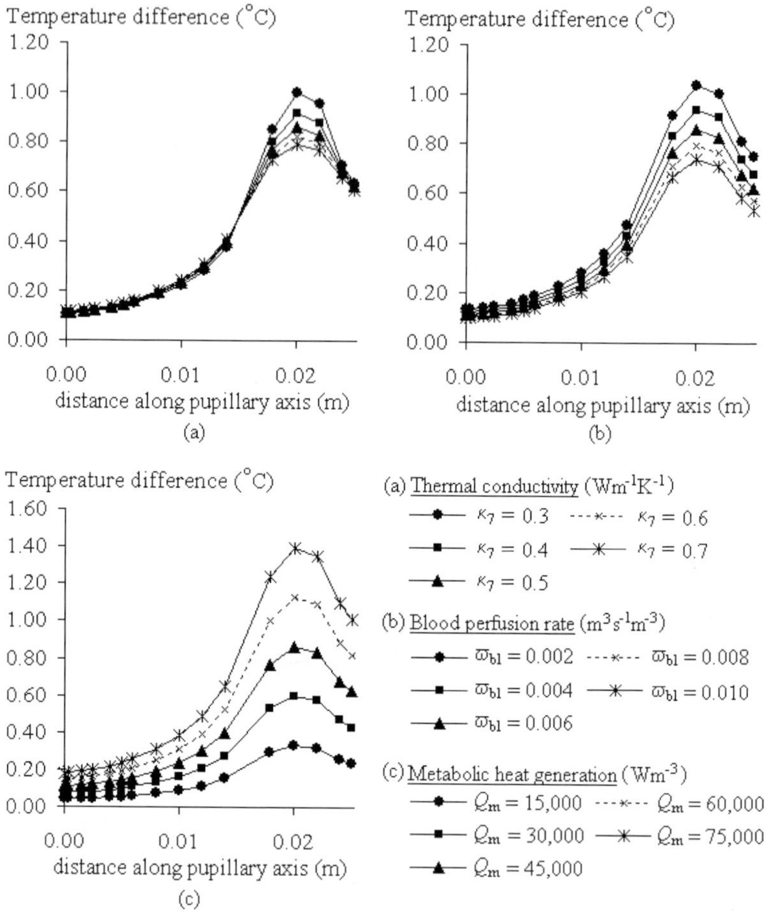

Fig. 7.11. Temperature variations along the pupillary axis of an eye with a category T3 eye tumor in Case 2.

the human eye are small due to the location of the eye tumor, which is positioned at the back of the eye. As distance from the eye tumor increases, the warming effect of the metabolic heat generated by the tumor diminishes.

Table 7.4 tabulates the magnitude of the temperature rise on the corneal surface due to the growth of the eye tumor in Case 2. As in Case 1, the results presented in Table 7.4 are obtained as control values of eye tumor thermal conductivity, blood perfusion rate, and metabolic heat generation. Contrary to the results obtained in Case 1, the largest increase in temperature in Case 2 occurs in the center of the corneal surface. This result is not surprising,

Table 7.4. Increase in temperature on the corneal surface in Case 2.

	Temperature increase (°C)			
		Tumor category		
Location	Normal	T1	T2	T3
Central (A)	34.274	0.036	0.091	0.121
Superior (B)	35.477	0.025	0.077	0.107
Inferior (C)	35.472	0.029	0.082	0.113
Nasal (D)	35.474	0.027	0.080	0.110
Temporal (E)	35.473	0.025	0.079	0.108
Average		0.028	0.082	0.112

since the center of the corneal surface is now closer to the eye tumor than the superior, inferior, nasal, and temporal points.

While the increase in temperature at the center of the corneal surface is the largest, no significant differences in the temperatures at the periphery of the corneal surface are found. This lack of varying temperatures indicates a lack of thermal asymmetry, which is to be expected, since the eye tumor is positioned in such a way that it is symmetrical about the pupillary axis. Consequently, the warming effects of the eye tumor are evenly dispersed into adjacent regions of the human eye, thus leading to a uniform increase in temperature at the corneal periphery.

Comparing the results in Table 7.3 with those in Table 7.4, we find that the eye tumor in Case 1 produces a temperature increase that is, on average, 2.4 times greater than the temperature increase in Case 2. This may be attributed to the greater distance between the eye tumor and the corneal surface in Case 2 when compared with the location of the eye tumor in Case 1 (see Fig. 7.4).

7.6.4. *Discussion*

From the numerical results presented in Secs. 7.6.2 and 7.6.3, we can see that the growth of a tumor inside the human eye produces a warmer ocular temperature distribution. For the location of the eye tumor in Case 1, a thermal asymmetry on the corneal surface is detected. Although simulations have not been carried out for the locations of an eye tumor that are at the negative y region of the human eye, results from our present study suggest that a temperature increase at the inferior of the corneal surface is expected to be the largest when the eye tumor grows at the negative y region of the human eye. Future studies may include the investigations of the ocular temperature changes for the locations of an eye tumor that are different from the two cases considered here.

In the present study, the model of the eye tumor is assumed to be a homogeneous region. This assumption, however, does not accurately depict the actual structure of tumors. In reality, a tumor may be divided into two zones, a necrotic zone at the center of the tumor and a regionally vascularized zone at the periphery.[46] Because of the tissue heterogeneity, the thermal properties of the tumor, particularly blood perfusion rate and metabolic heat generation may not be uniform throughout the entire tumor region. Instead, the vascularized periphery is expected to have a larger blood perfusion rate and metabolic heat generation than the necrotic center. Although it is not known whether such heterogeneity may cause any significant difference to the simulated results, the present model may be reworked in the future to address these problems, if sufficient information is available.

Investigations in the present study have been limited to choroidal melanoma. For eye tumors that grow on the iris (iris melanoma) and ciliary body (ciliary body melanoma), similar investigations may be performed by making minor modifications to the model developed in the present study. These areas are open for future studies.

To verify the accuracy of the model developed in this study, the numerical results are compared to the experimental observations reported by other researchers that can be found in the literature. Bourjat and Gautherie, who used IR thermography to observe the temperature profile of the corneal surface in patients with unilateral exophthalmos, found a marked orbital hyperthermy in the infected eye.[47] Bogdasarov *et al.* found that IR thermography is a feasible choice for the diagnosis of retinoblastoma in children.[48] Furthermore, they discovered that the intensity and the extent of the temperature increase inside the eye are related to the stages of tumor growth. Wittig *et al.*[49] observed warmer corneal surfaces and slight thermal asymmetries in patients with malignant uveal and conjunctival melanomas. The numerical predictions obtained from the present study are in agreement with the experimental results obtained by Bourjat and Gautherie,[47] Bogdasarov *et al.*[48] and Wittig *et al.*[49]

Along with the experimental results found by other researchers,[47–49] the numerical results obtained from the present study suggest that the growth of a tumor inside the human eye may be detected using IR thermography based on the thermal abnormality displayed on the corneal surface. From the results presented in Secs. 7.6.2 and 7.6.3, the average increase in the corneal surface temperature for category T1, T2, and T3 eye tumors are found, for Case 1, to be 0.066, 0.198, and 0.273°C, respectively and, for Case 2, to be 0.028, 0.082, and 0.112°C, respectively. Other than the category T1 eye tumor in Case 2, the increases in the corneal surface temperature are within the range of thermal sensitivity of a highly sensitive IR camera today.[50] The thermal asymmetry on the corneal surface may also indicate the location of tumor growth. For instance, a higher temperature at the superior of the corneal surface may imply that the tumor is growing at the positive y region inside the human eye (see Case 1).

It is worth noting that the corneal surface temperature is sensitive to factors such as the environmental condition (ambient temperature and ambient convection coefficient) and physiological variations among individuals (age, blood convection coefficient, and tear evaporation rate). According to Scott[36] and Ng and Ooi,[32] these factors have been found to produce changes in the corneal surface temperature that are of the same order of magnitude as those caused by the eye tumor. Therefore, to accurately measure the effects of the eye tumor on the corneal surface, it is important to ensure that changes in the corneal surface temperature recorded using an IR camera are solely caused by the eye tumor and are free from the effects of other factors that are classified as "noise." A parametric optimization based on ANOVA and the Taguchi method may be carried out to explain the precedence and the importance of the various factors that may affect the corneal surface temperature. These methods are presented next.

7.7. Parametric Optimization

The effects of eye tumor on the ocular temperature distribution have been demonstrated in the previous section. A tumor inside the human eye has been shown to cause an increase in the temperature of surrounding ocular tissue. Furthermore, an increase in temperature and a small degree of thermal asymmetry on the corneal surface have been predicted. We have also mentioned that factors pertaining to the environment and the physiological condition of the human eye can produce changes in the corneal surface temperature that are of the same order of magnitude as those caused by the eye tumor. In this section, a statistical analysis based on ANOVA is carried out to identify the precedence and importance of various factors that may affect the corneal surface temperature of an eye that is infected with a tumor. An optimization procedure based on the Taguchi method is then performed to determine the optimal setting of factors that maximizes the signal from the eye tumor, while isolating the effects of noise factors. The statistical study carried out in this section assumes that the use of IR thermography for detecting eye tumors is a well-established technique.

7.7.1. *Analysis of Variance*

ANOVA is a statistical tool that is often applied in the design of experiments and is generally used for deciding whether and which factor(s) or interaction(s) has a significant effect on the response of a system. In the context of the numerical investigation carried out in this chapter, ANOVA is used to determine the factors or interaction of factors that have dominant effects on the corneal surface temperature.

From the results obtained in Sec. 7.6, it is found that the metabolic heat generation, size, and location are among the factors of the eye tumor that determine the magnitude of temperature increase on the corneal surface. Effects of the tumor-blood perfusion rate and thermal conductivity are found to be small. From the results of the sensitivity analysis carried out by Ng and Ooi, the factors such as ambient temperature, ambient convection coefficient, blood convection coefficient, body temperature, and tear evaporation rate have been shown to affect the temperature on the corneal surface significantly.[32] The precedence and importance of each of these factors are investigated here.

We consider five factors: the ambient convection coefficient, the blood convection coefficient, the tear evaporation rate, the tumor size, and tumor metabolic heat generation, which we denote by A, B, C, D, and E, respectively. Effects of ambient and body temperature are not considered. It is assumed that the ambient temperature remains constant when the measurements of the corneal surface temperature are carried out. Likewise, the growth of an eye tumor is assumed to have no significant effect on the temperature of the human body. Although the location of the tumor growth has been found to play a major role in the thermal abnormality on the corneal surface, it is not considered in the present analysis due to the irregular shape of the choroidal surface, which makes a consistent establishment of distance between the eye tumor and the corneal surface difficult. In the present study, ANOVA is carried out only for the location of the eye tumor in Case 1.

A 2^n factorial design is used, where n represents the number of factors considered, i.e. five. Each of the five factors has two levels defined by low and high, which leads to 32 (2^5) runs. Selections of the low and high values of each factor are described as follows. According to Ng and Ooi,[32] a value of $10\,\mathrm{Wm^{-2}K^{-1}}$ has been selected to represent the value of the ambient convection coefficient at the control level, i.e. the condition depicting normal environmental condition. It is assumed that the value of the ambient convection coefficient may fluctuate as much as ±20% from the control as measurements is carried out. Values of 80 and 120% of the control, which are given by 8 and $12\,\mathrm{Wm^{-2}K^{-1}}$, respectively, are, thus, considered the low and high values of the ambient convection coefficient. The control value of the blood convection coefficient is given as $65\,\mathrm{Wm^{-2}K^{-1}}$, as suggested by Lagendijk[38] in 1982. More recently, Flyckt *et al.*[51] discovered that values between 250 and $300\,\mathrm{Wm^{-2}K^{-1}}$ better represent the blood convection coefficient inside the human eye. Based on this information, the values of $65\,\mathrm{Wm^{-2}K^{-1}}$ and $300\,\mathrm{Wm^{-2}K^{-1}}$ are selected to be the low and high values of the blood convection coefficient, respectively. The heat loss due to tear evaporation in a normal human eye is represented as $40\,\mathrm{Wm^{-2}}$ (see Scott[36]). According to Mishima and Maurice,[52] the tear evaporation rate of a normal human eye may range from 2.8 to $14.7\,\mathrm{mg{\cdot}hr^{-1}cm^{-2}}$, which corresponds to a heat loss of $20\text{–}100\,\mathrm{Wm^{-2}}$. The low and high values of the heat loss due to the tear evaporation rate are, thus, chosen to be 20 and $100\,\mathrm{Wm^{-2}}$, respectively.

Categories T1 and T3 eye tumors are chosen to represent the sizes of eye tumors at levels low and at high, respectively. The diameters of these tumors are given as 10 mm and 17 mm. In the analysis carried out in Sec. 7.6, the metabolic heat generation of the eye tumor has been assumed to have a value of between 15,000 and 80,000 Wm^{-3}. The range of metabolic heat generation is obtained using Eq. (7.6), where values of Λ between 1×10^6 and 5×10^6 $W{\cdot}day{\cdot}m^{-3}$, and an eye tumor doubling time, D_T of 63 days have been considered. According to Eskelin *et al.*[43] the doubling time of a typical eye tumor ranges from 34 to 220 days. Assuming the value of Λ is 3×10^6 $W{\cdot}day{\cdot}m^{-3}$, the range of metabolic heat generation corresponding to the range of eye tumor doubling time (34–220 days) is calculated as approximately 13,000–90,000 Wm^{-3}. Thus, the lower and upper values of this range are taken to be the low and high values of the eye tumor metabolic heat generation, respectively.

Table 7.5 summarizes the low and high values of each of the five factors considered. The upper case letters (A, B, ..., E) denote the factors while the lower case letters (a, b, ..., e) denote the runs. The following 32 runs are carried out: run 1, a, b, c, ..., ab, abc, ..., bcde, abcde. Run 1 corresponds to the time at which all factors are at low levels. For instance, the lower case letter appearing in a run would indicate that the letter's corresponding factor(s), represented in upper case letters, is at a high level. Run "ab" implies that the simulation is carried out for factors A and B at a high level, while factors C, D, and E are at a low level. Similarly, run "de" represents the simulation that is executed when factors A, B, and C are low, while factors D and E are high. In the 32 runs we carried out, the values of the other variables, such as thermal conductivity, ambient temperature, and body temperature are taken at the control level, such as stated by Ng and Ooi.[32] The thermal conductivity and the blood perfusion rate of the eye tumor are taken to be 0.5 $Wm^{-1}K^{-1}$ and

Table 7.5. Low and high values of the five factors considered in the ANOVA.

Factor	Description	Low	High
A	Ambient convection coefficient ($Wm^{-1}K^{-1}$)	8	12
B	Blood convection coefficient ($Wm^{-2}K^{-1}$)	65	300
C	Heat loss due to tear evaporation (Wm^{-2})	20	100
D	Tumor size (mm)	10	17
E	Tumor metabolic heat generation (Wm^{-3})	13,000	90,000

Table 7.6. Results obtained from the ANOVA.

Rank	Factor	[a]SS	[b]DOF	[c]MS = SS/DOF	Fo = MS/MSE
1	C	1.365	1	1.365	2089.17
2	B	1.099	1	1.099	1681.26
3	E	0.479	1	0.479	733.25
4	A	0.352	1	0.352	538.63
5	DE	0.177	1	0.177	271.20
6	D	0.162	1	0.162	247.12
7	BE	0.128	1	0.128	195.59
8	BD	0.103	1	0.103	157.29
9	BC	0.059	1	0.059	90.32
10	BDE	0.037	1	0.037	55.79
	Errors	0.014	21	[d]MSE = 0.0065	
Total DOF = 31(32 − 1)					

[a]SS: sum of squares; [b]DOF: degree of freedom; [c]MS: mean square; and [d]MSE: mean square of error.

$0.0063\ m^3s^{-1}m^{-3}$, respectively. Observations are made at the superior of the corneal surface (see Fig. 7.2) where temperature rise is expected to be the largest.

Results obtained from ANOVA are summarized in Table 7.6. The interactions of factors beyond the 10th rank are collectively treated as errors due to their small effects when clubbed together. From the results presented in Table 7.6, the tear evaporation rate, (C), appears to be the most dominant factor. This factor is followed by blood convection coefficient, (B), the metabolic heat generation of eye tumor, (E), and the ambient convection coefficient, (A). The combination of factors D and E, (DE), ranks fifth, while the size of the eye tumor, (D), ranks sixth. The effects of both the tear evaporation rate and the ambient convection coefficient are negative; implying that the larger values of A and C yield a cooler corneal surface. The fact that DE ranks higher than D is not surprising, since the size and metabolic heat generation of the eye tumor are mathematically related, whereby a larger tumor produces a greater amount of metabolic heat for the same amount of volumetric metabolic heat generation.

To determine the significance of the effect of each factor and interaction in Table 7.6, we compute the control limit $F_{\alpha,v1,v2}$, where α, ν_1, and ν_2 are the significance level, the numerator DOF, and the denominator DOF, respectively. The values of ν_1 and ν_2 are given as 1 and 21, respectively. For a significance level, α of 1%, the value of $F_{\alpha,v1,v2} = \mathrm{F}_{0.01,1,21}$ obtained from the F distribution chart is found to be 8.017. Since the values of F_o for all factors and interactions shown in Table 7.6 are greater than 8.017, we may conclude that these factors and interactions have significant effects on the corneal surface temperature.

7.7.2. *Taguchi Method*

From the results of ANOVA carried out in Sec. 7.7.1, the ambient convection coefficient, the blood convection coefficient, the tear evaporation rate, the tumor size, and the tumor metabolic heat generation have all been found to contribute significantly to the temperature changes at the superior of the corneal surface. Since our objective is to detect the eye tumor based on a warmer corneal surface temperature, it is essential that the IR camera captures only the signal originating from the eye tumor and is not affected by the other factors that are classified as noise.

In this section, we carry out an optimization procedure based on the Taguchi method to determine the optimal setting of factors that maximizes the signal from the factor of interest while minimizing effects from noise factors.[23] The Taguchi method is a statistical tool developed by Taguchi.[23] It is used primarily in the manufacturing sector but has recently been applied to various other fields, such as in biomedical engineering.[15] The Taguchi method discriminates between signal and noise and estimates the positive or negative effect of each factor in each alternative level.

An analysis based on the Taguchi method is carried out for all five factors, A, B, C, D, and E, which we have shown to be significant factors in ANOVA. Each factor is again assumed to have two levels, low and high, where their respective values are presented in Table 7.2. The signal of interest is defined by the effects from the factors pertaining to the eye tumor, including the size (D) and the metabolic heat generation (E), while effects from the ambient convection coefficient (A), blood convection coefficient (B), and tear evaporation rate (C) are categorized as noise.

One of the important variables in the Taguchi method is the signal to noise ratio (SNR), which reflects the variability in the response of a system caused by noise factors.[22] Three types of SNR are available, "larger is better," "nominal is best," and "smaller is better."[22] Since the signal of interest is defined by the increase in corneal surface temperature, it is decided that the "larger is better" SNR shall be used here. The "larger is better" SNR is mathematically given as:

$$\mathrm{SNR} = -10\log\left(\frac{1}{n}\sum_{i=1}^{i=n}\frac{1}{y_i^2}\right), \qquad (7.11)$$

where n is the number of occurrences and y_i is the ith response of the system defined by the temperature at the superior of the corneal surface.

Table 7.7 summarizes the results obtained from the Taguchi analysis. The values of T_{ave} represent the mean response of a particular factor at a given level. As shown in Table 7.7, the response of the system is found to increase with the size and metabolic heat generation of the eye tumor. The values of SNR in Table 7.7 suggest that the signal from the eye tumor is maximized when the size of the tumor becomes larger or when the metabolic heat generation of the tumor increases. This conclusion is, however, not definite as the combination of factors A, B, and C that causes the effects from the eye tumor to be insignificant has not been considered.

Table 7.7. Results of the Taguchi analysis.

	C	Low	High	Low	High	Low	High	Low	High	
	B	Low	Low	High	High	Low	Low	High	High	T_{ave}
	A	Low	Low	Low	Low	High	High	High	High	
D	E									
Low	Low	35.85	35.34	36.26	35.94	35.97	35.11	36.09	35.78	35.75
Low	High	36.01	35.50	36.30	35.98	35.75	35.26	36.13	35.82	35.54
High	Low	35.89	35.38	36.22	35.88	35.89	35.14	36.04	35.71	35.77
High	High	36.49	35.98	36.42	36.08	36.22	35.73	36.24	35.91	36.13
T_{ave}		36.06	35.55	36.30	35.97	35.86	35.31	36.13	35.80	
SNR		31.14	31.02	31.20	31.12	31.09	30.96	31.16	31.08	

Table 7.8. Effects and average $\overline{\text{SNR}}$ of various factors.

	$\overline{\text{SNR}}$		Effect	
Factor	Low	High	Low	High
A	31.12	31.07	35.97	35.78
B	31.05	31.34	35.70	36.05
C	31.15	31.04	36.09	35.66
D	31.08	31.11	35.79	35.95
E	31.07	31.12	35.76	35.99

For a more conclusive analysis, we calculate the effects and the average values of SNR ($\overline{\text{SNR}}$) of the various factors at both the low and high levels. These are presented in Table 7.8. The values of $\overline{\text{SNR}}$ of a particular factor are obtained by calculating the mean of the $\overline{\text{SNR}}$ values in Table 7.7 at the desired level. Similarly, the values of effects in Table 7.8 are obtained by averaging the values of T_{ave} in Table 7.7 for a particular factor at a given level. For instance, the value of $\overline{\text{SNR}}$ and the effect of factor A at low level are calculated from

$$\overline{\text{SNR}}_{\text{A,low}} = \frac{31.14 + 31.02 + 31.20 + 31.12}{4}$$

and

$$\text{Effect}_{\text{A,low}} = \frac{36.06 + 35.55 + 36.30 + 35.97}{4}, \text{ respectively.}$$

From Table 7.8, it is found that the combination of factors A, B, and C atlow, high, and low temperature produces effects that are stronger than the effects of factors D and E. This combination is undesirable since the signal that is captured comes primarily from factors A, B, and C, which have been classified as noise. The opposite combination is thus preferred where factors A, B, and C are at high, low, and high temperatures, respectively. At this combination, the effects are found to be smaller than the effects of factors D and E (see bold numbers). As a result, the contributions of the noise factors on the signal are minimized. A similar conclusion may also be derived from the values of $\overline{\text{SNR}}$ in Table 7.7.

7.7.3. *Discussion*

Based on the results from the ANOVA and the Taguchi method, the best setting to capture the signal from the eye tumor is at the high rate for the ambient convection coefficient and the tear evaporation and at the low rate for the blood convection coefficient. To increase the value of the ambient convection coefficient simply means to increase the convective heat loss from the corneal surface. This ideal level can be realized by keeping the ambient temperature low. To increase the evaporation rate of the tears, the eyelids of the patient can be kept open for a substantial period before measurement is taken. This allows a continuous evaporation of tears and prevents blinking that refreshes the layer of tears on the corneal surface. However, steps should be taken to ensure that the eyelids are not refrained from blinking for too long, as this would induce tearing that would cause severe changes in the temperature of the corneal surface. It is also possible to increase the tear evaporation rate by reducing the humidity of the surroundings.[53]

The blood convection coefficient depends on the blood flow inside the choroid.[38] In the human body, physically inactivity helps to maintain the blood flow at its basal level, which prevents any major fluctuations in the blood flow rate. Blood flow inside the human eye, however, has been shown to be auto-regulated,[54] meaning that the human eye is able to maintain an almost constant level of blood flow despite changes in the perfusion pressure which are usually associated with physical activities. According to Lovasik *et al.*[55] the choroidal blood flow increases by only 10% when an individual undergoes physical activity such as cycling. Kiss *et al.*[56] discovered that significant changes in the choroidal blood flow are only possible during heavy physical activity that increases the ocular perfusion pressure by nearly 70%. Although it may seem that the autoregulation of the ocular blood flow helps to maintain the blood convection coefficient at an almost constant level regardless of whether the patient is physically active or otherwise, it should be noted that the blood convection coefficient is the second most dominant factor that affects the corneal surface temperature (see Table 7.6). This factor should, therefore, be kept at its lowest (basal) level possible so that the interference on the signal is minimal.

It is also found that the signal from the eye tumor becomes stronger when the size of the eye tumor increases. Likewise, a larger metabolic heat generation of the eye tumor produces a greater response, thus leading to a stronger signal. From a clinical viewpoint, this would mean that the detection of the eye tumor is more efficient when the eye tumor is at later stages of growth. This may not be feasible since an early detection is crucial in the treatment of the eye tumor. This aspect requires further investigations before a definite conclusion can be drawn. One of the factors that has been excluded from this study is the location of the eye tumor. As we show in Sec. 7.1.6, when an eye tumor grows at the back of the human eye, changes in the corneal surface temperature are small and may not be effectively detected using an IR camera. For such cases, it may be possible to amplify the signal from the eye tumor by using the combination of factors (shown in Table 7.8) that minimizes the effect of noise.

Owing to the lack of similar studies, neither qualitative nor quantitative comparisons of the numerical results obtained in the present statistical analysis are possible. Nevertheless, if we compare the results from the present study with those obtained from the analysis carried out on breast tumor,[16] similarities in the setting of factors that minimize the effects of noise can be found. In breast tumor detection,[16] the signal from the tumor can be measured effectively by setting the noise factors to levels so that the heat loss from the surface of the breast is maximized while heat sources that originate from inside the breast that are not caused by the tumor are minimized. From our analysis, a similar pattern is observed where the best setting for detecting the eye tumor involves an increased heat loss from the corneal surface while heat from the ocular blood flow is kept as low as possible.

7.8. Concluding Remarks

We have successfully developed a 3D model of the human eye to investigate the effects of an eye tumor on the temperature distribution inside it. To account for the unknown thermal properties of the eye tumor, simulations have been performed for a range of values of eye tumor thermal conductivity, blood perfusion rate, and metabolic heat generation that have been derived from the thermal properties of tumors that grow at various parts of the human body. Despite the possible differences in the cellular structure between each type of tumor, we do not expect the actual thermal properties of the eye tumor to be significantly different from the range of values used in this study.

From the numerical results presented in this study, we have discovered that the presence of an eye tumor inside the human eye causes an increase in ocular temperature distribution. This increase in temperature distribution has been attributed to the metabolic heat generation of the tumor tissues. An increase in the corneal surface temperature and an asymmetrical thermal profile on the corneal surface has also been observed. The magnitudes of the temperature rise and the degree of thermal asymmetry have been suggested as potential indicators of an eye tumor inside the human eye.

When carrying out the parametric optimization in Sec. 7.7, we have assumed that the use of IR thermography for detecting eye tumors is a feasible choice. This assumption has not yet been fully verified although earlier experimental and numerical investigations appear to support it.[47–49] The most common method in diagnosing an eye tumor is through visual inspection using an ophthalmoscope. An ophthalmologist usually performs this method. The use of IR thermography as an alternative technique for detecting eye tumor is promising since it offers a quick and intelligent diagnostic tool. It may be used as a preliminary screening tool in mass screenings and ophthalmologists using the ophthalmoscope can verify any identification of tumors. This issue requires more investigation in the future, however.

One of the major setbacks that may impede any future investigations is the rare occurrence of an eye tumor within a given population. The American Cancer Society estimates that about 2,350 new cases of primary intraocular cancer were diagnosed in 2009.[57] This figure is smaller than the 180,000 cases of breast cancer diagnosed annually.[16] Nevertheless, eye tumors should not be taken lightly since vision loss from cancer is the second leading health concern.[58]

References

1. Ng, E.Y.K., Ung, L.N., Ng, F.C., and Sim, L.S.J. Statistical analysis of healthy and malignant breast thermography. *J Med Eng Tech* **25**:253–263, 2007.
2. Ng, E.Y.K., Chen, Y., and Ung, L.N. Computerized breast thermography: study of image segmentation and temperature cyclic variations. *J Med Eng Tech* **26**:12–16, 2001.
3. Harding, J.R. Investigating deep venous thrombosis with infrared imaging. *IEEE Eng Med Biol Mag* **17**:43–46, 1998.
4. Fushimi, H., Kubo, M., Inoue, T., Yamada, Y., Matsuyama, Y., and Kameyama, M. Peripheral vascular reactions to smoking-profound vasoconstriction by atherosclerosis. *Diabetes Res Clin Pract* **42**:29–34, 1998.

5. Ring, E.F.J. Progress in measurement of human body temperature. *IEEE Eng Med Biol Mag* **17**:19–24, 1998.
6. Ng, E.Y.K., Muljo, W., and Wong, B.S. Study of facial skin and aural temperature. *IEEE Eng Med Biol Mag* **25**:68–74, 2006.
7. Ng, E.Y.K. and Acharya, R.U. Remote-sensing infrared thermography. *IEEE Eng Med Biol Mag* **28**:76–83, 2009.
8. Morgan, P.B., Soh, M.P., Efron, N., and Tullo, A.B. Potential applications of ocular thermography. *Opt Vis Sci* **70**:568–576, 1993.
9. Purslow, C. and Wollsohn, J.S. Ocular surface temperature: a review. *Eye Contact Lens* **31**:117–123, 2005.
10. Chiang, H.K., Chen, C.Y., Cheng, H.Y., Hung, Y.C., Chen, K.H., and Chang, D.O. Development of infrared thermal imager for dry eye diagnosis. In: Longshore, R.E. and Sood A. (Eds.), *Infrared and Photoelectric Imager and Detector Devices* II, *Proc SPIE* **6294**:1–8, 2006.
11. Mapstone, R. Determinants of ocular temperature. *Br J Opthalmol* **52**:729–741, 1968.
12. Galassi, F., Giambene, B., Corvi, A., and Falaschi, G. Evaluation of ocular surface temperature and retrobulbar haemodynamics by infrared thermography and colour Doppler imaging in patients with glaucoma. *Br J Opthalmol* **91**:878–881, 2008.
13. Galassi, F., Giambene, B., Corvi, A., Falaschi, G., and Menchini, U. Retrobulbar hemodynamics and corneal surface temperature in glaucoma surgery. *Int Ophthalmol* **28**:399–405, 2007.
14. Ng, E.Y.K. and Sudharsan, N.M. Numerical computation as a tool to aid thermographic interpretation. *J Med Eng Tech* **25**:53–56, 2001.
15. Sudharsan, N.M. and Ng, E.Y.K. Parametric optimization for tumor identification: bioheat equation using ANOVA and the Taguchi method. *Proc Inst Mech Eng Part H, J Eng Med* **214**:505–512, 2001.
16. Gautherie, M., Quenneville, Y., and Gros, C.M. Metabolic heat production, growth rate and prognosis of early breast carcinomas. *Biomedicine* **22**:328–336, 1975.
17. Song, C.W., Lokshina, A., Rhee, J.G., Patten, M., and Levitt, S.H. Implication of blood flow in hyperthermic treatment of tumors. *IEEE Trans Biomed Eng BME* **31**:9–16, 1984.
18. Hu, L., Gupta, A., Gore, J.P., and Xu, L. Effect of forced convection on the skin thermal expression of breast cancer. *J Biomech Eng* **126**:204–211, 2004.
19. Deng, Z.S. and Liu, J. Modeling of multidimensional freezing problem during cryosurgery by the dual reciprocity boundary element method. *Eng Anal Bound Elem* **28**:97–108, 2004.
20. Deng, Z.S. and Liu, J. Monte Carlo method to solve multidimensional bioheat transfer problem. *Num Heat Transf* **42**:543–567, 2002.
21. Nardini, D. and Brebbia, C.A. A new approach in solid mechanics by an alternative boundary element procedure. *Int J Soil Dyn Earthquake Eng* **2**:228–233, 1983.
22. Lindman, H.R. *Analysis of Variance in Complex Experimental Designs*, W.H. Freeman & Co., San Francisco, 1974.
23. Taguchi, G. *Introduction to Quality Engineering*, Asian Productivity Organization, UNIPUB, 1986.

24. Romero, J.M., Finger, P.T., Rosen, R.B., and Iezzi, R. Three-dimensional ultrasound for the measurement of choroidal melanoma. *Arch Ophthalmol* **119**:1275–1282, 2001.
25. Kaiserman, I., Amer, R., Kaiserman, N., and Pe'er, J. Ultrasonographic tissue characteristics of mushroom-shaped uveal melanoma. *Cur Eye Res* **30**:171–177, 2005.
26. Melanoma, Choroidal, http://www.emedicine.com/oph/topic403.htm (last accessed: March 20, 2009).
27. Allen, D.C. *Histopathology Reporting: Guidelines for Surgical Cancer*, Second Edition, Springer, London, 2006.
28. Kujala, E. and Kivelä, T.O. Tumor, node, metastasis classification of malignant ciliary body and choroidal melanoma: evaluation of the 6th edition and future directions. *Ophthalmology* **112**:1135–144, 2005.
29. Charles, M.W. and Brown, N. Dimensions of the human eye relevant to radiation protection. *Phys Med Biol* **20**:202–218, 1975.
30. Fontana, S.T. and Brubaker, R.F. Volume and dimensions of the anterior chamber of the normal aging human eye. *Arch Ophthalmol* **98**:1803–1808, 1980.
31. Cicekli, U. Computational model for heat transfer in the human eye using the finite element method. M.Sc. Thesis, Department of Civil and Environmental Engineering, Louisiana State University, 2003.
32. Ng, E.Y.K. and Ooi, E.H. FEM simulation of the eye structure with bioheat analysis. *Comput Methods Programs Biomed* **82**:268–276, 2006.
33. Ooi, E.H., Ang, W.T., and Ng, E.Y.K. Bioheat transfer in the human eye: a boundary element approach. *Eng Anal Bound Elem* **31**:494–500, 2007.
34. Yoriyaz, H., Sanchez, A., and dos Santos, A. A new human eye model for ophthalmic brachytherapy dosimetry. *Radiat Prot Dosim* **115**:316–319, 2005.
35. Pennes, H.H. Analysis of tissue and arterial blood temperatures in the resting forearm. *J Appl Physiol* **1**:93, 1948.
36. Scott, J.A. A finite element model of heat transport in the human eye. *Phys Med Biol* **33**:227–241, 1988.
37. Emery, A.F., Kramar, P.O., Guy, A.W., and Lin, J.C. Microwave induced temperature rises in rabbit eyes in cataract research. *J Heat Transf* **97**:123–128, 1975.
38. Lagendijk, J.J.W. A mathematical model to calculate temperature distributions in human and rabbit eyes during hyperthermic treatment. *Phys Med Biol* **27**:1301–1311, 1982.
39. Incropera, F. and DeWitt, D.P. *Fundamentals of Heat and Mass Transfer*, Fifth Edition, John Wiley and Sons, New Jersey, 2002.
40. Jain, R. Analysis of heat transfer and temperature distributions in tissues during local and whole-body hyperthermia. In: Shitzer, A. and Eberhart, R. (Eds.), *Heat Transfer in Medicine and Biology: Analysis and Applications*, Volume 2, Plenum Press, New York, pp. 3–54, 1985.
41. Fieldman, H.J., Molls, M., and Vaupel, P. Clinical investigations on blood perfusion in human tumors. In: Molls, M. and Vaupel, P. (Eds.), *Blood Perfusion and Microenvironment of Human Tumors: Implications for Clinical Radiooncology*, Springer-Verlag, Berlin, pp. 47–62, 1998.
42. Collins, V.P., Loeffler, R.K., and Tivey, H. Observations on growth rates of human tumors. *Am J Roentgenol Radium Ther Nucl Med* **76**:988–1000, 1956.

43. Eskelin, S., Pyrhönen, S., Summanen, P., Hahka-Kemppinen, M., and Kivelä, T. Tumor doubling times in metastatic malignant melanoma of the uvea. *Ophthalmology* **107**:1443–1449, 2000.
44. Ang, W.T. *A Beginner's Course in Boundary Element Methods*, Universal Publishers, Florida, 2007.
45. Ooi, E.H., Ang, W.T., and Ng, E.Y.K. A boundary element model for investigating the effects of eye tumor on the temperature distribution inside the human eye. *Comp Biol Med* **39**:667–677, 2009.
46. Cravalho, E.G., Fox, L.R., and Kan, J.C. The application of the bioheat equation to the design of thermal protocols for local hyperthermia. In: Jain, R. and Gullino, P. (Eds.), *Thermal Characteristics of Tumor: Applications in Detection and Treatment*, Annals of the New York Academy of Sciences, 335, pp. 86–97, 1980.
47. Bourjat, P. and Gautherie, M. Unilateral exophthalmos investigated by infrared thermography. *Mod Probl Ophthalmol* **14**:278–285, 1975.
48. Bogdasarov, L., Lenskaia, O.P., and Belkin, B.M. Thermography in the diagnosis of retinoblastoma in children (article in Russian). *Med Radiol (Mosk.)* **30**:19–21,1985.
49. Wittig, I., Kohlmann, H., Lommatzsch, P.K., Krüger, L., and Herold, H. Static and dynamic infrared thermometry and thermography in malignant melanoma of the uvea and conjunctiva (article in German). *Klin Monatsbl Augenheilkd* **201**:317–321, 1992.
50. Ng, E.Y.K. A review of thermography as promising non-invasive detection modality for breast tumor. *Int J Therm Sci* **48**:849–859, 2009.
51. Flyckt, V.M.M., Raaymakers, B.W., and Lagendijk, J.J.W. Modeling the impact of blood flow on the temperature in the human eye and the orbit: fixed heat transfer coefficient versus the pennes bioheat model versus discrete blood vessels. *Phy Med Biol* **15**:5007–5021, 2006.
52. Mishima, S. and Maurice, D. Effects of normal evaporation rate on the eye. *Exp Eye Res* **1**:46–52, 1961.
53. McCulley, J.P., Aronowicz, J.D., Uchiyama, E., Shine, W.E., and Butovich, I.A. Correlations in a change in aqueous tear evaporation with a change in relative humidity and the impact. *Am J Ophthalmol* **141**:758–760, 2005.
54. Flammer, J., Orgul, S., Costab, V.P., Orzalesi, N., Krieglstein, G.K., Serra, L.M., Renard, J.P., and Stefánsson, E. The impact of ocular blood flow in glaucoma. *Prog Retinal Eye Res* **21**:359–393, 2002.
55. Lovasik, J., Kergoat, H., and Riva, C.E. Choroidal blood flow during exercise-induced changes in the ocular perfusion pressure. *Invest Ophthalmol Vis Sci* **44**:2126–2132, 2003.
56. Kiss, B., Dallinger, S., Polaf, K., Findl, O., Eichler, H.G., and Schmetterer, L. Ocular hemodynamics during isometric exercise. *Microvasc Res* **61**:1–13, 2001.
57. American Cancer Society, What Are the Key Statistics About Eye Cancer? http://www.cancer.org/docroot/CRI/content/CRI_2_4_1X_What_are_the_key_statistics_for_eye_cancer_74.asp?rnav=cri (last accessed: July 28, 2009).
58. Lee, S.B., Au-Eong, K.G., Saw, S.M., Chan, T., and Lee, H. Eye cancer incidence in Singapore. *Br J Ophthalmol* **84**:767–77, 2000.

Chapter 8

The Study of Ocular Surface Temperature by Infrared Thermography: The Principles, Methodologies, and Applications

Jen-Hong Tan*, Ng, E.Y.K.*, Rajendra Acharya, U.[†] and Chee, C.[‡]

Thermography is a technique that records the thermal pattern of a surface without in contact with the object of interest. Infrared (IR) thermography captures the thermal pattern by utilizing the IR radiation from any object. Thermal imaging is widely used today, and its role is increasingly important in biomedical field, as it can provide details on the variation and changes of temperature in vascular tissues. This technique has also been applied for the study of ocular surface temperature (OST) in ophthalmology. In this chapter, the principles, applications, methodologies, and the corresponding advantages with the use of this method in ophthalmology are elaborated and discussed. This chapter also presents several algorithms developed to acquire OST.

*School of Mechanical and Aerospace Engineering, College of Engineering, Nanyang Technological University, 50 Nanyang Avenue, Singapore 639798.

[†]Department of Electronics and Computer Engineering, Ngee Ann Polytechnic, Singapore.

[‡]Medical & Surgical Retinal Centre Department of Ophthalmology, National University Hospital, Singapore 119074.

8.1. Introduction to IR Thermography

IR thermography is a temperature-measuring technique that is neither intrusive nor contact required. Moreover, it does not alter surface temperature. Lawson brought this technology to modern medicine in 1956 and found that the temperature of the skin around breast carcinoma elevated compared to normal breast,[1] and hence determined the feasibility of using thermography to study breast lesions.[2] This thermography has undoubtedly revolutionized the way we measure temperature over the past 50 years.

This thermal imaging technique has numerous applications, for example, to investigate fluxes of convective heat, to comprehend fluid dynamics on complicated shapes,[3] and to monitor environmental issues.[4] In the medical field, this technology has been applied to manage neuropathic pain,[5,6] diagnose diseases such as breast cancer,[7,8] rheumatism,[9] fever,[10] skin lesion,[11] and impotence.[12] In ophthalmology, this thermal imaging technique currently are used to study and diagnose ocular diseases.[13]

Today, several ophthalmic imaging techniques are available to investigate ocular anterior anatomy and physiology such as confocal microscopy, slit lamp biomicroscopy, optical coherence tomography, etc. These techniques often provide detailed anatomical description, enable scientists to better understand and determine ocular diseases. Nevertheless, they are unable to detect certain pathological, physiological changes, especially those changes that are obscured or regions that are unreachable under anatomical examination but possible for thermography to capture.

Figure 8.1 shows an ocular thermal image of a normal subject. This form of a two-dimensional (2D) medical signal has brought in much research on ocular diseases such as unilateral exophthalmos,[14] dry eye,[15] glaucoma,[16] etc. Ocular thermography have also been applied in the diagnosis of vascular neuritis[17] and retinoblastoma in children.[18] Hence, it is possible for the OST to help diagnose ocular diseases.

Before the introduction of this technology to the field of ophthalmology, workers measured ocular temperature by invasive measuring techniques, such as needle probes. During measurement, OST is often lowered due to the application of topical anesthesia. The needle probe cools the ocular surface during its insertion into the eye.[19] This cooling leads to measurement error, especially when the penetration depth is below 40 mm. Furthermore,

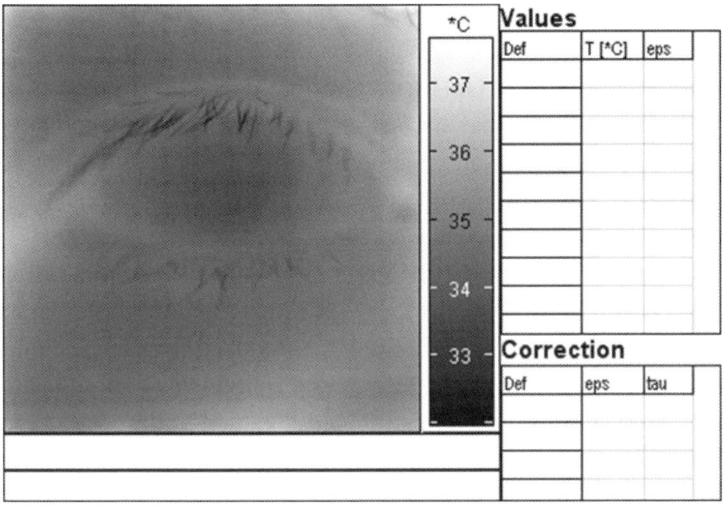

Fig. 8.1. Typical thermogram of normal eye.

this penetration can be traumatic, inducing greater ocular blood flow and thus altering OST.

With thermography, OST can be captured without the above drawbacks. It remotely measures temperature data of some surfaces and does not alter surface temperature. The acquired thermal data is of higher quality, and, hence, enables investigators to study the ocular surface with greater convenience, control, and accuracy.

8.2. Infrared Thermography and the Measured OST

Generally, IR thermography refers to the temperature measurement of some areas by capturing IR radiated from some surfaces. In ophthalmology, the temperature of the ocular surface is of the great interest. Acquisition of OST through an IR thermal imager is widely practiced, because it is noninvasive. However, as a thermal imager determines a surface temperature according to the amount of IR radiation detected, the opacity and the transmissive properties of various ocular tissues are important. It is possible for a thermal imager to receive radiation of an ocular tissue posterior to the ocular surface if tissues around anterior chamber are transparent to IR.

It has been discovered that ocular tissues have the absorption bands almost equal to the absorption bands of water, opaque to far IR.[20,21] The emissivity of water is 1. On IR spectral above 3 μm, it behaves similar to a black body. For ocular tissues, the radiated spectrum falls between 1 μm and 30 μm, with a maximum emission of 9 μm under normal circumstances (32°C). Hence, the spectrum emission of any ocular tissues is wholly absorbed by the tissues anterior to it.

The transmission between the cornea and tear film occurs differently. On the spectrum above 3 μm, water generally behaves like a black body. However, when the thickness of a water layer falls around 10 μm, the transmittance on spectrum 8 μm to 13 μm is determined to be around 30%; that value approximates to zero if the thickness value is 40 μm or above.[22,23] Tear film is determined to have a thickness around 40 μm.[24] The spectrum emitted by cornea is fully absorbed by tear film. Therefore, the OST measured in thermography is in fact the temperature of tear film.[23,25,26]

The above finding is verified by a recent study.[27] In this study, tear film stability was determined to be correlated with OST. Parameters such as corneal curvature, depth of anterior chamber, or central corneal thickness have no significant effect on OST.[27]

In clinical thermography, if one views a curved anatomical surface at an angle beyond 90°, the measured temperature will be 4°C less or more, since emissivity varies with angle of viewing. The amount of thermal radiation of other sources reflected from an anatomical surface also increases when the angle of viewing is greater, resulting in greater error in measured temperature. The resultant error is negligible when the angle of view falls within $\pi/4$. In practice, the error incurred due to the above factors is negligible for the recording of OST, since the angle of viewing on an eye with respect to thermography is assumed within $\pi/4$. As a result, the recorded OST can be seen as the temperature of tear film.

8.3. The Acquisition of OST

In general, an ocular thermogram is an image stored either in gray-scale or RGB format. RGB thermogram usually provides better visual representation of temperatures for manual inspection, and thermogram of gray-scale is useful for quantitative analysis.

The cornea-sclera boundary is indiscernible in ocular thermogram. Often in literature, the acquired OST refers to the temperature of a specific site or some defined region on anterior surface.[14,28–32] The approaches adopted by various researchers to study OST can be classified into three categories: manual measures, semi-automated methods, and fully automated methods.

8.3.1. *Manual Measures*

Efron *et al.* measured surface temperature in the way as illustrated in Fig. 8.2(a).[33] In total, 11 points are placed across the anterior surface, and the sixth point coincides with the estimated geometric center of the cornea. Horizontal temperature profile of OST was determined, expressed in terms of regression polynomial

$$\Delta T = ax^2 + bx + c,$$

where $a = 0.01°\text{C/mm}^2$, $b = 0.003°\text{C/mm}$, $c = 0.01°\text{C}$. x is the distance between the point of interest and the geometric centre of cornea.

Recently Tan *et al.*[34] developed a similar measure based on the method by Efron *et al.*[33] They determined the horizontal temperature profile and the vertical profile. Twenty points were used, as depicted in Fig. 8.2(b).[34] Purslow *et al.* devised a method that seems like an extended crosshair.[35] As illustrated in Fig. 8.3(a), they put up a structure of 23 points over the anterior ocular surface and recorded the corresponding thermal data.

The above-mentioned methods measured OST by the number of disjointed points. There were two similar methods, however recorded five thermal data points of various sites along the horizontal meridian.

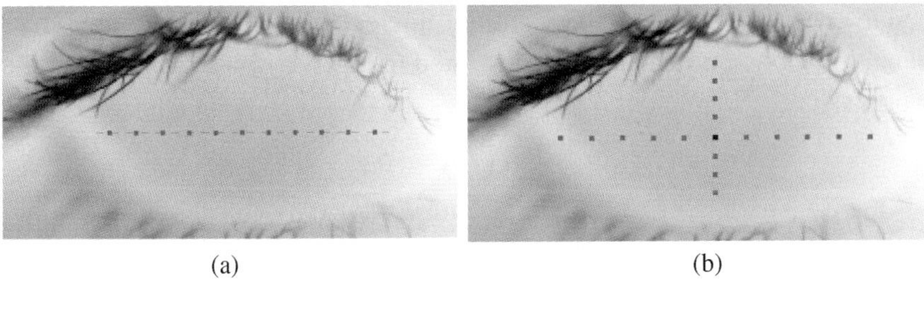

(a) (b)

Fig. 8.2. Methodology in the acquisition of OST by (a) Efron *et al.*[33] and (b) Tan *et al.*[34]

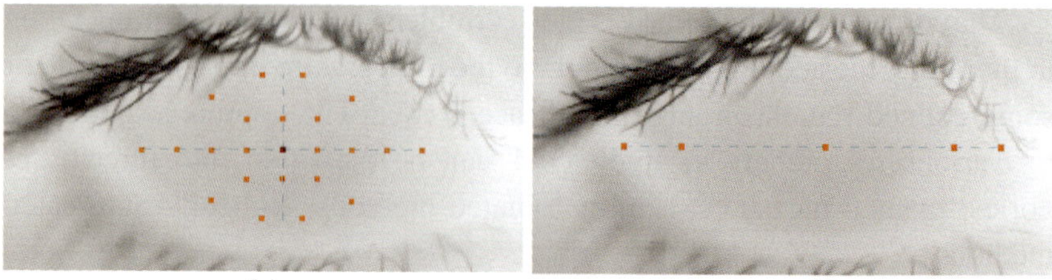

(a) (b)

Fig. 8.3. Methodology in the acquisition of OST by (a) Purslow *et al.*[35] and (b) Galassi *et al.*[36,37]

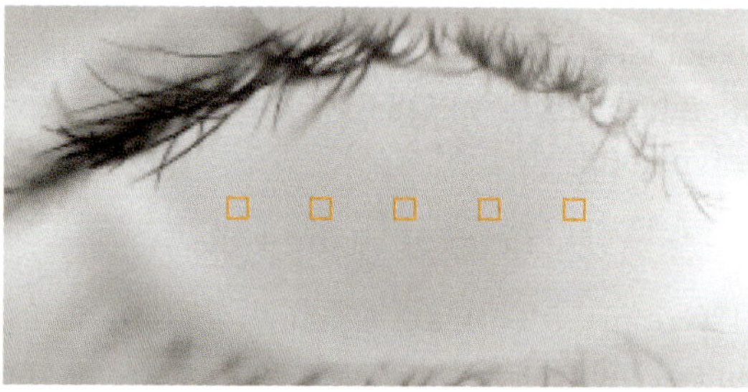

Fig. 8.4. Methodology by Morgan *et al.*[25,38,39]

Nevertheless, these schemes were slightly dissimilar to one another. Besides, instead of using points, research groups such as Morgan *et al.*[25,38,39] adopted squared boxes to acquire OST (as shown in Fig. 8.4).

8.3.2. *Semi-Automated and Fully Automated*

Acharya *et al.* developed a semi-automatic method to capture OST using a standard procedure.[40] This algorithm located the circular cornea, but the user had to crop a rectangle on the region of interest manually. The temperature deviation along cornea (TDC) (illustrated in Fig. 8.5) was calculated in the work as:

$$\mathrm{TDC} = \sum_{y} f(x, y) - f(x, y+1).$$

Snake algorithm, coupled with target-tracing function, was employed to achieve fully-automatic acquisition of OST,[41,42] as shown in Fig. 8.6. It was developed by Tan *et al.*, to enable localization of eye and cornea[41] without

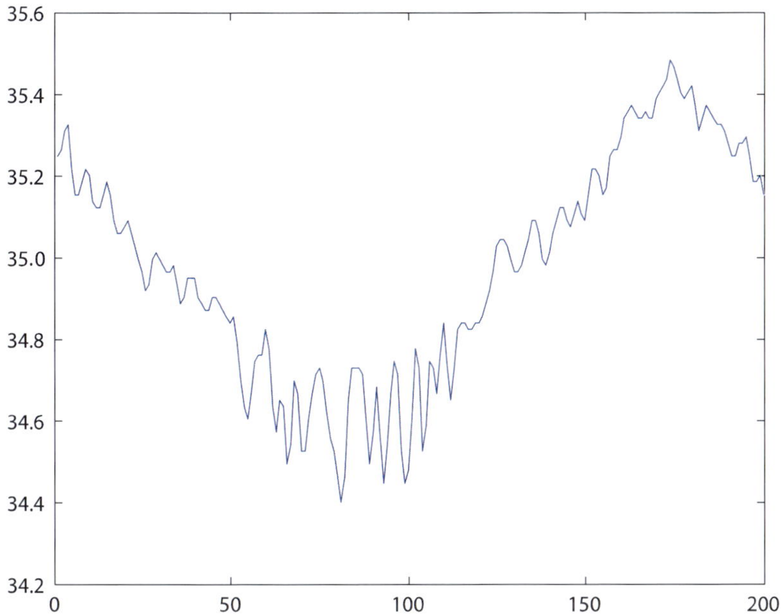

Fig. 8.5. Horizontal temperature profile across cornea.

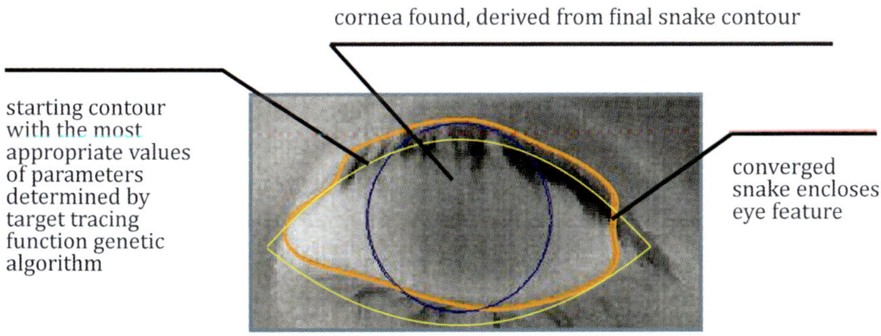

Fig. 8.6. The acquisition of OST by Tan *et al.*[41,42]

any manual intervention. The automated method derived the corneal radius and center based on the final snake points. The corneal center was equaled to the snake centroid. Corneal radius was obtained by a formula deduced on the basis of work in the past.[43]

8.4. Applications to Ocular Studies

OST is volatile and easily affected by factors other than changes in ocular physiology and pathology. The rise in environment temperature increases OST,[44,45] and an increase in air flow decreases OST.[46] Therefore, during data collection uniform temperature, air flow and humidity are maintained in most studies to minimize discrepancies in the final outcome.

Generally, investigations in thermal imaging by IR on OST can be classified into three categories: studies of ocular physiologies, studies of ocular diseases, and surgery of nonocular diseases. Previous research has indicated that the presence of ocular diseases may significantly alter surface temperature. Many investigators have determined that a large interocular difference in OST may indicate the presence of ocular diseases.[25]

8.4.1. *On Ocular Physiologies*

The IR thermal imager was introduced to the field of ophthalmology by Mapstone, who first utilized the Bofor IR camera system to obtain ocular thermogram and delineated thermographic patterns of normal, ischemic and hyperemic eyes.[29] Wachtmeister performed a similar study, and both normal and diseased eyes were investigated.[30] The affected eye was said to be warmer compared to normal counterpart in both anterior and posterior diseases.

In addition, there were researchers looking into the detection of ocular diseases based on thermal asymmetry.[47] They concluded that ocular diseases were present if the interocular differences in temperature were at least 0.5°C. But the finding was illustrated to be not correct.[47] Symmetry itself was determined to be insufficient to discriminate a diseased eye from a normal one.[47]

The introduction of a wide-field color-coded thermal imager to the field in the late 1980s has enabled research on OST with greater ease. Efron *et al.* employed the imager to determine the variation in temperature across the anterior surface.[33] The temperature apex was determined to be slightly inferior to the geometric center of cornea, and was surrounded by ellipsoidal isotherms.[33] Limbus was 0.45°C warmer than compared to the geometric center of cornea.[33]

In another work, the limbus was found to be 0.23°C ∼ 0.43°C warmer than that of the geometric center of cornea.[34] These authors recruited only Chinese origin subjects. Compared to Caucasian eyes, the tear volume and tear stability was found to be lower in Chinese eyes.[48–51] Their findings showed a lower temperature at limbus in brown eyes than in blue eyes.[34]

The above investigation also determined the horizontal and vertical OST profile. They concurred their findings on the horizontal profile with Efron *et al.*, and further illustrated that the vertical profile resembles a "spoon-shape."[34] It was proposed that, throughout a subjects' life, OST reduces by 0.01°C per year.[52] It was found that the interocular difference in OST was less than 0.62°C for 95% normal subjects.[25]

8.4.2. *On Ocular Diseases and Surgery*

Examination on exophthalmos was one of the earliest investigations into ocular diseases carried out using thermal imager by IR. Diseases such as dry eye and glaucoma were the focus in the later research. At first, OST of the dry eye was found to be high,[38] though it cools and evaporates faster,[53,54] the conjunctival hyperemia associated with it outweighs the cooling effect, leading an increase in surface temperature[38] and it correlates with another finding.[15] However, their findings were not correlated with the findings of Craig *et al.*[70] They showed that the temperature of corneal center in dry eyes was found lower. Besides, IR thermal imaging was also used to study the effect of acupuncture on dry eye.[55]

OST was determined to be correlated with ocular blood flow. Recently, Galassi *et al.* studied the OST of the patients who had primary open-angle glaucoma (POAG) and normal group using IR thermography.[46] The correlations between parameters such as OST, intraocular pressure, and retrobulbar hemodynamics were investigated together with color doppler imaging (CDI). The effect of retrobulbar hemodynamics on OST was highlighted in their experimental results.[36] In another study, the OST of central retinal vein occlusion (CRVO) patient was determined to be higher in comparison with normal subjects, and OST in ischemic CRVO eyes was lower than nonischemic ones.[56]

Besides the correlation between ocular blood flow and OST, the effects of two glaucoma surgeries were studied.[37] The temperature of corneal surface increased in both types of surgeries (deep sclerectomy and trabeculectomy) three month after operations, and postoperative changes in ophthalmic artery resistivity index were negatively correlated to corneal surface temperature for both types of surgeries.[37]

In addition to the studies on procedures for open-angle glaucoma, several cataract surgery procedures were compared and analyzed based on the phacoemulsification system in used. In these studies, the amount of heat generated during surgery,[57] the thermal impact on eye,[58] and the intraoperative thermal levels at the wound site[59] produced by different phacoemulsification systems were determined by the IR thermal imager.

8.5. Discussion

Thermometry was pioneered as a noninvasive method by Zeiss to measure ocular temperature.[60] However, Mapstone and Wachtmeister in fact brought IR thermal imaging to the field of ophthalmology. Since then, diagnostic criteria made on the observed asymmetry and anomalies were developed. However, these criteria were proved incorrect. The resolution of ocular thermogram was poor in the beginning, and the eye was not clearly discernible.

The rapid advancement in technology improved the resolution and, thus, provided a more detailed ocular thermogram. Such movement has enabled investigators to focus their concern on the thermographic pattern illustrated at anterior surface rather than at the orbito-ocular region. Besides, the horizontal temperature profile and the vertical profile have also been illustrated.

Investigations in the OST of ocular diseases by IR thermal imaging have been fruitful. Ocular diseases such as exophthalmos, dry eye, cataract, and glaucoma were investigated. Besides, the issues such as the OST in postherpetic neuralgia,[61] and the thermal impact by photorefractive keratectomy treatment[62] were also studied. However, for most of the investigations in ocular diseases, only mean temperatures on one or several sites (in this context, a site refers to a point or an area) over the anterior eye of both eyes were analyzed and compared.

In literature, investigators often employed first-order textural spatial statistics such as mean, standard deviation, or median over temperatures of a region to study ocular thermogram. These first-order spatial statistics alone are insufficient to delineate ocular thermographic pattern clearly. Other techniques in texture analysis, such as second-order spatial statistics, may do a better job.

Ocular thermal image does not have discernible corneal boundary, hence cornea is not easily located. This issue has been addressed in several investigations, as discussed in previous section. For the manual OST acquisition, the obtained values often do not represent the actual OST, and the corresponding accuracy is highly dependent on the area or the region covered. The semi-automated method requires human intervention to crop the eye. The fully automated method does not require human intervention and locate the eye and cornea automatically.

IR thermography does not emit ionizing radiation. It passively receives IR radiation. Unlike other ophthalmic imaging techniques, IR thermography poses no harm or induces no discomfort, as subject has no contact with the instrument and his/her eyes are not exposed to any electromagnetic and acoustic wave. It is capable of spotting pathological changes that are obscured under many other examinations, for example, mild inflammation can be overlooked under slit lamp biomicroscope examination as the changes are so subtle, however, enough to induce a measurable increase in OST.[34] These characteristics make it an excellent diagnostic tool, especially for the delicate human eye.

References

1. Lawson, R.N. Implications of surface temperature in the diagnosis of breast cancer. *Can Med Assoc J* **75**:309–310, 1956.
2. Lawson, R.N. Thermography: A new tool in the investigation of breast lesions. *Can Serv Med J* **13**:517, 1957.
3. Meola, C. and Carlomagno, G.M. Recent advances in the use of infrared thermography. *Meas Sci Technol* **15**:R27–R58, 2004.
4. Cehlin, M. *et al.* Visualization and measuring of air temperatures based on infrared thermography. *Proc. 7th Int. Conf. on Air Distribution in Rooms ROOMVENT*, Reading, UK, 2000.
5. Hooshmand, H. *et al.* Infrared thermal imaging as a tool in pain management an 11 year study: I. *Thermology International* **11**:53–65, 2001.
6. Hooshmand, H. *et al.* Infrared thermal imaging as a tool in pain management an 11 year study: II Clinical applications. *Thermology International* **11**:119–129, 2001.
7. Ng, E.Y.K. and Chen, Y. Segmentation of breast thermogram: improved boundary detection with modified snake algorithm. *J Mech Med Biol* **6**:123–136, 2006.

8. Ng, E.Y.K. A review of thermography as promising non-invasive detection modality for breast tumor. *Int J Therm Sci* **48**:849–859, 2008.
9. Ring, E.F.J. Progress in the measurement of human body temperature. *IEEE Engineering in Medicine Biology Magazine* **17**:19–24, 1998.
10. Ng, E.Y.K. and Rajendra, A.U. A review of remote-sensing infrared thermography for indoor mass blind fever screening in containing an epidemic. *IEEE Eng Med Biol Mag* **28**:76–83, 2008.
11. Kalili, T.K. and Gratt, B.M. Electronic thermography for the assessment of acute temporomandibular joint pain. *Compend Contin Educ Dent* **10**:979–983, 1996.
12. Merla, A. *et al.* Penile cutaneous temperature in systemic sclerosis: a thermal imaging study. *Int J Immunopathol Pharmacol* **20**:139–144, 2007.
13. Stefanie, P.B. *Thermotopography shows "enormous promise" for diagnosis and treatment of eye diseases.* Available: http://www.escrs.org/eurotimes/March2003/thermo.asp, 2008.
14. Bourjat, P. and Gautherie, M. Unilateral exophthalmos investigated by infrared thermography. *Mod Probl Ophthalmol* **14**:278–285, 1975.
15. Morgan, P.B. *et al.* Ocular surface cooling in dry eye: a pilot study. *J Br Contact Lens Assoc* **19**:7–10, 1996.
16. Tokuoka, S. *et al.* Cold response of skin and eye temperature in glaucoma. *Folia Ophthalmol Jpn* **41**:1159–1165, 1990.
17. Eremenko, A.I. Thermography in the diagnosis of vascular neuritis of the optic nerve. *Oftalmologicheskii Zhurnal* **4**:235–239, 1990.
18. Bogdasarov, I.B. *et al.* Thermography in the diagnosis of retinoblastoma in children. *Meditsinskaia radiologiia* **30**:19–21, 1985.
19. Rosenbluth, R.F. and Fatt, I. Temperature measurements in the eye. *Exp Eye Res* **25**:325–341, 1977.
20. Hartridge, H. and Hill, A.V. The transmission of infra-red rays by the media of the eye and the transmission of radiant energy by Crookes and other glasses. *Proc R Soc* **89**:58–76, 1915.
21. Mapstone, R. Measurement of corneal temperature. *Exp Eye Res* **7**:237–243, 1968.
22. Plyler, E.K. and Aquista, N. IR absorption of liquid water from 2 to 42 microns. *J Opt Soc Am* **44**:515, 1954.
23. Hamano, H. *et al.* Experiments in thermometry of the anterior portion of the eye wearing a contact lens by means of infra-red thermometer. *Contact Lens Anterior Eye* **13**:12–22, 1969.
24. Prydal, J.I. *et al.* Study of human precorneal tear film thickness and structure using laser interferometry. *Invest Ophthalmol Vis Sci* **33**:2006–2011, 1992.
25. Morgan, P.B. *et al.* Potential applications of ocular thermography. *Optom Vis Sci* **70**:568–576, 1993.
26. Fatt, I. and Chaston, J. Temperature of a contact lens on the eye. *Int Contact Lens Clin* **7**:195–198, 1980.
27. Purslow, C. and Wolffsohn, J.S. The relation between physical properties of the anterior eye and ocular surface temperature. *Optom Vis Sci* **84**:197–201, 2007.
28. Rosenstock, T. *et al.* Inflammation of the lacrimal drainage system — assessment by thermography. *Ophthalmic Surg* **14**:229–237, 1983.

29. Mapstone, R. Ocular thermography. *Br J Ophthalmol* **54**:751–754, 1970.
30. Wachtmeister, L. Thermography in the diagnosis of diseases of the eye and the appraisal of therapeutic effects: a preliminary report. *Acta Ophthalmol* **48**:945–958, 1970.
31. Murphy, P.J. *et al.* Corneal surface temperature change as the mode of stimulation of the non-contact corneal aesthesiometer. *Cornea* **18**:333–342, 1999.
32. Fielder, A.R. *et al.* Problems with corneal arcus. *Trans Ophthalmol Soc UK* **101**:22–26, 1981.
33. Efron, N. *et al.* Ocular surface temperature. *Curr Eye Res* **8**:901–906, 1989.
34. Tan, L. *et al.* Accuracy and sensitivity of the dynamic ocular thermography and inter-subjects ocular surface temperature (OST) in Chinese young adults. *Contact Lens Anterior Eye* **32**:78–83, 2009.
35. Purslow, C. *et al.* The effect of contact lens wear on dynamic ocular surface temperature. *Contact Lens Anterior Eye* **28**:29–36, 2005.
36. Galassi, F. *et al.* Evaluation of ocular surface temperature and retrobulbar haemodynamics by infrared thermography and colour Doppler imaging in patients with glaucoma. *Br J Ophthalmol* **91**:878–881, 2007.
37. Galassi, F. *et al.* Retrobulbar hemodynamics and corneal surface temperature in glaucoma surgery. *Int Ophthalmol* **28**:399–405, 2008.
38. Morgan, P.B. *et al.* Infrared thermography of the tear film in dry eye. *Eye* **9**:615–618, 1995.
39. Morgan, P.B. *et al.* Ocular temperature in carotid artery stenosis. *Optom Vis Sci* **72**:850–854, 1999.
40. Acharya, U.R. *et al.* Analysis of normal human eye with different age groups using infrared images. *J Med Sys* **33**:207–213, 2008.
41. Tan, J.H. *et al.* Automated detection of eye and cornea on infrared thermogram using snake and target tracing function coupled with genetic algorithm. *Quantitative Infrared Thermography International Journal*, 2009.
42. Tan, J.H. *et al.* Detection of eye and cornea on IR thermogram using genetic snake algorithm. *9th International Conference on Quantitative Infrared Thermography*, Krakow, Poland, pp. 143–150, 2008.
43. Yuille, A.L. *et al.* Feature extraction from faces using deformable templates. In: *Proceedings CVPR 89, IEEE Computer Society Conference*, San Diego, CA, USA, pp. 104–109, 1989.
44. Mapstone, R. Determinants of corneal temperature. *Br J Ophthalmol* **52**:729–741, 1968.
45. Morgan, P.B. Ocular thermography in health and disease, University of Manchester, Manchester, UK, 1994.
46. Freeman, R.D. and Fatt, I. Environmental influences on ocular temperature. *Invest Ophthalmol Vis Sci* **12**:596–602, 1973.
47. Alio, J. and Padron, M. Normal variations in the thermographic pattern of the orbitoocular regionocular region. *Diagn Imaging* **51**:93–98, 1982.
48. Cho, P. *et al.* Reliability of the tear break-up time technique of assessing tear stability and the locations of the tear break-up in Hong Kong Chinese. *Optom Vis Sci* **69**:879–885, 1992.
49. Cho, P. and Brown, B. Review of the TBUT technique and a closer look at the TBUT of HK-Chinese. *Optom Vis Sci* **70**:30–38, 1993.

50. Cho, P. and Yap, M. Age, gender, and tear break-up time. *Optom Vis Sci* **70**:828–831, 1993.
51. Patel, S. *et al.* Iris colour and the influence of local anaesthetics on precorneal tear film stability. *Acta Ophthalmol (Kbh)* **69**:387–392, 1991.
52. Morgan, P.B. *et al.* Corneal surface temperature decrease with Age. *Contact Lens Anterior Eye* **22**:11–13, 1999.
53. Mathers, W.D. *et al.* Ocular water evaporation and the dry eye: a new measuring device. *Cornea* **12**:335–340, 1993.
54. Rolando, M. *et al.* Increased tear evaporation in eyes with keratoconjunctivitis sicca. *Arch Ophthalmol* **101**:557–558, 1983.
55. Nepp, J. *et al.* The effect of acupuncture on the temperature of the ocular surface in conjunctivitis sicca measured by non-contact thermography: preliminary results. *Adv Exp Med Biol* **506**:723–726, 2002.
56. Sodi, A.A. *et al.* Ocular surface temperature in central retinal vein occlusion: preliminary data. *Eur J Ophthalmol* **17**:755–759, 2007.
57. Olson, M.D. and Miller, K.M. In-air thermal imaging comparison of Legacy Advantec, Millennium, and Sovereign WhiteStar phacoemulsification systems. *J Cataract Refract Surg* **31**:1640–1647, 2005.
58. Corvi, A. *et al.* Thermography used for analysis and comparison of different cataract surgery procedures based on phacoemulsification. *Physiol Meas* **27**:371–384, 2006.
59. Rose, A.D. and Kanade, V. Thermal imaging study comparing phacoemulsification with the Sovereign with WhiteStar system to the Legacy with AdvanTec and NeoSoniX system. *Am J Ophthalmol* **141**:322–326, 2006.
60. Zeiss, E. Über wärmestrahlungsmessungen an der lebenden menschlichen hornhaut. *Arch Augenheilkd* **102**:523–550, 1930.
61. Tullo, A.B. *et al.* Ocular and facial thermography in herpes zoster ophthalmicus and post-herpetic neuralgia. *Invest Ophthalmol Vis Sci* **37**:S49, 1996.
62. Betney, S. *et al.* Corneal temperature changes during photorefractive keratectomy. *Cornea* **16**:158–161, 1997.
63. Mori, A. *et al.* Use of high-speed, high-resolution thermography to evaluate the tear film layer. *Am J Ophthalmol* **124**:729–735, 1997.
64. Ng, E.Y.K., *et al.* Variations of ocular surface temperature with different age groups. In: Rajendra Acharya, U., Ng, E.Y.K., and Suri, J.S. (Eds.), *Image Modeling of Human Eye*, Chapter 17, Artech House, 2008.
65. Chiang, H.K. *et al.* Development of infrared thermal imager for dry eye diagnosis, In: *Proceedings of SPIE — The International Society for Optical Engineering*, San Diego, CA, USA, 2006.
66. Chang, T.-C. *et al.* Application of digital infrared thermal imaging in determining inflammatory state and follow-up effect of methylprednisolone pulse therapy in patients with Graves' ophthalmopathy. *Graefe's Arch Clin Exp Ophthalmol* **246**:45–49, 2008.

Chapter 9

Automated Microaneurysm Detection in Fluorescein Angiograms for Diabetic Retinopathy

Prerna Sethi* and Hilary W. Thompson†

Diabetic retinopathy (DR), an ocular complication of diabetes, is a leading cause of blindness in the Western world. DR is a progressive disease that has stages of severity, which are characterized by the number and types of lesions it produces on the retina. The symptoms of DR can drastically change the texture and appearance of the retina. Microaneurysms (MAs) are visible early and continue their presence as DR progresses. After the venous filling phase, fluorescein angiograms depict MAs as small, round, hyperfluorescent objects. Following this phase, manual MA counting can quantify the progress and stage of the disease. Computed MA detection can potentially be more rational than the manual process, which is prone to be tedious and time consuming, and introduces the potential for an inaccurate diagnosis. Hence, automated MA detection aids the early detection of DR, as it assists in quantifying MAs and differentiating them from other features.

Our objective in this chapter is to describe a comprehensive study on the current automated MA detection techniques. Several methods have

*Department of Health Informatics and Information Management, Adjunct Assistant Professor of Research, School of Biological Sciences, Louisiana Tech University, Ruston LA 71272.

†Professor of Biometry, LSUHSC School of Public Health, Adjunct Associate Professor of Ophthalmology, Biometry, and Neuroscience, Director, Clinical Trials and Biometry Unit, LSU Eye New Orleans, LA 70112.

been reported for automated MA detection using fluorescein angiography because of the modality's increased sensitivity as MAs are best seen on fluorescein angiograms.[1] These methods usually follow a sequence of computational processing steps. The first steps include image preprocessing for noise removal and contrast enhancement for discrimination between MAs and blood vessels. Next, image registration is performed to reduce the errors in alignment and scaling in the images so that the same eye images obtained at different visits are registered. Vasculature processing assists in the separation of vessel cross-section from the MA cross-section. Finally, classification is performed to discriminate the actual MAs based on the abnormality and severity of false detections, which are based on extracted features. The development of a fully automated system for MA detection and quantification will assist in prompt diagnosis at an early stage and prevent damage to the retina of DR patients.

9.1. Introduction

Diabetes is one of the most prevalent chronic diseases caused by an abnormal increase of glucose in the blood. Diabetes affects a large and varied portion of the population and is ranked the fifth deadliest disease in the United States. The International Diabetes Federation (IDF) indicates that diabetes currently affects 246 million people worldwide and is expected to affect 380 million by 2025. In terms of economic impact, the IDF projected that 232 billion dollars were spent globally for the treatment and prevention of diabetes and its associated complications. DR is an ocular complication of long-term diabetes that changes the tiny blood vessels in the retina by making them swell and causing them to leak fluid. DR is a progressive pathology and is one of the four main causes of problems associated with sight and is reported to be the number one cause of blindness in people younger than 50.[2,3] Due to its damaging effects and pathologic complications in sight and vision, DR has attracted the interest of researchers in the fields of biomedical computation and ophthalmology.

DR is a progressive disease that has stages of severity, which are characterized by the number and types of lesions it produces on the retina. The symptoms of DR can drastically change the texture appearance of the retina. The tiny blood vessels in the retina can cause leakage resulting in the formation of features such as MAs, cotton-wool spots, hemorrhages, or exudates.[4,5] Furthermore, the images of the disease can vary widely depending on the individual patient and the severity of the disease. Based on the number and type of lesions present, DR can be characterized into four stages[6]:

1. **Mild nonproliferative retinopathy** is the earliest stage of DR. During this stage, only MAs (small swellings in the retina's tiny blood vessels) occur.
2. **Moderate nonproliferative retinopathy** occurs as the disease progresses. During this stage, several MAs and retinal hemorrhages are present, along with some blood vessels, which may cause retinal blockage. Cotton-wool spots and some venous bleeding may become visible.
3. **Severe nonproliferative retinopathy** is the stage in which many blood vessels are blocked, depriving several areas of the retina of adequate blood supply. The following one or more symptoms may be present at this stage:[7]

    - more than 20 intraretinal hemorrhages in each of the four quadrants,
    - definite venous bleeding in two or more quadrants, and
    - prominent intraretinal microvascular abnormalities in one or more quadrants.

    Severe nonproliferative retinopathy carries a 50% chance of progression to proliferative retinopathy in one year.
4. **Proliferative retinopathy** is the advanced stage of DR, in which the signals sent by the retina for nourishment trigger the growth of new blood vessels. These newly developed blood vessels are abnormal and fragile and have thin walls. They grow along the retina and along the surface of the vitreous gel that fills inside the eye. The leakage of these blood vessels can cause severe vision loss and even blindness.

DR can be treated in many ways, depending on the stage of the disease and the specific problem that requires the fastest physician attention and care. However, the retinal surgeon relies on several tests to both monitor the progression of the disease and make decisions for the appropriate treatment. A comprehensive eye examination includes a visual acuity test, a dilated eye examination, and tonometry, combined with traditional retinal imaging methods such as the use of fundus images, fluorescein angiograms, and optical coherence tomography (OTC) to provide information about the circulatory system and the condition of the fundus background (retinal structure). Among those tests, intravenous fluorescein angiography (IVFA) has emerged as one of the most important diagnostic procedures. IVFA is also employed to diagnose other retinal vascular pathologies, such as age-related macular degeneration (AMD, a degenerative condition of the most sensitive area of the retina), retinal vessel occlusions, and many other problems related to retinal circulation.[8–10] Review of IVFA data requires individual, manual inspection of a series of images or of a single frame of color-filtered, monochrome views of the retinal circulation, where each individual frame is stored as a separate 35-mm slide or digital image. Currently, this image processing and analysis is performed manually, a time-consuming process that introduces the potential for inaccurate diagnosis.

MAs are visible early and continue their presence as DR progresses. Klien *et al.*[1] have reported a strong correlation between an increase in the number of MAs over time and the early development of DR. However, in the normal fundus photographs, MAs may be confused with small dot hemorrhages due to reduced visibility. The IVFA procedure enhances the visibility of these lesions. After the venous filling phase, fluorescein angiograms depict MAs as small, round, hyperfluorescent objects and edges better defined than those in the retinal fundus camera images, making the detection of MAs from fluorescein angiograms a preferred method over the fundus camera images. Following this phase, manual MA counting can quantify the progress and stage of the disease. Computed MA detection can potentially be more rational than the manual process, which is prone to be tedious and time consuming, and introduces the potential for an inaccurate diagnosis. For the reasons stated above, the manual counting of MAs is not used in current clinical practice. Automated MA detection aids the early detection of DR, as it assists in quantifying MAs and differentiating them from other features. Therefore, in this chapter, we will focus on this type of lesion.

Our objective in this chapter is to present a comprehensive study on the current automated MA detection techniques. Several methods have been reported for automated MA detection using fluorescein angiography because of the modality's increased sensitivity, as MAs are best seen

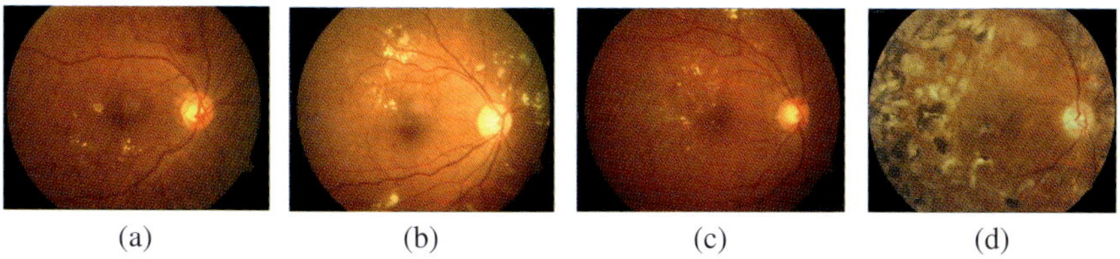

Fig. 9.1. Eyes at various stages of DR: (a) eye with mild nonproliferative, (b) eye with moderate nonproliferative, (c) eye with severe nonproliferative, and (d) eye with proliferative (Images obtained from Kasturba Medical Hospital, Manipal, India).

on fluorescein angiograms.[2] These methods usually follow a sequence of computational processing steps. The first steps include image preprocessing for noise removal and contrast enhancement for discrimination between MAs and blood vessels. Next, image registration is performed to reduce the errors in alignment and scaling in the images so that the same eye images obtained at different visits are registered. Vasculature processing assists in the separation of vessel cross-section from the MA cross-section. Finally, classification is performed to discriminate the actual MAs from the blood vessels, based on the abnormality and severity of false detections, which is based on extracted features. This chapter will cover a systematic description of these computational techniques.

9.1.1. *Preprocessing*

In many cases, a set of preprocessing steps is applied to a set of fluorescein angiographic images before the images are analyzed. The motivations for applying these steps is to reduce the unwanted variation in intensity or contrast that may occur due to the over or under exposure of the image and to separate MAs from the vessels, as they have the same intensity characteristics. It is essential to analyze the size, shape, and energy characteristics of the retinal objects in the image before segmenting for MAs. Further, the fluorescein angiograms are acquired during different visits for the same patient. The lapse of time between obtaining the images may result in discrepancies between the contrasts and provide an ineffective comparison. The common techniques in image processing include, but are not limited to, shade correction, Hough transform, and top-hat transform and edge detection. These techniques are applied to the image as a preprocessing step for the intensity and contrast enhancement. We briefly discuss these techniques, highlighting the related research in this domain.

9.1.1.1. *Shade correction*

Shade correction is a related technique that has been used to preprocess retinal images.[11–14] Shade correction is an approach that eliminates the background illumination that tends to be present in fundus or angiographic images. Cree *et al.* in Ref. [14] explains that the true image, $F(x, y)$ is an ideal image when there is a perfect optical system, the fundus is evenly illuminated by the flash, and the point (x, y) is defined as a position in the two-dimensional (2D) plane where all the imaging takes place. However, in practice, the fundus is often unevenly illuminated, or an optical disturbance eases the signal before it reaches the plane. Hence, the intensity of light $I(x, y)$ is defined as a product of the camera function, $C(x, y)$ and the image $F(x, y)$. This equation is represented below:

$$I = CF. \tag{9.1}$$

The background fluorescence due to the capillary network that extends through a large part of the retina and the choroid has an additive effect to the foreground features of the image. Therefore, the image F can be considered a sum of the background fluorescence, F_b and all the other features F_f of the image. For example,

$$F = F_f + F_b. \tag{9.2}$$

The illumination variation in the images is primarily caused by the following two factors: (i) the camera function, C, and (ii) the background fluorescence, F_b. Substituting Eq. (9.2) into Eq. (9.1), the image I can be defined as:

$$I = C(F_f + F_b). \tag{9.3}$$

Spencer *et al.*[15] have described a two-step procedure for preprocessing fluorescein angiograms. First, they normalized the images using a radiometric correction factor for the images to show the same range of pixel gray-levels and the same average gray-level. They then used a shade correction approach to remove the unnecessary changes in intensity that were incurred due to the photographic process and choroidal fluorescence.

9.1.1.2. *Hough transform*

Hough transform was proposed to identify straight lines in an image, but, later, the method has been extended to identify curves such as ellipses and circles. The purpose of Hough transform is to detect the parameterized curves in an image by groupings of edge points by object candidates and performing an explicit voting procedure over a set of parameterized image objects.[16] The mathematical representation behind a Hough transform is defined as follows. Consider a straight line in a normal form such as,

$$x \cos\theta + y \sin\theta = \rho. \tag{9.4}$$

This equation represents a line passing through the point (x, y) and is perpendicular to the line passing from the coordinates $(0, 0)$ and (ρ, θ), which is $(\rho \cos\theta, \rho \sin\theta)$ in rectangular space. To generate a Hough transform, a particular value of θ is chosen and is iterated through the angle defined by the granularity of θ. This process generates a curve in the rectangular representation of Hough transform (Fig. 9.2.).

As a preprocessing step to identify the MAs and the region of interest in the image Bernadres *et al.*[17] first segmented the image using two parallel lines to extract a circular area. They then applied the generalized and the modified Hough transform to detect the lines and the center and the radius of the circle. Azeem *et al.*[18] developed an approach to detect MAs based on circular Hough transform in fluorescein angiographic images. They were able to detect MAs due to the circular nature of the MAs and objects such

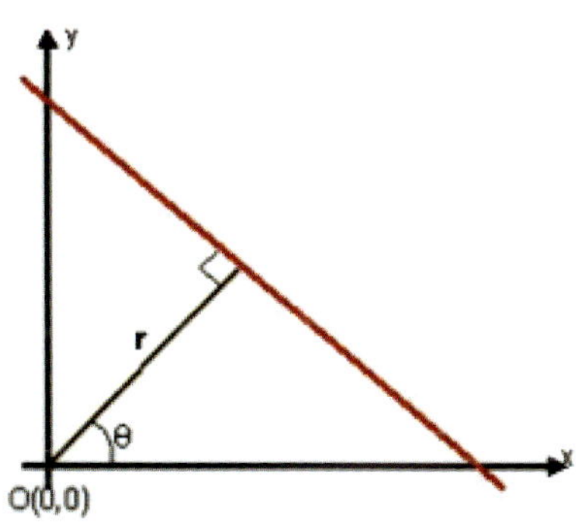

Fig. 9.2. Hough transform.

as vessel segments. However, circular Hough transform proved to be a computationally intensive procedure, almost making the procedure impractical to use.

9.1.1.3. *Top-hat transform*

Top-hat transform is a composite operation that extracts the bright objects from an uneven background by using morphological opening or closing operations. The mathematical formulation for a top-hat transform based on Refs. [19, 20] is described below.

Consider the gray-tone mathematical morphology operations. Let $f(x)$ be the values in the gray pixel levels in (x, u). The transformed image is defined by $f(x)$, where x is the set of pixels corresponding to the image. With the structuring element centered on x, the following transformation is defined as:

$$\text{Opening: } O^{\lambda}(X) = D^{\lambda}(E^{\lambda} f(X)) \quad \text{and} \tag{9.5}$$

$$\text{Closing: } C^{\lambda}(X) = E^{\lambda}(D^{\lambda} f(X)), \tag{9.6}$$

where $E^{\lambda} f(X)$ is defined as an erosion operation such that, $E^{\lambda} f(X) = \inf\{f(u){:}u \in \lambda_x\}$ and $D^{\lambda} f(X)$ is defined as a dilation operation, in which $D^{\lambda} f(X) = \sup\{f(u){:}u \in \lambda_x\}$. The opening and closing functions form the basis of top-hat transform. The opening function is usually used to separate distinct areas, and the closing operation is used to join them. The *peaks* and *valleys* can be extracted based on the size of the structuring element.[21]

Spencer *et al.*[13] used a bilinear top-hat transformation based on the linear structuring element to segment MAs from the arterioles and venules providing a robust discrimination between the linear and circular features. Mendonca *et al.*[22] applied the bilinear top-hat transform to reduce the detection of linear structures, such as retinal vessels.

9.1.2. *Image Segmentation*

Image segmentation is defined as a process that divides a digital image into nonoverlapping, disjoint regions by assigning labels to each pixel so that the pixels with the same label display certain common visual characteristics.[23] The process of image segmentation can be categorized into three approaches: the region approach, the boundary approach, and the edge approach.[24] We will briefly discuss each of these three approaches.

9.1.2.1. *The region approach*

Thresholding is one of the popular region approach techniques, in which all the pixels at or above the threshold gray level are assigned to an object while the pixels below the gray-level threshold fall on the other side of the object. Thresholding is computationally simple and easy to implement. Thresholding provides good results with the images where the objects have a uniform gray-level interior and the background is a contrasting but uniform gray level as well. Thresholding is further categorized into adaptive thresholding, global thresholding, and optimal threshold selection.

Particularly, in detecting MAs from the fluorescein angiographic images, Spencer *et al.*[13] performed an initial segmentation on the images by applying a bilinear top-hat transformation and matched filtering. They further applied thresholding to the processed image to convert the image into a binary image, which contained the candidate MAs. Mendonca *et al.*[22] applied thresholding by detecting the local maxima and considering those points as "seeds" that have an intensity value higher than a certain threshold value. A region growing procedure was then applied to limit the segments, which will be classified as MAs.

9.1.2.2. *The gradient-based method*

The boundary approach is a gradient-based segmentation method that attempts to segment the image into its interior and exterior points by initiating the process with a gradient magnitude image, which contains a single object on a contrasting background. The pixel with the highest gray level is identified as a boundary point. The three-by-three neighborhood centered around the first boundary point is searched, and the point with the highest gray level is chosen. This process is iterative and traces the maximum gradient boundary.

In gradient image thresholding, the image is thresholded at a low level to identify the object and the background below the threshold and the edge points above it. Increasing the threshold gradually causes the objects and the background to grow and when they converge, the points of contact define the boundary.

9.1.2.3. *Edge detection*

Edge detection is a feature-extraction technique that seeks to provide the information on the location of the regions in the image where the intensity changes sharply or discontinuities are detected. If a pixel lies on the boundary of the object, then the neighborhood pixels will show a region of transition. The two main characteristics of the edge detection operators are slope and direction. The points lying on an edge can be detected by: (1) detecting local maxima or minima of the first derivative or (2) detecting the zero crossing of the second derivative. Edge detection methods are classified broadly in three categories: first-order derivative (gradient), second-order derivative, and optimal edge detection.

9.1.2.3.1. The first-order derivative (gradient) methods

The first-order derivative methods have kernel operators that calculate the strength of the slope in the vertical or horizontal direction. Consequently, the edge strength is calculated as an aggregate of the different components of the slope. The Prewitt,[25] Roberts,[26] and Sobel[27] operators are classified as the first-order derivative methods.

The **Prewitt edge operator** measures the horizontal edge component using kernel K_x, and the vertical edge component using kernel, K_y. The gradient intensity of the current pixel is calculated by $|K_x| + |K_y|$. The Prewitt edge detector operator is easy to implement and is less computationally intensive than other methods. However, one of the major drawbacks of the Prewitt edge detector operator is its sensitivity to noise.[24]

The **Roberts edge operator** is a local differential operator used for detecting edges. The mathematical formulation is given as:

$$g(x, y) = \{[\sqrt{I(r, c)} - \sqrt{I(r+1, c+1]}^{2} + [\sqrt{I(r+1, c)} - \sqrt{I(r, c+1)}]^2\}^{1/2}, \quad (9.7)$$

where, $I(r, c)$ is the input image with the pixel coordinates (r, c). The Roberts edge operator marks the edge points in an image, providing no information about the orientation of the edge is available. The results have indicated that this method works best with binary images.[28]

The **Sobel edge operator** involves two convolution kernels denoted by $g(x)$ and $g(y)$, where

$$g(x) = \begin{bmatrix} -1 & 0 & 1 \\ -2 & 0 & 2 \\ -1 & 0 & 1 \end{bmatrix} \quad \text{and} \quad g(y) = \begin{bmatrix} -1 & -2 & -1 \\ 0 & 0 & 0 \\ 1 & 2 & 1 \end{bmatrix}. \tag{9.8}$$

The convolution kernels smooth the image and, hence, are less prone to noise. However, the edge localization is poor since it produces thicker edges.[27]

9.1.2.3.2. The second-order derivative methods

The second-order derivative methods search for zero crossings to find edges in an image, which are computed from the second-order derivative expression.[29,30] The Laplacian, Laplacian of Gaussian, and difference of Gaussian (DoG) methods are classified as second-order derivative methods.

The **Laplacian** of an input image I denoted by $f(x, y)$ is defined as,

$$f(x, y) = \frac{\partial^2 f(x, y)}{\partial x^2} + \frac{\partial^2 f(x, y)}{\partial y^2}. \tag{9.9}$$

The Laplacian operator is usually susceptible to noise and requires filtering. The **Laplacian of Gaussian** method is also known as the Marr-Hildreth edge detector. It is defined as,

$$LoG(x, y) = -\frac{1}{\pi\sigma^4}\left[1 - \frac{x^2 + y^2}{2\sigma^2}\right] e^{-\frac{x^2+y^2}{2\sigma^2}}. \tag{9.10}$$

The value of σ dictates the value of the Gaussian filter. The broader the Gaussian filter, the more smoothening is performed. However, the LoG is computationally intensive. The **DoG** can approximate the LoG. The LoG can be approximated by the difference of two Gaussians, DoG. The difference of the Gaussian also called the Mexican hat operator is defined as,

$$DoG(x, y) = \frac{e^{-\frac{x^2+y^2}{2\pi\sigma_1^2}}}{2\pi\sigma_1^2} - \frac{e^{-\frac{x^2+y^2}{2\pi\sigma_2^2}}}{2\pi\sigma_2^2}. \tag{9.11}$$

Here, the width of the edge can be adjusted by changing the values of σ_1 and σ_2.

9.1.2.3.3. The optimal edge detector

The **Canny edge detection**[31] is one of the most popular edge detection techniques. The algorithm utilizes an optimal edge detector based on a set of criteria to achieve the following optimization constraints:

- Achieve a good localization to mark the edges as closely as possible to the actual edges,
- Mark the edges only once when a single edge exists to minimize the number of responses to a single edge (to help detect the true negatives, that is the nonedges are not marked), and
- Maximize the signal-to-noise ratio in detecting the true positives.

According to Canny, the optimal filter that meets the above three criteria can be approximated as the first derivative of Gaussian function defined as,

$$G(x, y) = \frac{1}{2\pi\sigma^2} e^{\frac{x^2+y^2}{2\sigma^2}} \tag{9.12}$$

$$\frac{\partial G(x, y)}{\partial x} \alpha x e^{\frac{x^2+y^2}{2\sigma^2}} \quad \frac{\partial G(x, y)}{\partial y} \alpha x e^{\frac{x^2+y^2}{2\sigma^2}}. \tag{9.13}$$

Hafez *et al.*[32] have applied the Canny and Sobel edge detectors to find MAs in the fluorescein angiograms of the ocular fundus. They first detected the edges in the image using the Canny edge detector and then subtracted those segments that represented the vessel segments in the image. For the remaining objects, they calculated the edge threshold by computing the Sobel edge operator for each point in the object. The results obtained outperformed the Hough transform method in the computational time, as well as in detecting the number of false MAs.

We obtained a set of fluorescein angiographic images from the Louisiana State University Health Science Center, New Orleans and performed a series of preprocessing steps using a three-step preprocessing algorithm (Fig. 9.3.). First, we map low intensity pixels to high intensity values. The nonzero pixels of the resultant image correspond to the exudates or optic disc, and they are replaced by pixels with an average intensity value. This replacement aids in dismissing false alarms. Second, we smoothen the image for noise

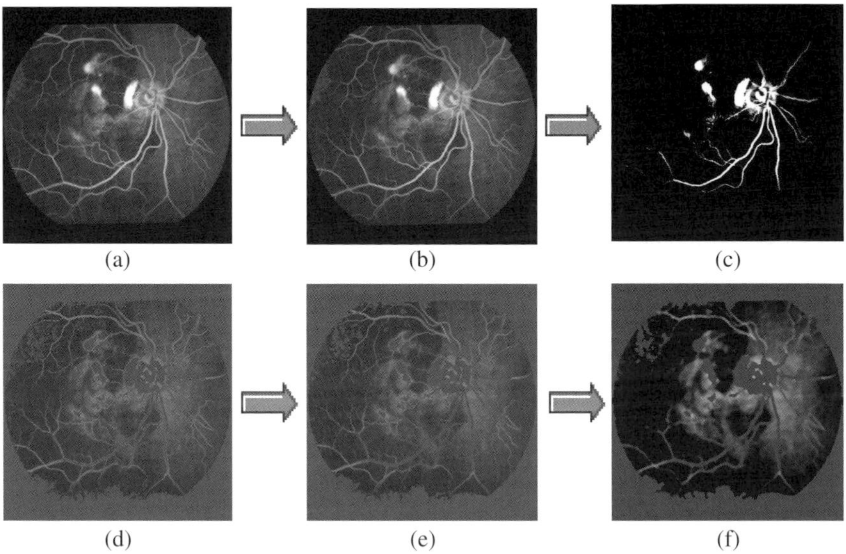

Fig. 9.3. Image preprocessing. (a) Original image, (b) Converted Grey-Scale Image, (c) Image Mapped to Different Intensities, (d) Identified Exudates Marked with Different Intensities, (e) Smoothed Image, (f) Contrast Enhanced Image.

removal. Finally, we perform contrast enhancement to aid in the sharpening of blood vessels for visual inspection. We then apply any of the above-mentioned techniques to segment the MAs from the other retinal objects.

9.2. Image Registration

The progression of DR is monitored by counting the MAs in the patients' images, which have been taken months apart. The digitization process causes the misalignment or mis-scaling of the images, and, hence, it becomes necessary to register the images obtained at different patient visits. It is important to align two or more images geometrically, so that the pixels in each image refer to the same physiological structure when the images are superimposed upon one another. A wide range of algorithms has been proposed for image registration based on the criteria, e.g. dimensionality, modality, and nature of transformation. An excellent survey on medical image registration can be found in Ref. [33]. In this chapter, we examine area-based or feature-based techniques for fluorescein angiographies (FAs). Numerous techniques are available for the registration of retinal images.[34–42] However, few techniques have been developed for the multimodal registration of fluorescein angiograms due to the nonlinear intensity differences between the modalities and the fluorescein angiograms and the low contrast in the late-phase images.[43] Here, we concentrate on the methods, which have been used in the past for the registration of fluorescein angiographic images.

A set of points from the sensed image are compared to the reference image on pixel intensities and optimized objective functions such as cross-correlation and mutual information are used in the area-based techniques. Cross-correlation is a commonly used technique[44–46] for image registration where a template is taken from one image and a similarity function is determined for each possible position of the template in the search image. This process results in a correlation surface, which peaks at the position of best match.[47] Ritter *et al.*[48] used mutual information as a similarity measure and a pyramid sampling-based reannealing search technique. The method seems to improve the speed and precision of the registration technique but does not always register the total surface of eye fundus images. Nagin *et al.*[45] extracted the blood vessels using edge enhancement and correlation techniques and performed a maximization of cross-correlation between the sections of the images. Ryan *et al.*[49] used the control point matching by computing the similarity transformation coefficients and calculating the Euclidean distance between the each similarity transformation coefficient and its nearest neighbor. This procedure is then followed by the expectation maximization algorithm with images being registered using the bilinear transformations for control point matching.

Dreo *et al.* use an intensity-based approach.[50] They present a comparative study of different optimization techniques and compare the performance on the sum of absolute difference objective function. Nunes *et al.*[51] propose a three-step multi-scale registration scheme, in which they propose an edge-preserving morphological filtering approach based on opening and closing with a linear rotating structuring element for the first step. In the second step, the morphological gradient of the filtered images are computed, and, eventually, the images are radiometrically corrected. Finally, a hierarchical approach is applied to register the images. Pinz *et al.*[52] propose an affine registration algorithm where the edges are detected at the boundaries of the vessels and are grouped by searching the other edges that satisfy the criteria such as opposite gradient direction and spatial proximity.

The motivations for using feature-based methods are to develop a correspondence between a number of points in the images and to use these points to obtain the parameters to register the two images. A Y-feature-based extraction method is proposed by Choe *et al.*[41] Choe *et al.* extract three bright vessels from the circular boundary of a candidate bifurcation location. The Y-feature (bifurcation) points are matched across modalities and through different phases of the circulation using the local maximization of mutual information. A random sample consensus (RANSAC) is then used for the pair-wise registration of images. Zana *et al.*[53] present an algorithm for the multimodal registration of retinal images based on the point correspondence of segmented vessels. The vascular tree using morphological segmentation is detected, and the bifurcation points are extracted. The matching for the bifurcation points is performed using a Bayesian Hough transform. Similarly, Laliberte *et al.*[38] study the similarity, affine and second-order polynomial transformation to match the bifurcation points.

Some algorithms have been developed in this area to register a red-free (RF) image with an FA image. In such cases, in order to extract the features, images from one of the modalities are transformed to simulate the appearance of vessels in the other modality. Another study by Matsopoulos *et al.*[54] extracts bifurcation points from the reference image, and the corresponding control points from the FA images are automatically generated using self-organizing maps (SOMs). Finally, the parameters of the affine transformation are estimated based on the corresponding points.

Based on all the approaches present in this section for image registration, only Refs. [41, 50] have validated their work on a complete set of fluorescein angiographic images.

9.3. Classification

After the steps of image preprocessing, image segmentation, and feature extraction, most of the computational methods attempt to discriminate the actual MAs based on the abnormality and severity of false detections based on extracted features. In the automated classification techniques, the algorithm is provided with a training data, which includes a series of observations to classify each candidate object as a MA or a falsely classified feature. The classifier is trained on this data set and is then subjected to a test data, which is a hold out data set where the values of the candidates are hidden. The statistical measures of specificity and sensitivity are used to measure the performance of a classification scheme. Sensitivity refers to the measure of true positives, which are correctly identified as MAs, while specificity refers to the measure of false negatives, which are correctly identified as some other object. The two measures are closely related to the type I and type II errors. Further, the receiver-operating characteristic (ROC) curve analysis is also commonly used to test the accuracy of a classifier in its ability to classify correctly the MAs. The ROC curve is a plot of sensitivity versus 1-specificity.

A number of studies[55–62] report the specificity and sensitivity in the detection of MAs for retinal images. Here, we discuss only those studies based on the IVFA images. Spencer *et al.*[13] compared the performance of their automated technique to that of the clinicians' response. The automated technique provided a sensitivity of 82%. However, there were few test images used in the experiment. Cree *et al.*[63] took a dataset of 68 trained and 20 test images and achieved a sensitivity of 82% at 5.7 false positives per image. Frame *et al.*[64] compared the classification results based on ROC analysis from a rule-based classifier to linear discriminant analysis and artificial neural network, and concluded that the performance of the rule-based classifier ($p = 0.92$) was better than the performance of the other two methods. Kamel *et al.*[65] use the learning vector quantization (LVQ) neural network to detect MAs from the fluorescein angiographic images based on a multi-stage training procedure. The multi-stage training procedure assisted in reducing the error rate to 0.75% when three stages of the network were used.

9.4. Automated, Integrated Image Analysis Systems

The algorithms developed over the past two decades and discussed in this chapter have contributed significantly to the development of integrated systems that encompass the ability to automate the vessel-tracing and feature-extraction procedures, automate image registration, and provide animation of images taken at different time intervals for an objective- and computer-assisted image analysis. In this section, we discuss some of the computational, integrated systems that have been developed to analyze retinal images. Tsai *et al.*[66] proposed an integrated image analysis tool called RIVERS (retinal image vessel extraction and registration system), which can be found at http://cgi-vision.cs.rpi.edu/cgi/RIVERS/. The tool encompasses the capabilities of automated vessel tracing, sub-pixel image registration, and the animation of images for ease of visualization. Cree *et al.*[67] developed an automated system for the detection of MAs, which incorporates the techniques for image preprocessing, registration of images, fovea detection, and identifying MAs and their turnover. Later, Cree *et al.* proposed another MA detector, the Waikato MA detector,[68] which is an automated system for the detection of MAs in retinal images that consists of the algorithms for shade correction, top-hat transform by morphological reconstruction (a Gaussian function template), and adaptive threshold determination. The MA detector was tested for 758 images, achieving the sensitivity and specificity of 85% and 95%, respectively. Jelinek *et al.*[69] utilized the Waikato MA detector and compared its performance with that of optometrists on a dataset of 543 images with no retinopathy and 215 images affected with retinopathy. They observed that the optometrists achieved a 97% sensitivity and 88% specificity in contrast to the MA detector, which achieved 85% sensitivity and 90% specificity in detecting the retinopathy.

9.5. Conclusion

In this chapter, we have attempted to summarize the automated techniques for the detection of MAs in fluorescein angiographic sequences. MAs are important lesions for detecting and determining a patient's stage of DR, since MAs are visible at an early stage and continue their presence in the later stages of the disease. An automated system for the detection of MAs encompasses the steps of image preprocessing, image segmentation, image registration, and image classification. A fully validated automated system can identify the retinal lesions and can provide significant advantages in cost and time for physicians and for early diagnosis of the disease.

References

1. Klien, R., Meuer, S., Moss, S., and Klien, B. The relationship of retinal microaneurysms counts to the 4-year progression of diabetic retinopathy. *Arch Ophthalmol* **107**:1780–1785, 1986.
2. Frilberg, T.R., Lace, J., Rosenstock, J., and Raskin, P. Retinal microaneurysm counts in diabetic retinopathy: color photography versus fluorescein angiography. *Can J Ophthalmol* **22**:226–229, 1987.
3. Klein, R., Klien, K., and Moss, S.E. Visual impairment in diabetes. *Ophthalmology* **91**:1–9, 1984.
4. Sjolie, A.K., Stephenson, J., Aldington, S., Kohner, E., Janka, H., Stevens, L., and Fuller, J. EURODIAB complications study group: retinopathy and vision loss in insulin-dependent diabetes in Europe. *Ophthalmology* **104**: 252–260, 1997.
5. Frank, R.N. Diabetic retinopathy. *Prog Retinal Eye Res* **14**:361–392, 1995.
6. Nayak, J., Bhat, P.S., Acharya, U.R., Lim, C.M., and Kagathi, M. Automated identification of different stages of diabetic retinopathy using digital fundus images. *J Med Syst* **32**:107–115, 2008.
7. http://www.icoph.org/standards/pdrdetail.html, retrieved November 9, 2009.
8. Arnold, C. *et al.* Fluorescein angiography in nonischemic optic disc edema. *Arch Ophthalmol* **114**(3): 293–298, 1996.
9. Laatikainen, L. and Blach, R.K. Behaviour of the iris vasculature in central retinal vein occlusion: a fluorescein angiographic study of the vascular response of the retina and the iris. *Br J Ophthalmol* **61**:272–277, 1977.
10. Cunha-Vaz, J.G. Diabetic retinopathy: surrogate outcomes for drug development for diabetic retinopathy. *Ophthalmologica* **214**:377–380, 2000.
11. Ward, N.P., Tomlinson, S., and Taylor, C.J. Image analysis and fundus photographs: the detection and measurement of exudates associated with diabetic retinopathy. *Ophthalmology* **96**:80–86, 1989.
12. Jagoe, R., Arnold, J., Blauth, C., Smith, P.L.C., Taylor, K.M., and Wootton, R. Retinal vessel circulation patterns visualized from a sequence of computer-aligned angiograms. *Invest Ophthal Vis Sci* **34**:2881–2887, 1993.
13. Spencer, T., Olson, J.A., McHardy, K.C., Sharp, P., and Forrester, J.V. An image-processing strategy for the segmentation and quantification of microaneurysms in fluorescein angiograms of the ocular fundus. *Comput Biomed Res* **29**:284–302, 1996.
14. Cree, M.J., Olson, J.A., McHardy, K.C., Sharp, P.F., and Forrester, J.V. The preprocessing of retinal images for the detection of fluorescein leakage. *Phys Med Biol* **44**:293–308, 1999.
15. Spencer, T., Philips, R.P., Sharp, P.F., and Forrester, J.V. Automated detection and quantification of microaneurysms in fluorescein angiograms. *Graefe's Arch Clin Exp Ophthalmol* **230**:36–41, 1992.
16. Shapiro, L.G. and Stockman, G.C. *Computer Vision*, Prentice Hall, 304, 2001.
17. Bernadres, R., Nunes, S., Pereira, I., Torrent, T., Rosa, A., Coelho, D., and Vaz, J. Computer-assisted microaneurysm turnover in the early stages of diabetic retinopathy. *Ophthalmologica* **223**:284–291, 2009.

18. Azeem, S.A., Hafez, M., and Auda, G. Using circular hough transform for detecting microaneurysms in fluorescein angiogram of the ocular fundus. In: *Proceedings of the International Conference on Industrial Electronics, Technology and Automation* (IETA2001), 2001.
19. Serra, J. *Image Analysis and Mathematical Morphology, Theoretical Advances*: Academic Press, London, 1982.
20. Serra, J. *Image Analysis and Mathematical Morphology*, 2. Academic Press, London, 1988.
21. Rodríguez, F., Maire, E., Courjault-Radé, P., and Darrozes, J. The black top hat function applied to a DEM: a tool to estimate recent incision in a mountainous watershed (Estibère Watershed, Central Pyrenees). *Geophysical Research Letters*, 2002.
22. Mendonca, M., Campilho, A.J., and Nunes, J.M. Automatic segmentation of microaneurysms in retinal angiograms of diabetic patients. In: *Proceedings of 10th International Conference on Image Analysis and Processing (ICIAP '99)*, pp. 728–733, Venice, Italy, September 1999, 1985.
23. Haralick, R.M. and Shapiro, L.G. Survey: image segmentation. *Comput Vis, Graph, Image Proc* **29**:100–132, 1985.
24. Prewitt, J. and Mendelsohn, M. The analysis of cell images. *Ann NY Acad Sci* **128**:1035–1053, 1996.
25. Roberts, L.G. Machine perception of three-dimensional solids. In: Tippett, J.G. (ed.), *Optical and Electro-Optical Information Processing*, MIT Press, Cambridge, MA, pp. 159–197, 1965.
26. Davis, L.S. A survey of edge detection techniques. *CGIP* **6**:492–501, 1997.
27. Neoh, H. and Hazanohuk, A. Adaptive edge detection for real-time video processing using FPGAs, CD. *Proceedings at the 2004 Global Signal Processing Expo (GSPx) and International Signal Processing Conference (ISPC)*, Santa Clara, California, September 27–30, 2004.
28. Castleman, K.R. *Digital Image Processing*, Prentice Hall, 1996.
29. Marr, D.C. and Hildreth, E. Theory of edge detection. *Proc R Soc Lond B* - **207**:187–217, 1980.
30. Sonka, M., Hlavac, V., and Boyle, R. *Image Processing, Analysis, and Machine Vision*, Second Edition, PWS Publishing, Pacific Grove, CA, 1999.
31. Canny, J. A computational approach to edge detection. *IEEE Trans Pattern Anal Mach Intell* **8**:679–714, 1986.
32. Hafez, M. and Azeem, S.A. Using adaptive edge technique for detecting microaneurysms in fluorescein angiograms of the ocular fundus. In: *Proceedings of IEEE MELECON* Cairo, Egypt, 2002.
33. Zitova, B. and Flusser, J. Image registration methods: a survey. *Image Vis Comput*, 977–1000, 2003.
34. Becker, D.E., Can, A., Tanenbaum, H.L., Turner, J.N., and Roysam, B. Image processing algorithms for retinal montage synthesis, mapping, and real-time location determination. *IEEE Trans Biomed Eng* **45**:105–118, 2003.
35. Can, A., Stewart, C., Roysam, B., and Tanenbaum, H. A feature-based, robust, hierarchical algorithm for registering pairs of images of the curved human retina. *IEEE Trans Pattern Anal Mach Intell* **24**:347–364, 2002.

36. Belongie, S., Malik, J., and Puzicha, J. Shape matching and object recognition using shape contexts. *IEEE Trans Pattern Anal* **24**:509–522, 2002.
37. Lloret, D., Serrat, J., Lopez, A., Soler, A., and J. Villanueva. Retinal image registration using creases as anatomical landmarks. *Proc IEEE Int Conf Image Process* **3**:203–207, 2000.
38. Laliberté, F., Gagnon, L., and Sheng, Y. Registration and fusion of retinal images — an evaluation study. *IEEE Trans Med Imaging* **22**:661–673, 2003.
39. Ege, B., Dahl, T., Sndergaard, T., Larsen, O., Bek, T., and Hejlesen, O. Automatic registration of ocular fundus images. In: *Work Comp Assisted Fundus Image Analysis*, 2000.
40. Heneghan, C., Maguire, P., Ryan, N., and de Chazal, P. Retinal image registration using control points. In: *IEEE Int Symp Biomed Imaging* 349–352, Washington, DC, USA, 2002.
41. Choe, T.E. and Cohen, I. Registration of multimodal ?ourescein images sequence of the retina. In: *Proc ICCV*, 2005.
42. Mendonca, A., Campilho, J., and Nunes, A. A new similarity criterion for retinal image registration. In: *Proceedings IEEE International Conference on Image Processing*, pp. 696–700, 1994.
43. Tsai, C.L., Chun, Y.L., Gehua, Y., and Lin, K.L. The edge-driven dual-bootstrap iterative closest point algorithm for registration of multimodal fluorescein angiogram sequence. *SPIE Symposium on Medical Imaging*, 2008.
44. Glazer, F., Raynolds, G., and Annanan, P. Scene matching by hierarchical correlation. In: *Proc IEEE Comput Soc Conf Computer Vision and Pattern Recognition*, p. 432, 1983.
45. Nagin, P., Schwartz, B., and Reynolds, G. Measurement of fluorescein angiogram of the optic disc and retina using computerised image analysis. *Ophthalmology* **92**:547, 1982.
46. Corona, E. and Mitra, S. A fast algorithm for registration of individual frames and information recovery in fluorescein angiography video image analysis. *Proceedings of Fifth IEEE Southwest Symposium on Image Analysis and Interpretation*, 2002.
47. Byrne, J.P., Undrill, P.E., and Phillips, R.P. Feature-based image registration using parallel computing methods. In: *Proceedings of the First Conference on Visualization in Biomedical Computing*, pp. 304–310, 1990.
48. Ritter, N., Owens, R., Yogesan, K., and Saarloos, P. The application of mutual information to the registration of stereo and temporal images of the retina **1342**:67–76, 1997.
49. Ryan, N., Heneghan C., and Chazal P. Registration of digital retinal images using landmark correspondence by expectation maximization. *Image Vis Comput* **22**:883–898, 2004.
50. Dreo, J., Nunsea, J.C., and Siarry P. Robust rigid registration of retinal angiograms through optimization. *Comp Med Imaging Graph* **30**:453–463, 2006.
51. Nunes, J.C., Bouaoune, Y., Delechelle, E., and Bunel, P. A multiscale elastic registration scheme for retinal angiograms. *CVIU* **95**:129–149, 2004.
52. Pinz, A., Bernogger, S., Datlinger, P., and Kruger, A. Mapping the human retina. *IEEE Trans Med Imaging* **17**: 606–620, 1998.

53. Zana, F. and Klein J.C. A multimodal registration algorithm of eye fundus images using vessels detection and Hough transform. *IEEE Trans Med Imaging* **18**:419–428, 1999.
54. Matsopoulos, G.K., Mouravliansky, D.K.K., and Nikita, K.S. Automatic retinal image registration scheme using global optimization techniques. *IEEE Trans Inf Technol Biomed* **3**:47–60, 1999.
55. Yang, G., Gagnon, L., Wang S., and Boucher M.C. Algorithm for detecting microaneurysms in low-resolution color retinal images. *Vision Interface*, 265–271, 2001.
56. Hipwell, J.H., Strachan, F., Olson, J.A., McHardy, K.C., Sharp P.F., and Forrester, J.V. Automated detection of microaneurysms in digital red free photographs: a diabetic retinopathy screening tool. *Diabetic Med* **17**:588–594, 2000.
57. Streeter, L. and Cree, M.J. Microaneurysm detection in colour fundus images. In: *Proceedings of Image and Vision Computing New Zealand*, pp. 280–285, 2003.
58. Pallawala, P.M.D.S., Hsu, W., Lee, M.L., and Goh, S.S. Automated microaneurysm segmentation and detection using generalized eigenvectors. *Seventh IEEE Workshops on Application of Computer Vision (WACV/MOTION'05)* — Volume 1, *WACV-MOTION* **1**:322–327, 2005.
59. Fleming, A.D., Philip, S., Goatman, K.A., Olson, J.A., and Sharp, P.F. Automated microaneurysm detection using local contrast normalization and local vessel detection. *IEEE Trans Med Imaging* **25**:1223–1232, 2006.
60. Narasimha-Iyer, H., Can, A., Roysam, B., Stewart, C.V., Tanenbaum, H.L., Majerovics, A., and Singh, H. Robust detection and classification of longitudinal changes in color retinal fundus images for monitoring diabetic retinopathy. *IEEE Trans Biomed Eng* **53**:1084–98, 2006.
61. Wong, L.Y., Acharya, R.U., Venkatesh, Y.V., Chee, C., Min, L.C., and Ng, E.Y.K. Identification of different stages of diabetic retinopathy using retinal optical images. *Inf Sci* **178**:106–121, 2008.
62. Quellec, G., Lamard, M., Josselin, P.M., Cazuguel, G., Cochener, B., and Roux, C. Optimal wavelet transform for the detection of microaneurysms in retina photographs. *IEEE Trans Med Imaging* **27**:1230–1241, 2008.
63. Cree, M.J, Olson, J.A, McHardy, K.C, Sharp, P.F., and Forrester, J.V. A fully automated comparative microaneurysm digital detection system. *Eye* **11**:622–628, 1997.
64. Frame, A.J., Undrill, P.E., Cree, M.J., Olson, J.A., McHardy, K.C., Sharp, P.F., and Forrester, J.V. A comparison of computer based classification methods applied to the detection of microaneurysms in ophthalmic fluorescein angiograms. *Comput Biol Med* **28**:225–238, 1988.
65. Kamel, M., Belkassim, S., Mendonca, A.M., and Campilho, A. A neural network approach for the automatic detection of microaneurysms in retinal angiograms. In: *the Proceedings of International Joint Conference on Neural Networks* **4**:2695–2699, 2001.
66. Tsai, C.L., Madore, B., Leotta, M.J., Sofka, M., Yang, G., Majerovics, A., Tanenbaum, H.L., Stewart, C.V., and Roysam, B. Automated retinal image analysis over the internet. *IEEE Trans Inf Technol Biomed* **12**:480–487, 2008.
67. Cree, M.J. The Waikato Microaneurysm Detector, A technical report. Available at: http://roc.healthcare.uiowa.edu/results/documentation/waikato.pdf.

68. Cree, M.J., Olson, J.A., McHardy, K.C., Sharp, P.F., and Forrester, J.V. Automated microaneurysm detection. *Proc IEEE Int Conf Image Processing* **29**:284–302.
69. Jelinek, H., Cree, M.J., Worsley, D., Luckie, A., and Nixon, P. An automated microaneurysm detector as a tool for identification of diabetic retinopathy in rural optometric practice. *Clin Exp Optometry* **89**:299–305, 2006.

Chapter 10

Computer-Aided Diagnosis of Diabetic Retinopathy Stages Using Digital Fundus Images

Rajendra Acharya, U.*, Oliver Faust*, Sumeet Dua*,†, Seah Jia Hong*, Tan Swee Yang*, Pui San Lai* and Kityee Choo*

10.1. Introduction to Diabetic Retinopathy

Diabetes is a condition in which an individual's blood sugar-level exceeds the normal range. Prolonged diabetes damages small blood vessels in the retina resulting in diabetic retinopathy (DR). Early stage DR can go undetected, because it often progresses through the first stage without causing debilitating symptoms. Damage only occurs in the later stages. Hence, regular cost-effective eye screening for diabetic subjects is beneficial. This chapter documents a system that can automatically screen massive numbers of the images of varying DR patients at different stages. In this work, 120 digital fundus images were analyzed and classified into three groups: images that showed normal retinas, images that showed nonproliferative DR (NPDR) (mild, moderate, and severe DR), and images that showed proliferative DR (PDR). Features such as hemorrhages, exudates, blood vessels, and textures were extracted from the retinal images. These features were programmed into an artificial neural network (ANN) for automated classification. The proposed system delivers an average classification accuracy of 96.67%, sensitivity of 100%, and specificity of 95%.

*Department of Electronics and Computer Engineering, Ngee Ann Polytechnic, Singapore.
†Department of Computer Science, College of Engineering and Science, Louisiana Tech University, Louisiana, USA.

DR is one of the most diagnosed causes of vision loss around the world.[1] More specifically, DR represents an end-organ response to a systemic disease, and is a significant risk factor for vision loss and blindness. Individuals with diabetes are 25 times more likely to become blind than individuals without diabetes[2] because the retina is the innermost layer of the eye, the earliest disease-related changes can be seen in it.

The retina is a thin layer of tissue that covers the inner section of the eye. The tissue is made of chemical photo detector cells. These detector cells consist of two common types: rods and cones. Rods function in dim light, providing black-and-white vision; cones aid in daytime vision and color perception. A third, much rarer type of photoreceptor, the photosensitive ganglion cell, is important for receiving responses to bright daylight. These detector cells are not equally dense within the retina area. This uneven denseness is considered sampling with an unequal (spatial) sampling interval. The result of the detection process is an electrical signal, which is sent to the brain via the optic nerve. Figure 10.1 shows the structure of the retina.

The optic disc is an area of the eye where the optic nerves leave the eye, leading to the brain; therefore, there are no detector cells in this area. This lack of detector cells causes a break in the visual field, called a "blind spot." The macula is an oval spot located in the center of retina and is responsible for the acuteness of vision. The fovea is located in the center of macula. It helps to discriminate colors and has the highest density of photo detector cells.

Experts differentiate three DR stages. The following list provides a brief description of these stages.[3] In the mild NPDR stage, at least one

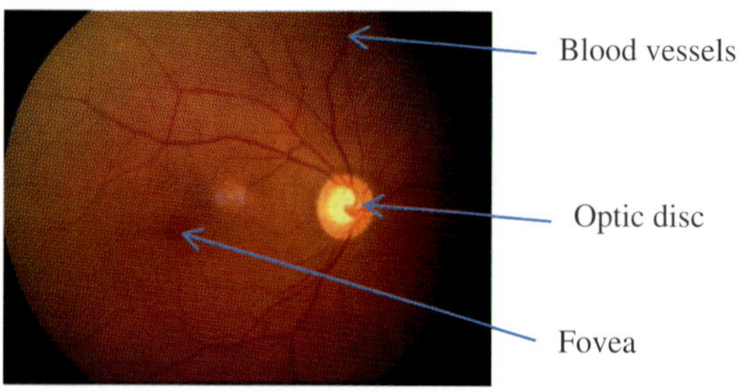

Fig. 10.1. Structure of normal retina.

microaneurysm (MA) is present. The MAs that are present may or may not contain hard exudates, cotton wool spots, or hemorrhages. In moderate NPDR, more MAs and retinal hemorrhages exist in the retina. Cotton wool spots and a slight venous beading may also be present. In severe NPDR, more MAs, retinal hemorrhages, and hard exudates appear. In PDR, large areas of the retina are starving. Because of diet changes caused by diabetes, the blood vessel tree is unable to provide enough nourishment to the retina. The signals sent by the retina for nourishment pave the way for the growth of new, fragile blood vessels, which may leak. Such leakage can lead to severe vision loss and blindness.

DR is a common cause of vision loss among working-class people in developed countries.[4] Laser photocoagulation may slow down the disease progression if the disease is detected early. Digital fundus images, from diabetes subjects, should be taken every year, and these images should be examined in order to avoid the progression of the disease by administering the right treatment at the right time.[5] The use of digital image-processing and data-mining techniques for automatic DR stage detection in the area of screening DR have increased with improved technology.[6]

Computer-aided diagnosis systems assist physicians in detecting abnormalities in retina fundus images.[7] Hayashi *et al.* proposed a method that can identify blood vessel intersections and, at the same time, detect abnormal widths in blood vessels.[7] The method was tested using 450 fundus images.

Z. Xiaohui and O. Chutatape presented three-step approaches for the detection and classification of bright lesions fundus images.[8] The steps include local contrast enhancement (pre-processing stage), improved fuzzy C-means, and hierarchical support vector machine (SVM) classification. Their results show that it is possible to classify bright nonlesion areas, exudates, and cotton-wool spots.

A neural network was used to detect diabetic features, namely, vessels, exudates, and hemorrhages in fundus images and compare the network performance to an ophthalmologist screening.[9] Blood vessels, exudates, and hemorrhages were detected with accuracy rates of 91.7%, 93.1%, and 73.8%, respectively. The system can aid healthcare professionals during screenings of diabetic patients.

Niemeijer *et al.* presented a method to detect red lesions based on the hybrid approach.[10] They were able to detect the red lesions with a sensitivity of 100% at a specificity of 87%. This method detects red lesions better than several other automatic systems and as well as human experts.

Li *et al.* proposed a new method to measure the severity of retinal arteriolar narrowing the arteriolar-to-venular diameter ratio.[11] The blood vessels were detected using a combined Kalman filter and Gaussian filter. Their results indicate 97.1% accuracy for identifying vessel starting points, and an accuracy of 99.2% for tracking retinal vessels.

DR may advance from mild to severe NPDR. Vallabha *et al.* proposed a novel method to automatically detect and classify vascular abnormalities in DR.[12] This method detects vascular abnormalities using scale and orientation selective Gabor filter banks. The proposed method classifies fundus images into mild and severe classes.

In our work, we have extracted four features using image-processing techniques and input are fed to a neural network classifier for automatic classification. Figure 10.2 shows an overview diagram of the proposed scheme. The layout of the chapter is as follows: In Sec. 10.2, we address both data acquisition process and preprocessing of the raw images. In Sec. 10.3, we present the feature extractions, using image processing and in Sec. 10.4, we

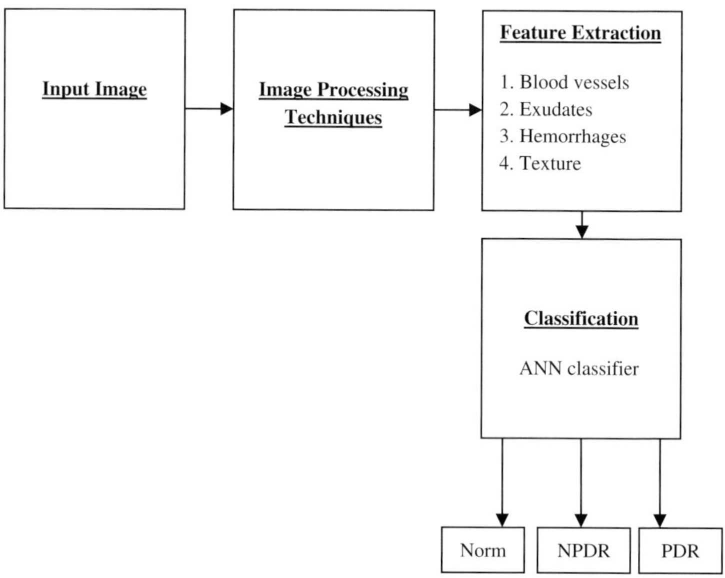

Fig. 10.2. Proposed system.

present the neural network classifier used. In Sec. 10.5, we show the results of the proposed method. In Secs. 10.6 and 10.7, we discuss the results and conclusion.

10.2. Data Acquisition

For this chapter, we have studied 120 retinal photographs of eyes with mild NPDR, moderate NPDR, and severe NPDR, PDR along with the images of normal eyes (data provided by the Kasturba Medical Hospital, Manipal, India). Images, taken by a fundus camera interfaced to a computer, were stored as 24-bit JPEG images of 256 × 256 pixels. The photographs in each group are described in 10.3. Figure 10.3 shows the optical images that belong to the normal eye and to eyes in the moderate, severe, and Proliferative DR stages.

Preprocessing was performed to remove nonuniform background, which may have been caused by nonuniform illumination or variations in the eye pigment color, from the fundus images. Adaptive histogram equalization was performed to solve this problem.[13] This method computes several

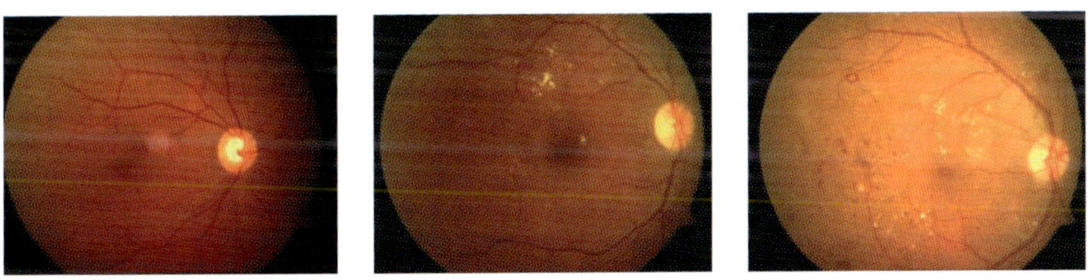

(a) (b) (c)

Fig. 10.3. Fundus images: (a) normal, (b) NPDR, and (c) PDR.

Table 10.1. Range of age and number of subjects in each group.

Types	Normal	Mild NPDR	Moderate NPDR	Severe NPDR	PDR
Age (years)	32±8	62±15	58±13	45±12	70±10
Number of Subjects	40	14	13	13	40

histograms, each of which correspond to a distinct section of the image, and uses them to redistribute the pixel values of the image.

10.3. Feature Extraction

Features, namely (i) blood vessels, (ii) exudates, and (iii) hemorrhages, were extracted from the fundus images. A brief description of these features is given below.

10.3.1. *Blood Vessel Detection*

The number of blood vessels, which nourish the retina, is one of the features that define the different DR stages. In this work, we used various morphological image-processing methods to identify the blood vessels. Figure 10.4 is a flowchart illustrating the blood vessel detection algorithm. In our experiments, the green channel of the fundus RGB image was used to detect blood vessels.[14] The image was converted to grayscale and subjected to an image inversion operation. The adaptive histogram operation was used to

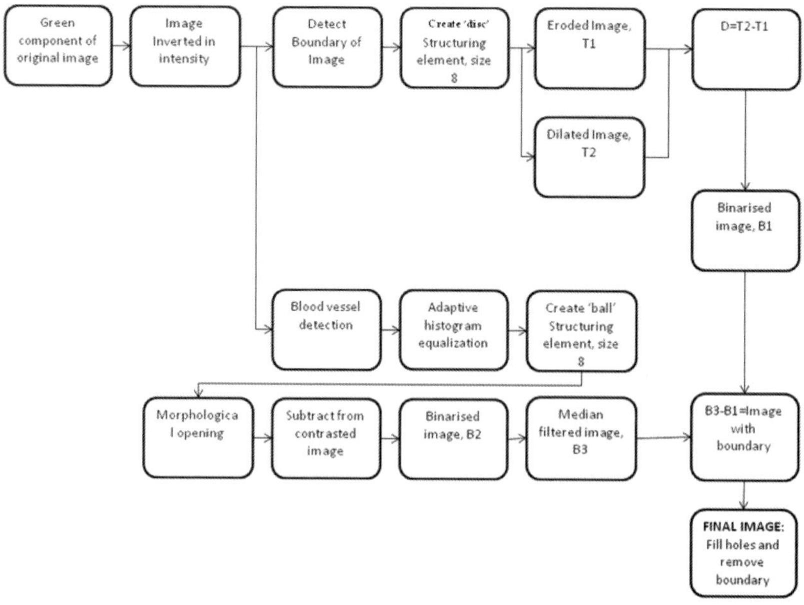

Fig. 10.4. Block diagram of the blood vessel detection.

improve the image contrast. The morphological "opening" operation was implemented, using the "ball" structuring element, to highlight the blood vessels and smoothen the image background.[13] After that, the inverted image was subtracted from the enhanced image. In the resulting image, the blood vessels showed higher contrast than the background.

The resulting image was binarized with a threshold value of 0.099, and, then, median filtering was performed to remove noise. To extract blood vessels from fundus images, a border was created and all remaining noise in the image was eliminated using the MatLab function "imfill". The images that showed only borders were subtracted from the inverted image, effectively removing the borders. Next, the image was inverted again to obtain an image depicting only blood vessels.

10.3.2. *Exudates Detection*

The green component of the RGB image was used for exudates detection.[14–16] Figure 10.5 shows the flowchart for this process. Dilation and erosion techniques using octagon and disc-shaped structuring elements were used to detect the exudates. The "Closing" operation was implemented using an octagon-shaped structuring element to obtain a better contrast image.[13] The resultant image contained both optic disc and exudates. Furthermore, the image gray levels are comparable with the exudates.

In order to remove artifacts from the image, neighborhood operation was performed by columns. The result of this operation was an image, which contained only optic disc and exudates. It is common for exudates to have irregular shapes. A disc-shaped structuring element was used to solve this problem (irregular exudates) and, then, the resultant image was binarized with a threshold value of 0.7 to highlight only the exudates. In most fundus images, the optic disc had the highest pixel value and occupied approximately one-seventh of image. A mask of 80×80 was used at the highest pixel location to remove the optic disc. Finally, the opening operation was performed using a disc-shaped structuring element to obtain only exudates.

10.3.3. *Hemorrhages Detection*

Figure 10.6 shows the two steps involved in hemorrhage detection: (1) detection of blood vessels alone and (2) detection of both blood vessels and

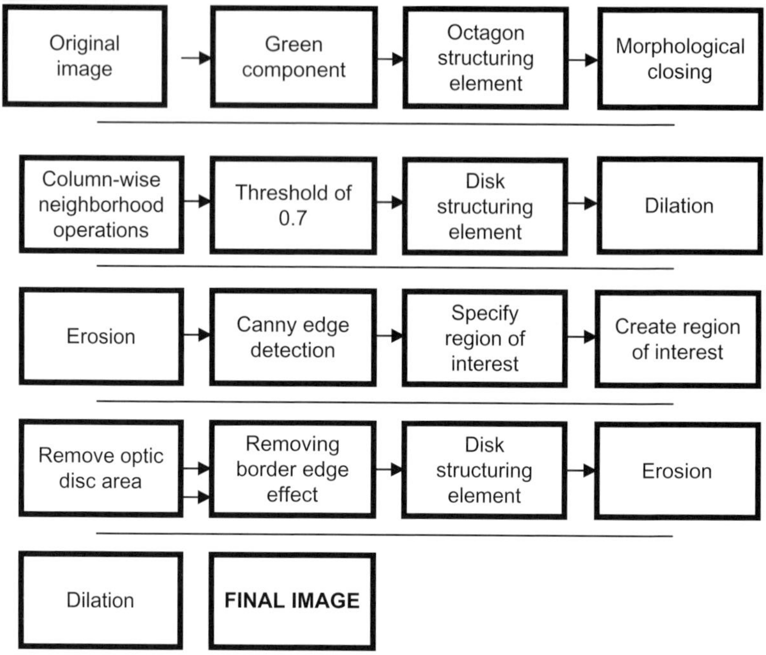

Fig. 10.5. Block diagram of the exudates detection.

hemorrhages. Once blood vessels and hemorrhages were detected, the image with blood vessels alone was subtracted from the image with blood vessels and hemorrhages to obtain only the hemorrhages. The red channel of the RGB image was used to detect hemorrhages.[14] An inversion operation was performed on the RED channel of the image. Next, the image was subjected to adaptive histogram equalization to increase the image contrast. Two "ball" shaped structuring elements, of sizes 6 and 25, were used to detect blood vessels without hemorrhages and blood vessels with hemorrhages.[14] We have divided the detection of blood vessels without hemorrhages and the detection of blood vessels with hemorrhages into two steps, as given below.

For the detection of blood vessels without hemorrhages, a small "ball"-shaped structuring element of size 10 was used, and the image was subjected to both erosion and dilation. Next, the image was enhanced using adaptive histogram equalization. Finally, the resulting image was subjected to image enhancement again to highlight only the blood vessels.

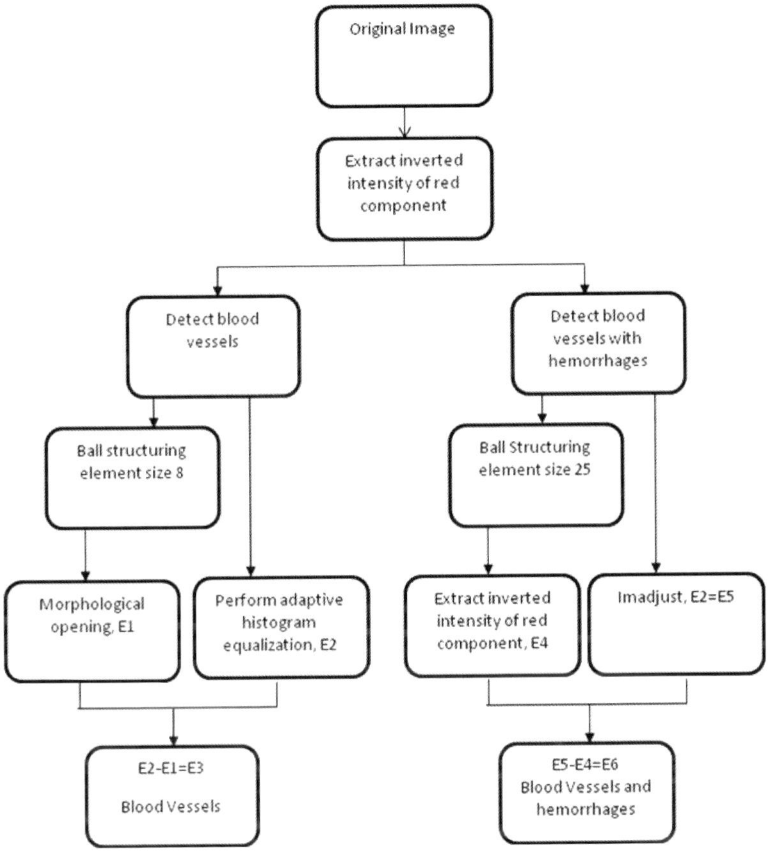

Fig. 10.6. Block diagram of the hemorrhages detection.

To detect both blood vessels and hemorrhages, we used a "ball"-shaped structuring element of size 45, because hemorrhages are typically large. First, the image was subjected to erosion and dilation using this "ball"-shaped structure element. Then, the image was enhanced to increase the image contrast. Next, this image was dilated and, then subtracted from the enhanced image and subjected to histogram equalization, again to obtain the image with blood vessels and hemorrhages.

The image that did not contain hemorrhages was subtracted from the image with blood vessels and hemorrhages to obtain an image with hemorrhages and noise. The resulting image was binarized with a threshold value of 0.25, and then it was subjected to median and Wiener filtering to remove the artifacts. Finally, the optic disc was removed (explained in detection of exudates) to obtain only hemorrhages.

10.3.4. *Contrast*

For an image, represented by a function $f(x, y)$ having N discrete gray levels, we defined the spatial gray level dependency matrix $P(d, \Phi)$ for each d and Φ, as given by:

$$P(d, \Phi) = \begin{vmatrix} p_{0,0} & p_{0,1} & \cdot & \cdot & p_{0,N-1} \\ p_{1,0} & p_{1,1} & \cdot & \cdot & p_{1,N-1} \\ \cdot & \cdot & \cdot & \cdot & \cdot \\ \cdot & \cdot & \cdot & \cdot & \cdot \\ p_{N-1,0} & p_{N-1,1} & \cdot & \cdot & p_{N-1,N-1} \end{vmatrix}, \qquad (10.1)$$

where

$$p_{i,j} = \frac{\text{number of pixel pairs with intensity (i, j)}}{\text{total number of pairs considered}}.$$

In this equation, the term $p_{i,j}$ is the relative number gray of a level pair (i, j) when pixels are separated by the distance d along the angle Φ. Finally, each element is normalized using the total number of occurrences to get co-occurrence matrix P.

Contrast is a texture parameter that indicates the amount of intensity variation in the image. This variation is given by[17]:

$$\Phi_2 = \sum_{i=0}^{N-1} \sum_{j=0}^{N-1} (i - j)^2 \, p_{i,j}. \qquad (10.2)$$

$p_{i,j}$ is the elements of the co-occurrence matrix shown in Eq. (10.1). The contrast is 0 for a constant image.

10.4. Classifier Used

In this work, we have used a backpropagation algorithm (BPA) to classify images into the three main DR stages. We provide a brief description of this algorithm in the following section.

10.4.1. *Backpropagation Algorithm*

BPA is a supervised learning technique used to train ANNs[18] and it is widely used to feed-forward networks. Sigmoid transfer function in the ANN makes it nonlinear in nature.

The BPA algorithm is an iterative gradient algorithm used to reduce the mean square error between actual and desired outputs. This algorithm is also known as "the generalized delta rule."[18] The neurons in layers, between the input and output layers, are called hidden nodes, and they do not directly interact with the environment. Using BPA, the weights associated with the hidden layers are updated continuously so that actual output and desired output match with least error.

To address the problem stated above, we experimented between one to four hidden layers. Based on the experimental results, we chose a neural network with one hidden layer that had five neurons, because this configuration gave the best classification result. A learning constant $\eta = 0.9$ was chosen to control the step size, while the mean square error was defined as 0.001.

Figure 10.7 shows the neural network classifier configuration that was used in this work. The output layer had two neurons, implying four possible

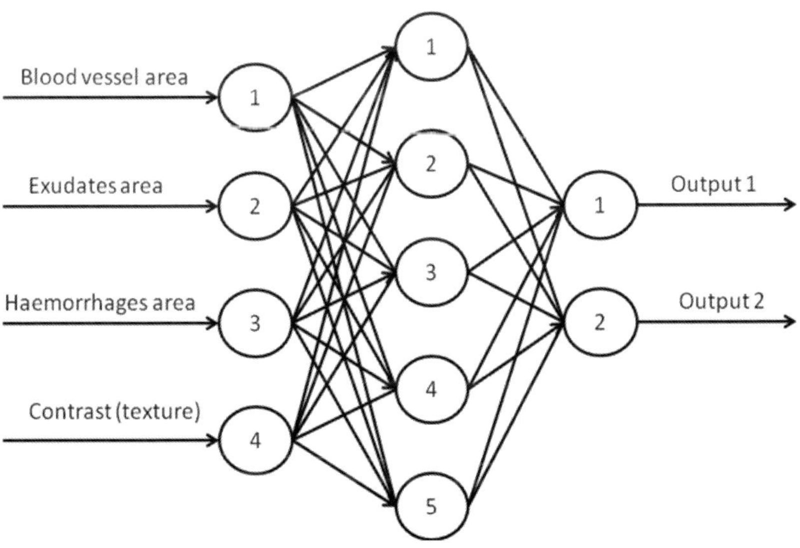

Fig. 10.7. Three-layer feed-forward neural network.

classes at the output. However, the network was trained to identify only three classes (normal, NPDR, and PDR) given by the decoded binary outputs of [00, 10, 01], respectively. Values of the blood vessels, exudates, hemorrhages, and contrast of the image were computed and fed as input to the classifier.

10.5. Results

Table 10.2 shows the analysis of variance (ANOVA) results of features extracted for normal, NPDR, and PDR images. These results show that the blood vessel area is more for NPDR and less for PDR. Exudates are absent for normal and present for NPDR. The hemorrhage area is larger for PDR, and the contrast is high for normal fundus images. All features are clinically significant ($p < 0.0001$).

Table 10.3 shows the classification results of the ANN classifier. We have used 90 images for training and 30 for testing. Our proposed system

Table 10.2. Results of area of blood vessels, exudates, hemorrhages, and contrast.

Features	PDR	NPDR	Normal	p-value
Blood vessel area	51,171 ± 20,652	343,975 ± 14,890	325,000 ± 20,278	$p < 0.0001$
Exudate area	8,148 ± 3,149	8,776 ± 2,106	—	$p < 0.0001$
Hemorrhage area	5,103 ± 4,329	2,213 ± 2,107	—	$p < 0.0001$
Contrast	0.0698 ± 0.0212	0.0691 ± 0.0182	0.0867 ± 0.0136	$p < 0.0001$

Table 10.3. Results of automatic classification.

Class	No. of data used for training	No. of data used for testing	Overall percentage of success
Normal	30	10	100.00
PDR	30	10	90.00
Non-PDR	30	10	100.00
Average			96.67

classifies all normal and NPDR images correctly, and PDR images are classified correctly with an accuracy of 90%. The average classification accuracy is 96.67%. Table 10.4 shows the results of sensitivity, specificity, and positive predictive accuracy for the proposed system. The table shows

Table 10.4. Results of sensitivity, specificity, and positive predictive accuracy for the proposed system.

True positive	True negative	False positive	False negative	Positive predictive value	Sensitivity	Specificity
10	19	1	0	90.91%	100%	95%

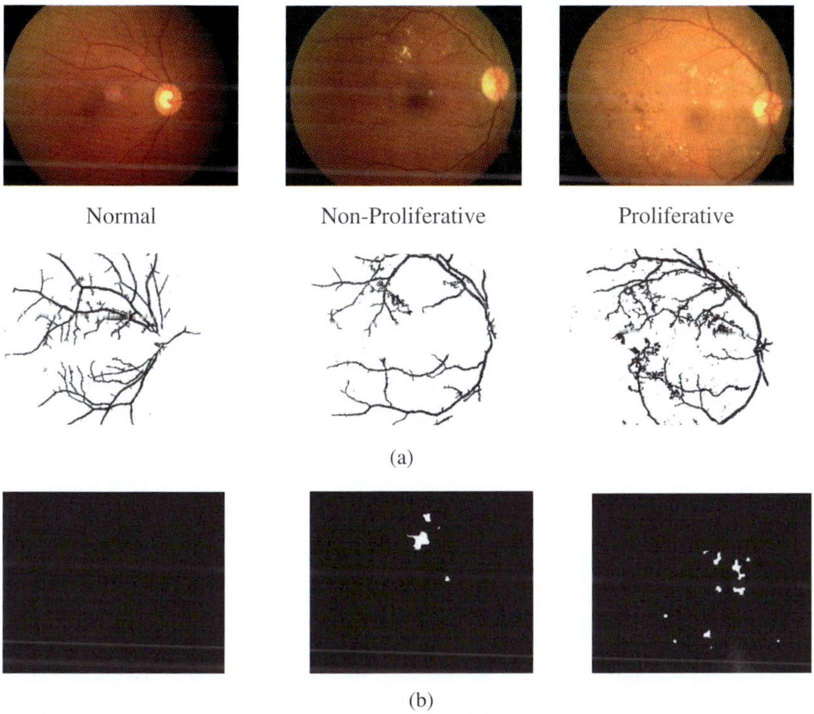

Fig. 10.8. Results of (a) blood vessel detection and (b) exudate detection for normal, NPDR, and PDR images using image processing.

that sensitivity and specificity of the proposed system are 100% and 95%, respectively.

Figure 10.8 (a–c) show the results of blood vessel detection, exudates detection, and hemorrhage detection for normal, NPDR, and PDR images. These figures show that there are fewer blood vessels for normal and more for NPDR and PDR stages. Exudates and hemorrhages do not exist in normal but they are present for NPDR and PDR stages.

10.6. Discussion

A computer diagnostic system was developed to detect three early lesions: hemorrhage MAs, hard exudates, and cotton-wool spots, and to classify NPDR based on these three types of lesions using 361 images.[19] The correct diagnosis rates, between computer system and reading center, for determining each lesion were 82.6%, and 88.3% for hemorrhages and MAs, hard exudates, and cotton-wool spots, respectively. The results from the proposed classification system were comparable to those provided by human experts, and can be used as a clinical aid to physicians for screening, diagnosing, and detecting NPDR.

A decision support system for the early diagnosis of DR by detecting the presence of MAs was developed by Kahai *et al.*[20] Their results show that their support system was able to achieve sensitivity and specificity of 100% and 67%, respectively.

Four retinal conditions: normal retina, moderate NPDR, severe NPDR, and PDR were automatically classified using the area and perimeter of blood vessels of red, green, and blue layers of the images and a neural network classifier.[21] Their proposed method demonstrated an accuracy of more than 80%, sensitivity, and specificity of more than 90% and 100%, respectively.

Using blood vessels, exudates, and texture parameters as well as feed-forward neural network, three classes: normal, NPDR, and PDR were automatically classified.[15] They identified the unknown class with an accuracy of 93%, and a sensitivity and specificity of 90% and 100%, respectively.

A low-cost screening method to identify normal and abnormal fundus images, based on exudates and lesions, was proposed.[22] The classification was performed with a statistical classifier and a local-window-based verification strategy. Their method identified all retinal images with exudates with 100% accuracy and normal images as normal with 70% accuracy.

The normal, as well as mild, moderate, severe NPDR, and PDR classes were automatically classified using higher order spectra features and the SVM classifier.[23] Acharya *et al.* demonstrated sensitivity and specificity of 82% and 88%, respectively using 300 digital fundus images.

Features, namely, blood vessels, exudates, MAs, and hemorrhages were extracted and assembled into an input vector before being fed to an SVM classifier. This classifier had to identify five groups: normal retina, mild NPDR, moderate NPDR, severe NPDR, and PDR.[16] They identified the unknown class with sensitivity and specificity of 82% and 86%, respectively.

In our present work, four features, namely, blood vessels, exudates, hemorrhages, and textures were used. These parameters were presented to a neural network for automated classification. Our results show an accuracy of 96.6%, sensitivity of 100%, and specificity of 95%. The classification accuracy can be further improved by using better features, more diverse fundus images taken under good lighting conditions, and better classifiers.

10.7. Conclusion

DR is a progressive eye disease caused by prolonged diabetes. It can result in vision loss, if not detected at an early stage. In this work, we have proposed an automated system to identify normal, NPDR, and PDR fundus images using image-processing and data-mining techniques. The proposed system can identify normal, NPDR, and PDR accurately with an accuracy of 96.67%, sensitivity of 100% and specificity of 95%. Our results show that the system can help ophthalmologists to automatically identify early stage DR. Additionally, ophthalmologists can use this system as an adjunct tool for screening.

References

1. Klein, R., Klein, B.E., and Moss, S.E. Visual impairment in diabetes. *Ophthalmology* **91**:1–9, 1984.
2. Eye Disease: http://visioneyedoctor.com/eye-disease/page-4.htm.
3. Frank, R.N. Diabetic retinopathy. *Prog Retin Eye Res* **14**:361–392, 1995.

4. Ong, G.L., Ripley, L.G., Newsom, R.S., Cooper, M., and Casswell, A.G. Screening for sight-threatening diabetic retinopathy: comparison of fundus photography with automated color contrast threshold test. *Am J Ophthalmol* **137**:445–452, 2004.
5. Fong, D.S., Aiello, L., Gardner, T.W., King, G.L., Blankenship, G., Cavallerano, J.D., Ferris, F.L., and Klein, R. Diabetic retinopathy. *Diabetes Care* **26**:226–229, 2003.
6. http://reseau-ophdiat.aphp.fr/Document/Doc/confliverpool.pdf.
7. Hayashi, J., Kunieda, T., Cole, J., Soga, R., Hatanaka, Y., Lu, M., Hara, T., and Fujita, H. A development of computer-aided diagnosis system using fundus images. In: *Proceeding of the Seventh International Conference on Virtual Systems and MultiMedia (VSMM 2001)*, pp. 429–438, 2001.
8. Xiaohui, Z. and Chutatape, O. Detection and classification of bright lesions in colour fundus images. *Int Conf Image Process* **1**:139–142, 2004.
9. Gardner, G., Keating, D., Williamson, T., and Elliott, A. Automatic detection of diabetic retinopathy using an artificial neural network: a screening tool. *Br J Ophthalmol* **80**:940–944, 1996.
10. Niemeijer, M., van Ginneken, B., Staal, J., Suttorp-Schulten, M., and Abramoff, M. Automatic detection of red lesions in digital color fundus photographs. *IEEE Trans Med Imaging* **24**:584–592, 2005.
11. Li, H., Hsu, W., Lee, M.L., and Wong, T.Y. Automated grading of retinal vessel caliber. *IEEE Trans Biomed Eng* **52**:1352–1355, 2005.
12. Vallabha, D., Dorairaj, R., Namuduri, K.R., and Thompson, H. Automated detection and classification of vascular abnormalities in diabetic retinopathy. *Thirty-Eighth Asilomar Conference on Signals, Systems and Computers*, 2004.
13. Gonzalez, R.C. and Woods, R.E. *Digital Image Processing*, Prentice Hall, New Jersey, Second Edition, 2001.
14. Acharya, U.R., Lim, C.M., Ng, E.Y.K., Chee, C., and Tamura, T. Computer based detection of diabetes retinopathy stages using digital fundus images. *J Eng Med* **223**(H5):545–553, 2009.
15. Nayak, J., Bhat, P.S., Acharya, U.R., Lim, C.M., and Kagathi, M. Automated identification of different stages of diabetic retinopathy using digital fundus images. *J Med Sys* **32**:107–115, 2008.
16. Sinthanayothin, C., Boyce, J., and Williamson, C.T. Automated localization of the optic disk, fovea, and retinal blood vessels from digital colour fundus images. *Br J Ophthalmol* **38**:902–910, 1999.
17. Randle, V. and Engler, O. *Introduction to Texture Analysis: Macrotexture, Microtexture and Orientation Mapping*, RCR Press, 2000.
18. Yegnanarayana, B. *Artificial Neural Networks*, Prentice-Hall of India, New Delhi, 1999.
19. Samuel, C.L., Elisa, T.L., Yiming, W., Ronald, K., Ronald, M.K., and Ann, W. Computer classification of a nonproliferative diabetic retinopathy. *Arch Ophthalmol* **123**:759–764, 2005.
20. Kahai, P., Namuduri, K.R., and Thompson, H. A decision support framework for automated screening of diabetic retinopathy. *Int J Biomed Imaging*: 1–8, 2006.
21. Wong, L.Y., Acharya, U.R., Venkatesh, Y.V., Chee, C., Lim, C.M., and Ng, E.Y.K. Identification of different stages of diabetic retinopathy using retinal optical images. *Inf Sci* 178, 2008

22. Wang, H., Hsu, W., Goh, K., and Lee, M. An effective approach to detect lesions in colour retinal images. In: *Proceedings of the IEEE Conference on Computer Vision and Pattern Recognition*, pp. 181–187, 2000.
23. Acharya, U.R., Chua, K.C., Ng, E.Y.K., Wei, W., and Chee, C. Application of higher order spectra for the identification of diabetes retinopathy stages. *J Med Sys* **32**:481–488, 2008.

Chapter 11

Reliable Transmission of Retinal Fundus Images with Patient Information Using Encryption, Watermarking, and Error Control Codes

Myagmarbayar Nergui*, Sripati Acharya, U.*, Rajendra Acharya, U.†, Wenwei Yu‡ and Sumeet Dua§

Today, digital media are important to many aspects of entertainment, business, and medicine. Many service and sales providers, broadcast companies, wireless cellular phone service providers, and entertainment electronics companies use digital signal processing (DSP) techniques to improve their quality of service (QoS). Ophthalmologists and other eye care clinicians can leverage this trend, especially with eye images, which can be transmitted along with patient information. Patient diagnosis and image information are worked on and stored digitally. Presently, patient diagnosis (text) and retinal fundus image information compose two forms data and are read as different files. Hence, a patient diagnosis may be matched to the wrong information. This interchanging of patient information can be prevented by embedding patient diagnosis and retinal fundus image

*Department of Electronics & Communication, National Institute of Technology Karnataka, Surathkal, India.
†Department of ECE, Ngee Ann Polytechnic, Singapore.
‡Graduate School of Medical System Engineering, Chiba University, Japan.
§Department of Computer Science, Louisiana Tech University, Ruston, LA 71272, USA.

information so that they can be separated without detectable damaging either file. In this chapter, we explain digital watermarking, a data-hiding technique that can embed patient diagnosis information within retinal fundus images. Additionally, we explain our digital watermarking and compression techniques, cryptographic protocols, and error correcting codes in this chapter. We have compared the performances of different types of error control codes (ECCs) with that of our technique. Watermarking helps to secure patient records from being compromised and assures that the information will remain intact even in noisy channels.

11.1. Related Studies

In hospitals and healthcare centers, today, patient diagnosis information and retinal fundus images constitute a large component of the data-generated information. The volume of data processed and stored in hospital and healthcare centers is increasing as the usage of electronic diagnostic instruments increases. Patient diagnosis must be coordinated with retinal fundus images so that patient and image information can be securely and accurately maintained. It is common for hospital and clinic admission staff to record patient information in text files. The patient diagnosis file is then modified during each examination. The diagnosis (text) information and its corresponding retinal fundus images are preserved under an admission number or name. The current practice is to store the retinal fundus images individually in a standard file format. In this format, a lost file identification tag can make it difficult to match the image to the correct patient. Furthermore, patient records stored as standard files are not protected. Unauthorized staff may gain access to the information. Therefore, the highest priority is to secure the information. Digital imaging and communication (DICOM) is the main file format for the medical images transmission in clinic centers.[1,2] The DICOM image format employs a header containing key words, attributes, and details of the image, ensuring that the patient image is not erroneously distinguished from patient data. Almost hospitals and healthcare centers have data-management systems that are grounded on the DICOM standard.[1,2] The DICOM standard has some disadvantages. For example, it does not protect patient diagnosis information from being viewed by unauthorized users, nor does it defend information integrity as it passes through imperfect transmission mediums or is stored in imperfect media. We have proposed a method for solving these problems using the DICOM standard in this chapter. We suggest using cryptographic protocols (to prevent security attacks) and using ECCs to preserve data security.

Digital watermarking[3–10] can be used to conceal statistic data into image. For intellectual property authentication and copyright protection, many applications use the digital watermarking techniques to protect rights. Applying this technique, text data, such as patient diagnosis history, can be concealed in the image.[11–14] In order to disguise it, the information is embedded into a cover data, most likely unobjectionable audio, image, or video. The media is then protected against distortion potentially caused by watermarking. Hence, digital watermarking techniques can be applied to conceal patient text data into image data, creating a complex data comprising both image and embedded text data that is hidden in the image. Cryptographic algorithms and ECCs process this complex block of data to secure the information.

Two types of domain methods, spatial domains and frequency domains, are watermarking techniques. The first watermarking method, spatial domains, is composed of least significant bit (LSB) insertion.[4,5,7,8] This method works that the LSB of each pixel value of the image data is substituted by one bit of the ASCII character set presenting the data to be embedded into image, so that it is placed into the LSB of the consecutive pixels in the image. It is possible to embed 192 kb text information data into a 256×256 digital color image. Since difference between original image and embedded image is so small, and is only one bit, the human eye cannot detect the distortion between the watermarked (embedded) image and the original image. In the second watermarking method, frequency domains,[5,7,9] the watermarking is performed in the transform domain. The redundant frequency content, which is related to the image, is substituted with characters from the text file. The substitution allows images to be watermarked using the discrete Fourier transforms (DFT), the discrete cosine transforms (DCT), or the discrete wavelet transforms (DWT).[14] Because diagnosis data often comprise confidential and private data, a cryptographic algorithm, advanced encryption standard (AES),[15,16] is applied to protect patient data. This algorithm is robust and safe against attacks by hackers.

Compression techniques are used to compress data so that it requires less bandwidth during a data transmission and less disk space for storage. Using compression schemes for communication equipment, such as modems, bridges, and routers, improves the throughput over standard channels. Image compression is a kind of data compression technique applied to digital images to cut down the redundancy of the image data for storing or transmitting in an efficient form.

Information passing through a communication channel can be affected due to nonqualities in the transmission medium and receiver system. A transmitted signal reduces in strength, as it passes through a physical channel. This strength reduction is caused by energy absorption and spreads during image transmission, introducing noise as it does so. In satellite communication terminology, this phenomenon is called "free space loss" and can exceed 200 dB on a 14/12-GHz frequency geosynchronous satellite channel.[17] The fading fluctuates with the geometry of the surrounding environments, the speed of the vehicle, and the terrain in mobile radio applications. A radio signal transmitted can be detoured by surrounding environments, such as buildings, mountains, and plants onto one of several paths. Various types of the signals may reflect to each other positively or negatively, as these signals may have different wavelengths and different degrees of fading, depending on the location of the receiver. An active receiver will receive a series of energy peaks and troughs, which reflect the frequency of occurrence. The number of peaks and troughs depends on agent speed. These types of communication channels are called "multi-path fading channels." Therefore, we conclude that transmission channels are not perfect and can often receive corrupted data.

In 1948, Shannon[18] argued that the right pre-coding of information could decrease errors made by a noisy channel to an expected level without sacrificing information, as long as the information does not exceed the channel capacity. ECC was issued to determine the veracity of this theorem. Since the 1950s, ECC has been used to defend the information integrity in communication channels or in storage. ECC has become a substantial component of the telecommunications revolution, and has improved computer systems, the internet, digital recordings, and space exploration.[19] All CDs,

CD-ROMs, hard disks, and DVDs employ ECCs to defend the data encoded in them. ECC techniques are employed in the data link layer to enhance the reliability of the transmission channel.

The patient data is encrypted (applying the AES) and hidden in the retinal fundus image. The complex block of data composing patient text data and image information is encoded by using ECC algorithms. We have employed Huffman codes, Bose-Chaudhuri-Hocquenghem (BCH) codes, Reed-Solomon (RS) codes, convolutional codes, and Turbo codes in our experiment. For measuring the improvement supplied by this system of coded data transmission, we calculated the bit error rate (BER) and the symbol error rate (SER) at the receiving system, for the error control coded and uncoded transmission pass through an AWGN channel by employing Monte-Carlo techniques. We obtained and plotted the BER values against the various levels of SNR. The coding gain (CG), a quantity of the decrease in the SNR required to accomplish a particular BER compared to uncoded case, is determined from this plot. Hence, the CG measures the improvement of information integrity supplied by using ECC.

The chapter is organized as follows. Section 11.2 is outlined the encryption algorithm employed for enhancing protection of data and the digital watermarking technique employed for data concealing characteristics. In Sec. 11.3, we outlined compression techniques. In Sec. 11.4, we explained the prominent characteristics of the ECC and interleavers techniques employed in this study. In Sec. 11.5, we showed our simulation results that measure the efficiency of these systems. In Sec. 11.6, we discussed our experiment. Finally, we presented our conclusions in Sec. 11.7.

Figure 11.1a, b shows the proposed transmitter and receiver block diagrams.

11.2. Encryption and Watermarking of Text Information

11.2.1. *Encryption*

For storage, information is encrypted and hidden in image data used for digital watermarking to improve the security. John Daemen and Vincent Rijmen designed an algorithm that can withstand attacks and hacking

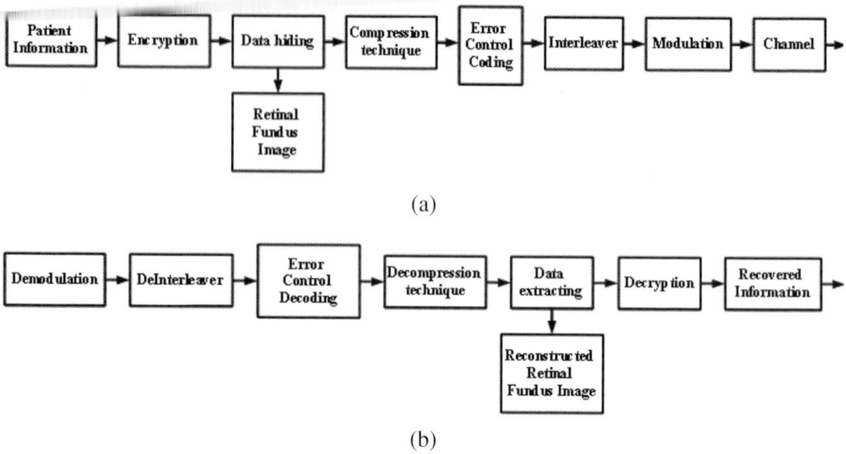

Fig. 11.1. (a) Proposed transmitter block diagram and (b) proposed receiver block diagram.

attempts, the AES or the Rijndeal algorithm, and used it to encrypt text data.[14–16]

During the encryption process, data changes from plain text to cipher text, which is unintelligible to the human eye. Applying digital watermarking techniques, encrypted data is hidden into image data, and created watermarked image data. The watermarked image data is encoded with a suitable ECC algorithm. The secure image is passed through the communication channel. The received data (corrupted by noise) is corrected by ECC algorithms, and then patient data is extracted from the watermarked image and recovered into its original form using decryption process of AES. Three types of AES are capable of applying cryptographic keys of 128-, 192-, and 256-bits length to encrypt and decrypt data into 128-bits blocks.

For AES, the Rijndeal algorithm uses the following operating combinations: security, performance, efficiency, implementation, and flexibility. This algorithm has a quick setup time, and its agility is well suited for our purposes. The Rijndeal algorithm requires low memory, making it well suited for conditions in which space is limited. The amount of security that the Rijndeal algorithm offers is contingent on the computational power and time required to decrypt the cipher. In this chapter, we have applied the AES algorithm with a cryptographic 256-bit key to encrypt and decrypt our patient information.

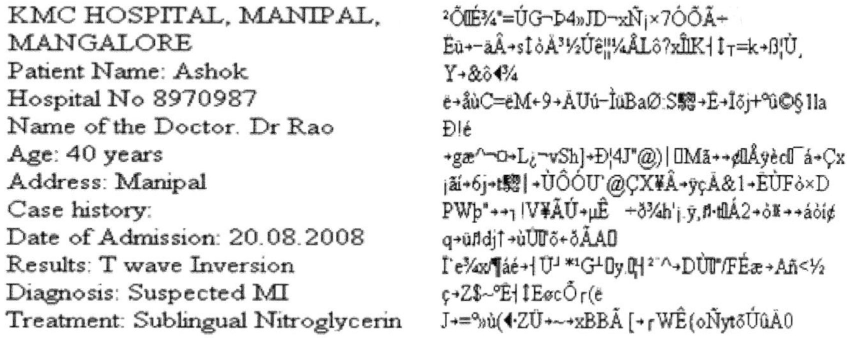

Fig. 11.2. (a) Original patient diagnosis data and (b) encrypted patient data.

We show the original patient text and the encrypted patient text, respectively in Figs. 11.2a and 11.2b.

11.2.2. *Digital Watermarking Technique*

During digital watermarking, data is hidden within a digital signal, or cover data, which may be audio, video, or pictorial. Whenever the watermarked signal is copied, the hidden data is also conveyed to the copied file.

Digital watermarking can be divided by visible or invisible. For visible watermarking, the embedded data is seeable in the watermarked video or pictorial record. Generally, the embedded data can be any logo or text data that recognizes the proprietor of the cover data. For invisible watermarking, the embedded data is inserted into an audio stream or a pictorial or videographic record, but cannot be detected by human eyes. Invisible watermarking is applied in intellectual property right protection systems, such as steganography, which is the writing of hidden messages and an application of digital watermarking.[3] Digital watermarking hides message data in cover data without distorting the information. This watermarking technique can compress storage or transmit text information, even in image and video records. We have applied LSB insertion watermarking to hide patient information in retinal fundus images.

We show an original retinal fundus image in Fig. 11.3a, and a watermarked (embedded) retinal fundus image in Fig. 11.3b. No distortion in the

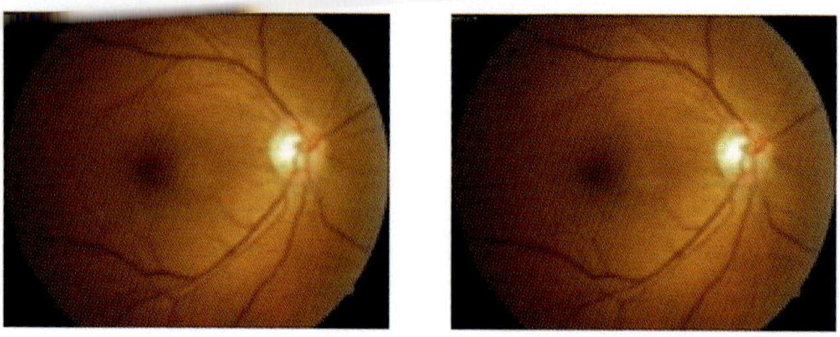

(a) (b)

Fig. 11.3. (a) An original retinal fundus image and (b) a watermarked/embedded retinal fundus image.

watermarked retinal fundus image of Fig. 11.3b is visible to the naked eye. Every pixel of the retinal fundus image comprising color data (red, green, blue) is represented by 24 bits.

11.3. Compression Technique

Compression techniques compress data so that it can travel on less bandwidth and can be stored using less memory. Compression schemes improve throughput over standard channels, and is commonly performed on digital medical images to reduce redundancy. Most compression algorithms use the repetition contained in data. For instance, a character set that includes letters, digits, and punctuation is normally composed of a seven-bit ASCII code, but a compression algorithm can use a three-bit code to represent the eight most common letters. The compression ratio is equal to the number of bits before compression over the number of bits after compression.

Two important compression concepts are lossless and lossy compression, as explained below.

In **lossless compression**, data is compressed without any loss of data, and all the data is available for use. Lossless compression is necessary for many medical images. Lossless compression is also known as entropy coding since it uses statistics/decomposition techniques to remove redundancy, and is applied to medical imaging.

The following techniques are included in lossless compression:

- Run length encoding,
- Huffman encoding,

- Lempel-Ziv-Welch (LZW) coding, and
- Arithmetic coding.

Lossy compression standards are JPEG (Joint Photographic Experts Group), MPEG (Motion Picture Experts Group), and MP3 (MPEG-1, Layer 3). Some loss of information is acceptable in the lossy compression technique. In these cases, it is acceptable to lose some data information to create smaller graphics files. The loss could be applied as a color resolution or as a graphic detail. For instance, high-resolution information points could be lost whenever a picture is displayed on a low-resolution instrument or equipment. The loss is also satisfactory in voice, audio, and video compression, depending on the desired quality. Lossy algorithms supply higher compression ratios than lossless algorithms do. Lossy algorithms are used for applications that can work well with lower quality reconstructed images. Using this scheme, the decompressed image is not identical to the original image, but is reasonably close to it. The loss of information in the image is unperceivable to the human eye.

Lossy compression can allow for compression ratios of between 100:1 and 200:1, relying upon the type of data being compressed. However, the lossless compression ratios commonly accomplish only a 2:1-compression ratio.

The following techniques are included in lossy compression:

- Transformation coding,
- Vector quantization,
- Fractal coding,
- Block truncation coding, and
- Subband coding.

Lossless compression standards are Gzip, Unix compress, zip, GIF, and Morse code.

The lossless image compression occurs in two stages: decorrelation and entropy coding. Decorrelation is a run-length coding technique, which removes spatial redundancy or inter-pixel redundancy. The second stage takes out coding redundancy. This stage includes Huffman coding, LZW, and arithmetic coding.

In this chapter, we used the lossless compression technique, Huffman coding, because of using it for watermarked image, which embeds patient text information into retinal fundus images, should recover information without loss. Here, we talk about Huffman coding briefly.

11.3.1. *Huffman Coding*

During his Ph.D. work, David A. Huffman developed the Huffman coding technique as a class assignment; the class was information theory and was taught by Professor Robert M. Fano at Massachusetts Institute of Technology (MIT). In 1952, Huffman published a paper entitled "A Method for the Construction of Minimum-Redundancy Codes" based on this work. Codes produced using this technique are called Huffman codes. Not all string and numeric characters occur with the same frequency. However, all numerical characters are allocated the same amount of space. The size of one string and numerical character is one byte.

A definition of Huffman coding is an ensemble where X is a set of (x, A_x, P_x), x is a value of a random variable, A_x is a set of possible values for x, $A_x = \{a_1, a_2, \ldots, a_i\}$, P_x is a probability for each value, $P_x = \{p_1, p_2, \ldots, p_i\}$, where $P(x) = P(x = a_i) = p_i$, $p_i > 0$.

The Huffman coding algorithm makes optimal symbol codes. In Huffman coding, probable characters are compared with short and less probable characters using codes. Every code can be uniquely decoded to create a binary, or Huffman tree, generated from the exact frequencies of the text. A decoder can use the Huffman tree to decode the string by following the paths according to the string and adding each new character that it finds.

11.4. Error Control Coding

In practice, communication lines, or transmission channels, and digital recordings introduce errors in data transmitted through them. Thus, communication lines and digital recordings are not perfect. ECCs are usually applied to detect and correct the errors influenced by imperfect transmission mediums and digital recording. Working principle of the ECC is adding redundant data to transmitted messages. This redundant data can find and make correction the errors added into the data message. Hence, using this special characteristic of ECC strategies, original data can be retrieved from corrupted data, which was introduced by errors passing through imperfect communication channels or digital recording. ECC techniques are currently utilized in securing deep space communications in addition to storage media. ECC schemes also are applied to many of the communication appliances for ensuring information integrity. The operations of ECC are executed in the data-link layer to enhance the reliability of the physical transmission mediums.

Transmission mediums and digital recording can cause errors. Thus, the categories of ECC address different types of damages.[17,19–23]

ECCs are sorted as block codes or trellis codes, depending on their structure. BCH codes and RS codes are examples of block codes, and convolutional codes and turbo codes are examples of trellis codes. In this study, we have applied Hamming codes, BCH codes, RS codes, convolutional codes, and turbo codes to improve the quality of watermarked retinal fundus images while transmitting through communication channel and storing at storage media. Their abilities are measured by comparing the BER and SER as the SNR after applying the ECC. The BER/SER is determined to be the ratio of the number of bits/symbols received in error to the total number of bits/symbols passed through the channel in a given time slot. Hence, the main role of the ECC is to protect information from corruption induced by the imperfect transmission medium and imperfect digital recording. In this study, the AWGN channel model is used because of its simplicity. Then also, the signal strength variation is different while transmitting through channels where the SNR alters randomly with time. This variation in signal strength is a feature of fading channels. We have shown the performance of the Hamming code, the RS code, and the BCH code over such channels in Fig. 11.7, and the performance of the convolutional and the turbo codes, which had code rates of 0.333 and 0.5 over the AWGN channel in Figs. 11.8 and 11.10.

11.4.1. *Hamming Codes*

In 1950, Richard Hamming described a class of error correction code, which has become popular in usage. Hamming codes are considered an extension of simple parity codes that are applied in multiple parity bits. Every parity bit is determined over a subset of the information bits in a message. Hamming codes was first applied for long distance telephone use to control errors.

Hamming codes are $[n, k]$ linear error correcting block codes. Hamming codes are commonly expressed as a function of a single integer, where $n-k$ is the number of parity or check bits. There are many Hamming codes, namely, (7, 4), (15, 11), and (31, 26), etc.

Here, we briefly explain the Hamming Code (15, 11).

A codeword length is $n = 2^m - 1 = 2^4 - 1 = 15$ bits,
The number of information bits is $k = 2^m - m - 1 = 2^4 - 4 - 1 = 11$ bits,
The number of parity bits is $m = n - k = 15 - 11 = 4$ bits,
The minimum distance of this code is $n - k = 7 - 4 = 3$-bit, and
The error correcting capability is $t = (d_{\min} - 1)/2 = (3 - 1)/2 = 1$.

For block code, we convert a sequence of input bits s with length K, into a sequence of transmitted bits t with length N, to add redundancy bits to transmitted bits. In a linear error correcting block code, the redundancy N–K bits are linear functions of the input K bits and are called parity-check bits. Hamming code is linear and can be written using the following matrix. The transmitted codeword t is found from the input sequence s by a linear operation, $t = [G] \cdot s$, where $[G]$ s the generator matrix of the Hamming code.

The Hamming codes are cyclic code and perfect code. The decoding process of the Hamming codes, which determines the codeword, is sent if errors occur in transmission and storage, and is easy to use with the simple form of the parity check matrix of these codes.

For the Hamming (15, 11) code, the parity check matrix equals $[H] = [I_{n-k} \vdots P^T]$. I_{n-k} is $[(n-k) \times (n-k)]$ identity matrix. $[P]$ is the parity check matrix.

$$[H] = \begin{bmatrix} 1 & 0 & 0 & 0 & 1 & 0 & 0 & 1 & 1 & 0 & 1 & 0 & 1 & 1 & 1 \\ 0 & 1 & 0 & 0 & 1 & 1 & 0 & 1 & 0 & 1 & 1 & 1 & 1 & 0 & 0 \\ 0 & 0 & 1 & 0 & 0 & 1 & 1 & 0 & 1 & 0 & 1 & 1 & 1 & 1 & 0 \\ 0 & 0 & 0 & 1 & 0 & 0 & 1 & 1 & 0 & 1 & 0 & 1 & 1 & 1 & 1 \end{bmatrix}$$

The Generator Matrix:

$[G] = [P \vdots I_k]$ I_n is $[n \times n]$ identity matrix. P is parity check matrix.

$$[G] = \begin{bmatrix} 1 & 1 & 0 & 0 & 1 & 0 & 0 & 0 & 0 & 0 & 0 & 0 & 0 & 0 & 0 \\ 0 & 1 & 1 & 0 & 0 & 1 & 0 & 0 & 0 & 0 & 0 & 0 & 0 & 0 & 0 \\ 0 & 0 & 1 & 1 & 0 & 0 & 1 & 0 & 0 & 0 & 0 & 0 & 0 & 0 & 0 \\ 1 & 1 & 0 & 1 & 0 & 0 & 0 & 1 & 0 & 0 & 0 & 0 & 0 & 0 & 0 \\ 1 & 0 & 1 & 0 & 0 & 0 & 0 & 0 & 1 & 0 & 0 & 0 & 0 & 0 & 0 \\ 0 & 1 & 0 & 1 & 0 & 0 & 0 & 0 & 0 & 1 & 0 & 0 & 0 & 0 & 0 \\ 1 & 1 & 1 & 0 & 0 & 0 & 0 & 0 & 0 & 0 & 1 & 0 & 0 & 0 & 0 \\ 0 & 1 & 1 & 1 & 0 & 0 & 0 & 0 & 0 & 0 & 0 & 1 & 0 & 0 & 0 \\ 1 & 1 & 1 & 1 & 0 & 0 & 0 & 0 & 0 & 0 & 0 & 0 & 1 & 0 & 0 \\ 1 & 0 & 1 & 1 & 0 & 0 & 0 & 0 & 0 & 0 & 0 & 0 & 0 & 1 & 0 \\ 1 & 0 & 0 & 1 & 0 & 0 & 0 & 0 & 0 & 0 & 0 & 0 & 0 & 0 & 1 \end{bmatrix}$$

The above two matrices are $[G]$ and $[H]$ in systematic form. For linear block codes, $[G]$ and $[H]$ must satisfy $[G] \cdot [H]^T = 0$, an all-zeros matrix.

Hence, the right side of the $[H]$ matrix is all nonzero n-tuples. The left side of the matrix is the $(n - k)$-identity matrix. Thus, $[G]$ can be found from $[H]$ by taking the transpose of the right side of $[H]$ with the identity k. The right side of $[G]$ is an identity matrix. The (15, 11) Hamming code has a code rate $= 11/15 = 0.7333$.

11.4.2. *BCH Codes*

BCH codes represent a large class of multiple random error-correcting codes. A. Hocquenghem first discovered BCH codes in 1959, and R. C. Bose and D. K. Ray-Chaudhuri discovered BCH codes independently in 1960.[23] These researchers brought out the codes, but not the decoding algorithms.

BCH codes are played the significant and powerful role to linear block codes, which are cyclic codes with a variety of parameters. The most common BCH codes are characterized as follows.

For any integer $m \geq 3$ and $t < 2^{m-1}$;
Codeword length $n = 2^m - 1$,
Parity check bits $n - k \leq mt$, and
Minimum distance $d_{\min} \geq 2t + 1$,

where m is the number of parity bits and t is number of errors that can be corrected. t or less than t random errors can be corrected over a span of $2^m - 1$ bit positions. This ECC is called a t-error-correcting BCH code. Extension of BCH codes was the nonbinary codes, which is using symbols from the Galois field $GF(q)$ by Gorenstein and Zieler.[23]

In 1960, Peterson investigated the decoding algorithm of binary BCH codes.[23] Since this publication, Berlekamp, Massey, Chien, Forney, and many others have improved Peterson's algorithm.[23]

The Hamming single error-correcting codes are BCH codes. BCH codes perform well with many code parameters, block lengths, and code rates. For example, in a single error-correcting (15, 11) BCH code with a code rate of 0.733(11/15), $m = 4$, $t = 1$; $n = 2^m - 1 = 2^4 - 1 = 15$; $n - k \leq mt = 4 \cdot 1 = 4$; $k = n - mt = 15 - 4 = 11$; and $d_{\min} \geq 2 \cdot 1 + 1 = 3$.

A BCH code is represented by a polynomial expression over a finite field and has an especially selected generator polynomial.

Let α be a primitive element in $GF(2^m)$. For $1 \leq i \leq t$, let $\phi_{2i-1}(x)$ be the minimum polynomial of the field element α^{2i-1}. The degree of $\phi_{2i-1}(x)$ is m.

The generator polynomial $g(x)$ of t error-correcting primitive BCH codes of length $2^m - 1$ is given as $g(x) = LCM\{\phi_1(x), \phi_3(x), \ldots, \phi_{2t-1}(x)\}$. The degree of $g(x)$ is $m \cdot t$ or less. Thus, the number of parity-check bits of the code $n - k$ is at most $m \cdot t$. For instance, for the (15, 11) BCH code, minimum polynomial of the field element α:

$$\begin{aligned}\phi_{2t-1}(x) &= \phi_{2\cdot 1-1}(x) = \phi_1(x) = (x - \alpha)(x - \alpha^2)(x - \alpha^4)(x - \alpha^8)\\ &= x^4 + x + 1.\end{aligned}$$

The generator polynomial $g(x)$ of a single error-correcting primitive BCH code of length $2^m - 1 = 2^4 - 1 = 15$ is given by $g(x) = LCM\{\phi_1(x)\} = x^4 + x + 1$.

11.4.3. *Convolutional Codes*

In a communication system, a convolutional code transforms each b-bit information symbol to be converted into an r-bit symbol. Code rate is calculated by ratio of the transmitted bit to the received bit (b/r). The last bits of convolutional code are the transformation bits that are created from the last l information bits, where l is calculated by $b+1$ and called the constraint length of the code.

Convolutional codes are generated by shift register, while transmitted information bits passing through linear finite state circuit. Additional combinatorial logic performs a modulo-two addition. In convolutional coding, an information frame and the previous m information frames are encoded into a single codeword frame. Hence, successive frames are coupled by the encoding procedure. Codes obtained this way are called tree codes, and tree codes with additional properties of linearity and time invariance are called convolutional codes.

In Fig. 11.4a, we have shown a code rate 1/2 (b/r) encoder with a constraint length of 3. An input bit $b1$ of an input sequence is supplied by the leftmost register. The output r bits of encoder are generated by two generator polynomials created by shift registers. Working principle is that

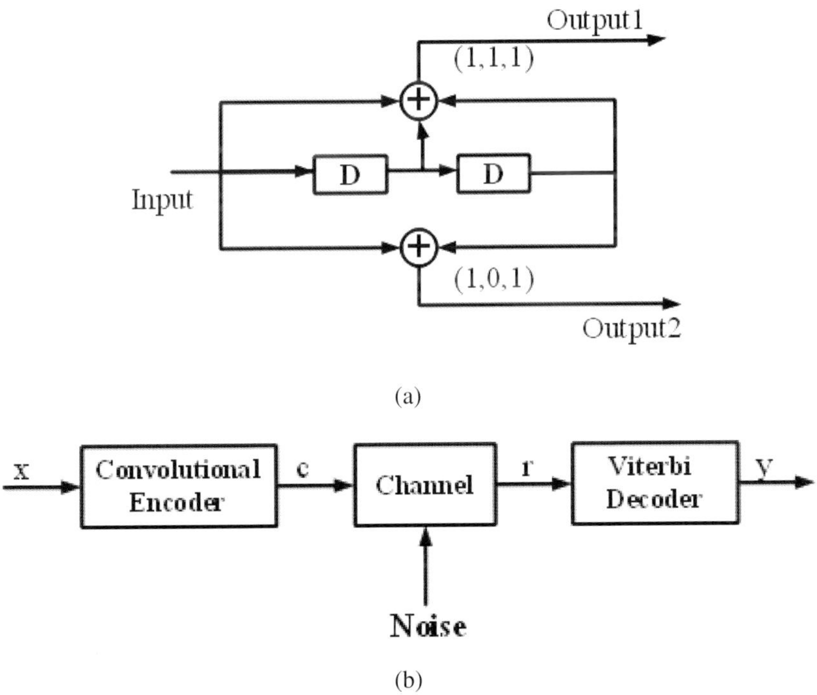

Fig. 11.4. (a) Rate 1/2 non-recursive and non-systematic convolutional encoder with constraint length 3 and (b) convolutional coding system.

the bit shifts all register values to the right ($b1$ moves to $b0$, $b0$ moves to $b-1$, and so on) and waits for the next input bit, and output bits are added by module 2. The encoder will work until all registers are on zero state, even there are no input bits. In Fig. 11.4a, we show the generator polynomials, $G1 = (1, 1, 1)$ and $G2 = (1, 0, 1)$.

The state transition diagram and the trellis structure are also used for their description. At the receiver end, the decoding strategy for convolutional codes is the Viterbi algorithm, which is used to decode the convolutional code. In Fig. 11.4b, we have shown a convolutional coding system. First, the input x bits are encoded into c bits using the convolutional encoder. Then, the c bits are passed through a noisy channel, and the r bits are received. Then, the received r bits are decoded by the Viterbi decoder. The Viterbi algorithm calculates maximum likelihood (ML) estimation on the estimated code bits y from the received r bits, so that it maximizes the probability $p(r/y)$ that r bits are received conditioned on the estimated code y bits. y bits must be one of the allowable code bits.

11.4.4. *RS Codes*[14]

RS codes are block-based error-correcting codes and it is used widely in the applications of communication systems and digital recording. A block of information representing symbols enters into the RS encoder, which generates codeword symbols for adding redundant symbols to input symbols. These redundant symbols created from the input symbols. The codeword symbol is modulated to be efficiently transmitted or stored. The receiver receives the weakened and corrupted codeword transmitted through noisy channel, and passed it through the RS decoder. The decoder detects and corrects errors inserted into codeword during transmission and storage. RS codes have been widely used, especially as maximum distance separable (MDS) codes.[17,19–24]

An RS code is defined as an (n, k) RS code over $GF(q^m)$, meaning that the encoder encodes k data symbols of m bits each to an n symbols codeword. In this case, n–k parity symbols of m bits each are added. The MDS property implies that for a given value of n and k, RS codes have the largest possible minimum distance $d_{\min}$. RS codes are subcategory of linear systematic block codes and use symbol containing of m bits (m is a natural number and $m \geq 2$). For any given value of m, there are $2^m - 1$ symbols with m bit RS codeword created. For instance, if $m = 8$, each symbol has 8-bit, each codeword has $2^8 - 1 = 255$ symbols, and the operation generated over $GF(2^8)$. Some prominent properties of RS codes are Refs. [17,19–24]

For the RS codeword $n = q^m - 1$ constructed from k data symbols (each symbol has m bits), the minimum distance of the RS code is calculated by $d_{\min} = n - k + 1$. Then also, the code rate of RS code is defined by a ratio of number of data symbols to number of codeword symbols, that is k/n, and the error correcting capability is $t = \lfloor (n - k)/2 \rfloor$.

11.4.5. *Turbo Codes*[14]

Shannon's noisy channel coding theorem[18] established that it is possible to communicate information over a noisy channel with a small probability of error as long as the rate of information transfer is less than the channel capacity C. The channel capacity depends on the signal to noise ratio (SNR) and bandwidth of the channel. Furthermore, Shannon computed a lower boundary on the minimal value of SNR expected for supporting an appropriate code rate with an arbitrarily small probability of error. This small probability of error is known as the Shannon limit. An ECC is considered powerful if it performs as well as or nearly as well as the Shannon code.

The development of turbo codes in 1993[25] helped to improve code designs so that they were able to perform nearly as well as Shannon's channel coding theorem. Traditional code designs, such as BCH and Reed Muller, or algebraic code designs, have algebraic or topological structures, which ensure good distance properties and decoding algorithms for the code. Turbo codes possess random properties that have enough structure for an effective iterative decoding process as originally envisaged by Shannon. Iterative decoder of turbo codes applied any code rate and larger than 10^4 bits can accomplish a lower BER at 10^{-5} SNRs within 1 dB of Shannon limit. Turbo codes are consisting of two or more simple component codes along with an interleaver. We can find some research examples of turbo decoders. For instance, soft-in–soft-out (SISO) decoders for each component code are applied in an iterative manner.[26] In this study, simple turbo code is applied to assure data retains integrity, as it passes through imperfect transmission media. We have shown the outcomes found by applying turbo codes in this study in Fig. 11.10.

11.5. Results

After patient information is encrypted by using AES, it is then hidden in the retinal fundus image using digital watermarking; the image with hidden patient information is compressed using the lossless compression technique. In this study, Huffman coding technique is applied for lossless compression technique, which is given a 0.736-compression ratio. Then, the compressed watermarked image is changed over a bitstream and is encoded by employing suitable ECCs. In this chapter, we assume that the channel modulation is the binary phase shift keying (BPSK). Then modulated signals are passed through an assumed imperfect transmission media, which is corrupted by AWGN. The ultimate goal of our study is to quantify the character of the recovered text data and quality of rebuilt retinal image. The main quantity is that the BER is measured as a function of the SNR after decoding the obtained retinal image corrupted by the communication channel noise. Figures 11.5 (a–i) show us to visually estimate the difference between the images (also patient information) as retrieved over an AWGN channel (9-dB SNR) and the recovered image (also patient information) after being worked out by the (15, 11) Hamming code, (15, 11) BCH code, and (15, 11) RS code. From these figures, we can see that the quality of images and text data has improved significantly after applying ECCs. In this study, five types of ECC schemes are applied for the robustness and reliability of the transmission system. That is the Hamming code, the BCH code, the RS code, the convolutional code, and turbo codes. Then, we have compared the performance of Hamming code (15, 11), the BCH code (15, 11) and the RS (15, 11) code.

We show the original retinal fundus image in Fig. 11.5a, the original patient text information that has to be hidden into retinal image in Fig. 11.5b, the image after hiding text data in Fig. 11.5c, respectively. The watermarked image (image plus text) corrupted by noise with 9-dB SNR as received at the receiver without ECC is shown in Fig. 11.5d. The reconstructed image

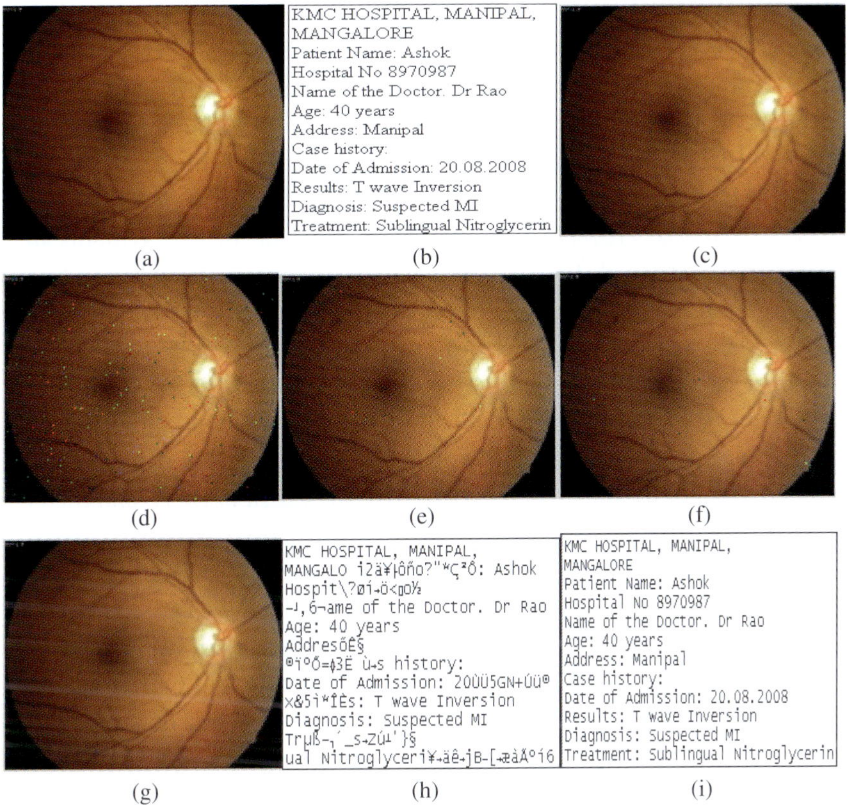

Fig. 11.5. (a) An original retinal fundus image, (b) original patient text information, (c) the retinal fundus image after hidden text data, (d) the watermarked retinal fundus image (image + text) corrupted by noise with 9-dB SNR without using ECC, (e) the reconstructed retinal fundus image using the (15, 11) Hamming code for 9-dB SNR, (f) the reconstructed retinal fundus image using the (15, 11) BCH code for 9-dB SNR, (g) the reconstructed retinal fundus image using the (15, 11) RS code for 9-dB SNR, (h) the recovered patient information corrupted by noise with 9-dB SNR without using ECC, and (i) the recovered patient information using RS code for 9-dB SNR.

using the (15, 11) Hamming code (9-dB SNR) is shown in Fig. 11.5e. The reconstructed image using the (15, 11) BCH code (9-dB SNR) is shown in Fig. 11.5f. The reconstructed image using the (15, 11) RS code (9-dB SNR) is shown in Fig. 11.5g. The recovered patient information corrupted by noise with 9-dB SNR as received at the receiver without ECC is shown in Fig. 11.5h. The recovered patient information using RS code (9-dB SNR) is shown in Fig. 11.5i. In this chapter, we used the (15, 11) Hamming code the (15, 11) BCH code, and the (15, 11) RS code over $GF(2^4)$. These three codes have same code rate, which is $11/15 = 0.7333$. Nevertheless, their error-correcting capabilities are different. In this study, we wanted to compare with the results of these codes. The (15, 11) Hamming code and (15, 11) BCH code can correct up to one bit in error within a 15-bit length codeword.

However, the (15, 11) RS code can correct up to two symbols in a 15-symbol codeword. In a fading channel, errors tend to occur in bursts rather than in random patterns. Interleaving, the rearranging of the symbol sequence, increases the burst error-correcting capability of an ECC. A deinterleaver is the inverse process of interleaver. A block interleaver is very simple and widely applied in the communication systems. In this chapter, in order to improve the capability of error correction, a block interleaver is applied along with the (15, 11) Hamming, the (15, 11) BCH, and the (15, 11) RS codes.

The plots in Fig. 11.6 show the compared result of performance of the (15, 11) Hamming code, the (15, 11) BCH code, and the (15, 11) RS code on transmission of retinal fundus image. From this result, we can observe that this RS code performs better than the Hamming and BCH codes, and has a good CG with 4 dB at a BER of 5×10^{-6}. From this plot, the BER of uncoded case will reach to roughly 10^{-3} at SNR of 9 dB. This raised BER indicates almost one bit in error for every 10^3 bits passed through. Hence, using the RS code leads to increased information integrity. The RS code results are similar to those found in literature. Wicker has given a plot of the probability of the decoder error (essentially the SER) as a function of SNR for a (31, 27) two error-correcting ECC.[23] It is observed that the SER is approximately 10^{-5} at an SNR of 8 dB.

We use a Huffman coding compression technique after the watermarking technique. Huffman coding is given a 0.736-compression ratio. We use Hamming, BCH, and RS codes with a code rate of 0.733 as error-correcting codes. This compression ratio is similar to the code rate used in ECC, meaning that data size after using ECCs is equal to original image size. We save the memory space of storage media and transmission in channel.

We check the performance of these codes over time variant channels. We assume that this channel can change suddenly from a good state (channel introduces few errors) to a bad state (channel introduces many errors). The

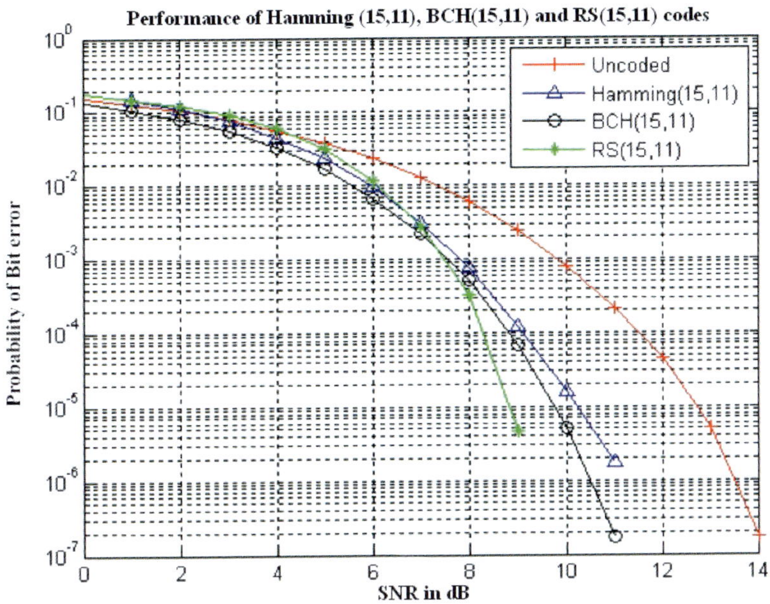

Fig. 11.6. Compared result of (15, 11) Hamming code, (15, 11) BCH code, and (15, 11) RS code performances with block interleaver (The retinal fundus image passed through noisy channel which is corrupted by AWGN channel.).

bad state errors are related to burst errors in fading channels. For insuring the performances of Hamming, BCH, and RS codes over such channels, we randomly alter the SNR on the communication channel. Such channel SNR values are altered over 10 values roaming amongst 10 dB (a large amount of SNR) and 0 dB (a small amount of SNR). The performances of the Hamming, BCH, and RS codes are recorded in detail of the BER plot in Fig. 11.7. We find an improvement of about 5 dB at a BER of 10^{-5} for RS code.

We can see that (15, 11) RS code is more beneficial fitted for burst errors correction than (15, 11) Hamming code and (15, 11) BCH code.

Figure 11.8 shows the performance of convolutional code with a code rate of 0.5. This convolutional code performance is better than the Hamming and BCH code performances but less than the RS code. From these results, we can conclude that the RS code is a better error-correcting code capability; error occurs in random and burst.

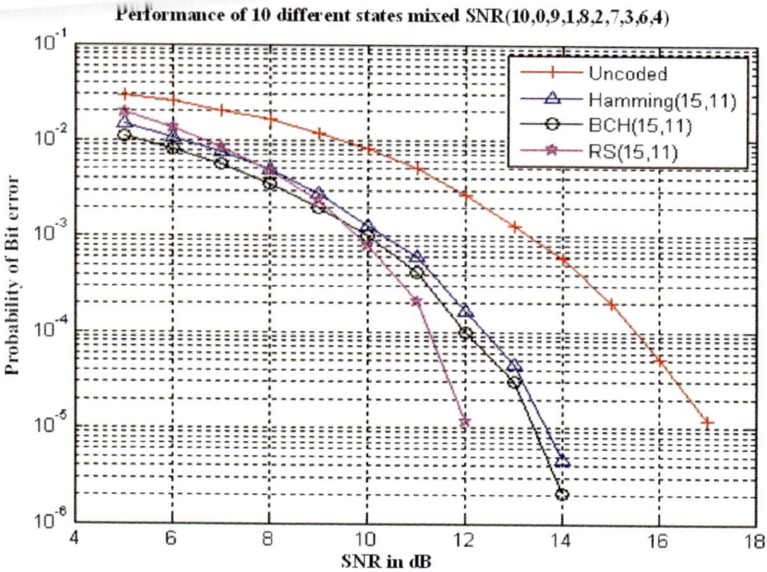

Fig. 11.7. Performance of (15, 11) Hamming code, (15, 11) RS code, and (15, 11) BCH code on the transmission of retinal fundus images through burst error channels that are corrupted by (various SNR levels) burst noise.

11.5.1. *Using Turbo Codes for Transmission of Retinal Fundus Image*

After getting the simulation results of the (15, 11) Hamming code, the (15, 11) BCH code, the (15, 11) RS code, and the convolutional code, we have experimented with a turbo code. The turbo code encoder used is shown in Fig. 11.9a. This turbo encoder consists of two recursive systematic convolutional (RSC) encoders, which are connected in parallel, an interleaver, which is placed before the second RSC encoder, and puncturing block. Input bits are encoded by both RSC encoders. The RSC encoder 1 encodes the input bits in their original order, while the RSC encoder 2 encodes the input bits in interleaved order. We can change the code rate of the turbo code by using puncturing scheme.[21] We have shown the performance of the original rate one-third turbo code and the punctured rate one-half turbo code.

Figure 11.9b shows an RSC encoder, which is used in the turbo encoder. u_k is an input bit and c_k is the coded output bit.

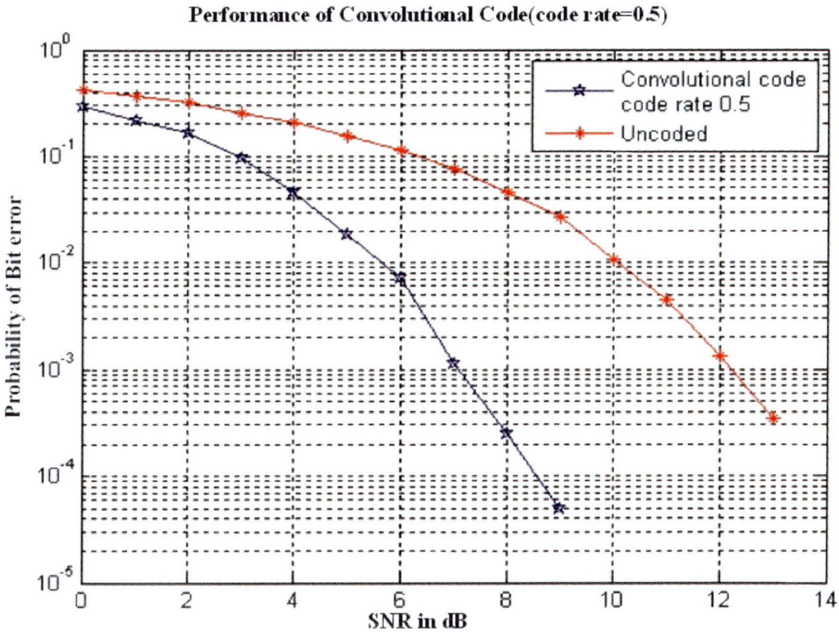

Fig. 11.8. Performance of convolutional code with a code rate 0.5 on transmission of retinal fundus images through AWGN channel.

A recursive systematic convolution encoder consists of linear finite-state shift registers and module 2 adder. The interleaver plays a main role in the turbo codes. In this chapter, a pseudo random interleaver is applied, because its performance is better in Ref. [27].

Mainly, puncturing and multiplexing is used to enhance a given code rate, while deleting bits. Figure 11.9c shows the turbo decoder block diagram. Turbo decoder consists of demultiplexer, two MAP decoders, interleavers, deinterleaver and decision maker. Turbo decoding begins with the formation of *a posteriori* probabilities (APPs) for each data bit. MAP (maximum *a posteriori*) decoder determines most likely information bit.

A logarithmic ratio of the APP of u_k, conditioned on the received signal y, is defined as:

$$L(u_k) \triangleq \log\left[\frac{P(u_k = 1/y_1^N)}{P(u_k = 0/y_1^N)}\right] \tag{11.1}$$

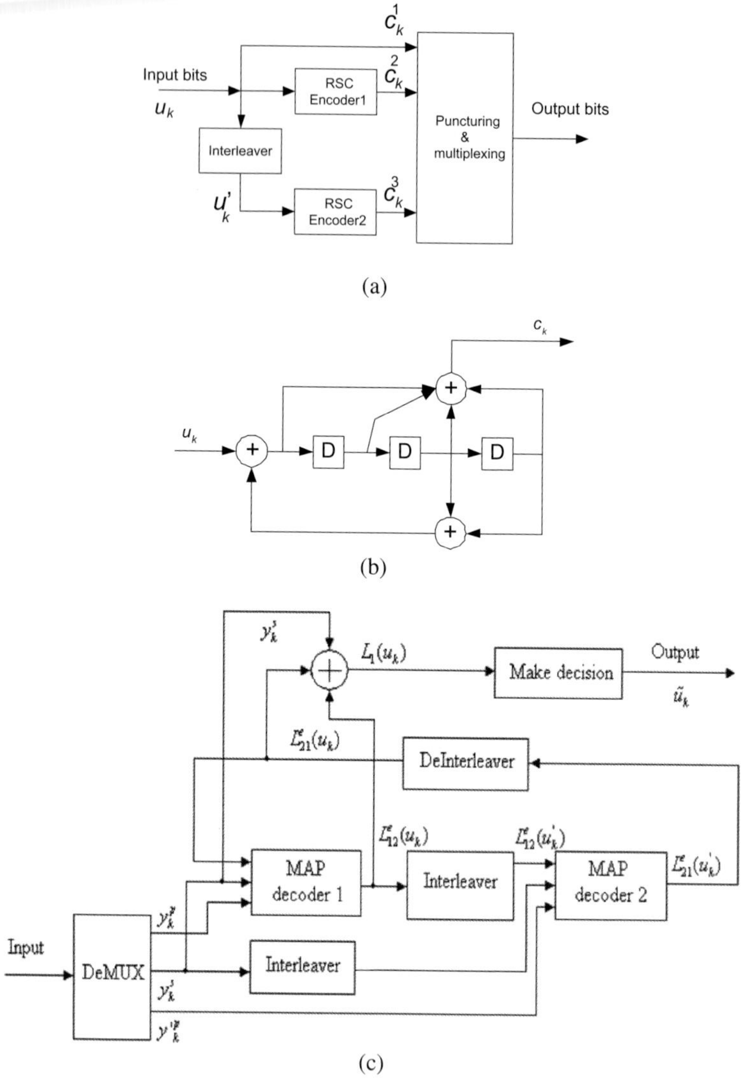

Fig. 11.9. (a) Turbo code encoder, (b) RSC encoder, and (c) block diagram of a turbo decoder.[26]

The decoding decision of $\tilde{u}_k$ is made based on the sign of $L(u_k)$, i.e.

$$\tilde{u}_k = \text{sign}[L(u_k)]. \tag{11.2}$$

$L(u_k)$ is computed by three terms, L_apriori, L_channel, and $L^e(u_k)$. L_apriori is *a priori* information based on the input bit u_k at time k. The previous decoder provides it.

$$\begin{aligned} L(u_k) &= [L^e(u_k) + Lc \cdot y_k^{1,s}] + \log \frac{\sum_{u^+} \tilde{\alpha}_{k-1}(s') \cdot \tilde{\beta}_k(s) \cdot \gamma_k^e(s',s)}{\sum_{u^-} \tilde{\alpha}_{k-1}(s') \cdot \tilde{\beta}_k(s) \cdot \gamma_k^e(s',s)} \\ &= L\_\text{apriori} + L\_\text{channel} + L^e(u_k), \end{aligned} \tag{11.3}$$

where L_apriori and L_channel denote $L^e(u_k)$ and $Lc \cdot y_k^{1,s}$, respectively. $\sum_{u^+}()$ is the summation of all the possible transition branch pairs (s_{k-1}, s_k) at time k given input $u_k = 1$, and $\sum_{u^-}()$ is the summation over all the possible transition branch pairs (s_{k-1}, s_k) at time k, given input $u_k = 0$. Lc is the channel reliable factor; its computation is given as the following,

$$Lc = \frac{4 \cdot A \cdot \text{SNR\_}b}{p}, \tag{11.4}$$

where A is a fading amplitude, equal to one for an AWGN channel, $\text{SNR\_}b$ is the bit SNR ($\frac{E_b}{N_0}$), and p denotes $1/r_c$, r_c is code rate of the turbo encoder.

$L^e(u_k)$ is extrinsic information based on all parity and systematic information except the systematic value at time k. It can be passed to a subsequent decoder and is computed using the following equations:

$$L^e(u_k) \triangleq \log \frac{\sum_{u^+} \tilde{\alpha}_{k-1}(s') \cdot \gamma_k^e(s',s) \cdot \tilde{\beta}_k(s)}{\sum_{u^-} \tilde{\alpha}_{k-1}(s') \cdot \gamma_k^e(s',s) \cdot \tilde{\beta}_k(s)}, \tag{11.5}$$

where

$$\gamma^e(s',s) = \exp\left[\sum_{i=2}^{q} \left(Lc \cdot \frac{1}{2} \cdot y_k^{i,p} \cdot c_k^i\right)\right]. \tag{11.6}$$

$\tilde{\alpha}_k(s)$, $\tilde{\beta}_{k-1}(s')$ can be computed recursively, with initial conditions, as described below:

$$\begin{aligned} \tilde{\alpha}_k(s) &= \frac{\sum_{s'} \tilde{\alpha}_{k-1}(s') \cdot \gamma_k(s',s)}{\sum_s \sum_{s'} \tilde{\alpha}_{k-1}(s') \cdot \gamma_k(s',s)}, \\ \tilde{\alpha}_0(s) &= \begin{cases} 1 & \text{if } s = 1 \\ 0 & \text{otherwise} \end{cases}. \end{aligned} \tag{11.7}$$

$$\tilde{\beta}_{k-1}(s') = \frac{\sum_s \tilde{\beta}_k(s) \cdot \gamma_k(s', s)}{\sum_s \sum_{s'} \tilde{\alpha}_{k-2}(s') \cdot \gamma_{k-1}(s', s)},$$
$$\tilde{\beta}_N = \begin{cases} 1 & \text{if } s = 1 \\ 0 & \text{otherwise} \end{cases}. \tag{11.8}$$

$$\gamma_k(s', s) \propto \exp\left[\frac{1}{2} \cdot L^e(u_k) \cdot u_k + Lc \cdot \frac{1}{2} \cdot y_k^{1,s} \cdot c_k^1\right]$$
$$\times \exp\left[\sum_{i=2}^{q} \left(Lc \cdot \frac{1}{2} \cdot y_k^{i,p} \cdot c_k^i\right)\right]. \tag{11.9}$$

For example, at any given iteration, decoder 1 $L_1(u_k)$ is computed as:

$$L_1(u_k) = Lc \cdot y_k^{1,s} + L_{21}^e(u_k) + L_{12}^e(u_k),$$
$$\tilde{u}_k = \text{sign}[L_1(u_k)], \tag{11.10}$$

where $L_1(u_k)$ is given in Eq. (11.3). $L_{21}^e(u_k)$ is extrinsic information for decoder 1, derived from decoder 2, and $L_{12}^e(u_k)$ is the third term in Eq. (11.3), which is used as the extrinsic information for decoder 2 derived from decoder 1. The decoders share the information with each other. The value of $L_1(u_k)$ decides the degree of the reliability of $\tilde{u}_k$.

In this chapter, we have used the turbo code with a code rate of 1/2 and 1/3. The rate 1/2 code is derived by puncturing both the parity and data bits. In the decoding algorithm, we also reinserted parity bits and data bits are in the parity and data streams to replace those that were deleted by puncturing.

The plots in Figs. 11.10a and b show the turbo codes performance with code rate of 1/2 and 1/3 on the transmission of retinal fundus images through AWGN channel. These codes perform better than RS codes. Nevertheless, this advantage is obtained at the cost of greater decoder complexity and a larger decoding delay.

11.6. Discussion

Assuring the privacy and authenticity of patient information is the most important aspect of medical storage. Watermarking is adjusted here to hide patient text information into retinal fundus images. Color image pixel is usually represented by 24 bits. LSB of retinal fundus image is substituted

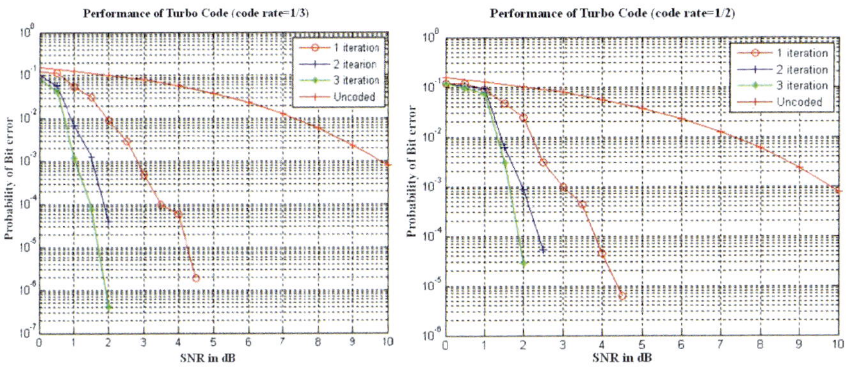

Fig. 11.10. The turbo code performance (a) with code rate 1/3 and (b) with code rate 1/2.

by the ASCII character bit of the patient text information. Hence, the LSB is "sacrificed" for hiding patient information, and the resultant loss is usually satisfactory. The role of the AES algorithm increases the security and privacy of patient information. The compression technique is employed to compress the watermarked retinal fundus image. Thus, the size of the watermarked retinal image is reduced before transmission over channels and storing to storage media. The retinal fundus image passing through channel and storing to storage media can be corrupted by imperfect physical channel and storage in imperfect digital recording or storage media. Hence, various types of ECCs or error correction techniques are utilized to produce more robust data managing and reliable transmission and storage against to imperfect transmission.

The effect of additive white Gaussian noise channel and fading channel (burst error noisy channel) on watermarked patient information was analyzed applying Monte-Carlo simulation techniques.[14] The authors of Ref. [14] have compared the performance of the (15, 11) RS code and the turbo code with a code rate of 0.5 and 0.33 to correct the error bits corrupted by imperfect channel. In Ref. [14], the turbo code was superior to the RS code in AWGN channel. Ultimate goal of this chapter is to show that the performance of different types of ECCs to reduce the effect of channel-caused errors significantly and then compared them. We studied comparisons of the (15, 11) Hamming code, (15, 11) the BCH code, and the (15, 11) RS code (different error-correcting codes with same code rate). We also studied comparisons of the convolutional code and the turbo code. Computation time of different types of ECCs is different to each other.

We showed below the comparison of computation time of different types of error-correcting codes.

$$t_{\text{Hamming}} < t_{\text{BCH}} < t_{\text{Convolutional}} < t_{\text{Reed-Solomon}} < t_{\text{Turbo Code}}.$$

From results, we observed that Hamming code is the simplest code and has the least computation time, but the error-correcting capability is not good when compared to others. The turbo code takes larger time for computation. We also showed below the comparison of the performance of different types of error-correcting codes.

$$\text{Perf}_{\text{Hamming}} < \text{Perf}_{\text{BCH}} < \text{Perf}_{\text{Convolutional}} < \text{Perf}_{\text{Reed-Solomon}} < \text{Perf}_{\text{Turbo Code}}.$$

From above result, the turbo code executes the best among these five codes. However, the complexity of the decoding algorithm of turbo codes outcomes large calculation time. If the protection given by the RS code is enough, then it may apply in many practical applications.

In this work, we assume that data-embedding techniques combined with ECCs will be widespread in many applications, such as the transmission, storage, and image authentication in medical and law enforcement in the future.

11.7. Conclusion

Hiding patient information, such as text files, demographical data, and physiological signals, into retinal fundus images or pictures is demonstrated to aid efficient transmission and storage. First, the patient text data is encrypted. Then, watermarking is performed in the retinal fundus image. After using a watermarking technique, a compression technique is used. The technique of proposed transmission and receiver is tried out for different channel conditions, which are AWGN and burst error (various SNR levels), and is tested with the Hamming, BCH, RS, convolutional, and turbo codes. These procedures can protect against errors made by imperfect digital recording or storage media and imperfect communication channel. In conclusion, the performance of turbo codes shows the best result, but it has the large calculation time of decoding procedure. So that, the turbo code for the real-time recovery of image and text data may be impossible in all applications. The RS (15, 11) code performs reasonably well, with small calculation time of decoding procedure and may apply to many practical applications.

References

1. Digital Imaging and Communications in Medicine (DICOM). National Electrical Manufacturers Association, Rosslyn, Virginia USA, DICOM Committee, 2001.
2. Sergio, S.F., Marina, S.R., Ramon, A.M., Marcelo, S., Nivaldo, B., Gustavo, H.M.B.M. *et al.* Managing medical images and clinical information. InCor's experience. *IEEE Trans Inf Technol Biomed* **11**:17–24, 2007.
3. Stefan, K. and Fabien, A.P.P. *Information Hiding Techniques for Steganography and Digital Watermarking*. Artek House, Boston, 2000.
4. Shoemaker, C. Independent Study. Hidden bits: a survey of techniques for digital watermarking. EER-290 Prof Rudko, Spring (2002).
5. Chung, Y.Y. and Wong, M.T. Implementation of digital watermarking system. Dig. Tech. Pap. *IEEE Int Conf Consum Electron* 214–215, 2003.
6. Giakoumaki, A., Pavlopoulos, S., and Koutsouris, D. Secure and efficient health data management through multiple watermarking on medical images. *Med Biol Eng Comput* **44**:619–631, 2006.
7. Gonzales, R.C., Woods, R.E., and Eddins, S.L. *Digital Image Processing Using MATLAB*. Prentice Hall, 2004.
8. Gonzales, R.C. and Woods, R.E. *Digital Image Processing*, Second Edition, Prentice Hall, 2001.
9. Elliott, M. and Schuette, B. Digital image watermarking. *ECE 533 Image Processing*, University of Wisconsin-Madison, 2006.
10. Miyazaki, A. Digital watermarking protection technique for multimedia. *Tech Rep IEICE* **102**(41):61–66, 2002.
11. Acharya, U.R., Deepthi, A., Bhat, P.S., and Niranjan, U.C. Compact storage of medical images with patient information. *IEEE Trans Inf Technol Biomed* **5**:320–323, 2001.
12. Acharya, U.R., Bhat, P.S., Kumar, S., and Min, L.C. Transmission and storage of medical images with patient information. *Comput Biol Med* **33**:303–310, 2003.
13. Nayak, J., Bhat, P.S., Acharya, U.R., and Kumar, M.S. *Efficient Storage and Transmission of Digital Fundus Images with Patient Information Using Reversible Watermarking Technique and Error Control Codes*. Springer Science + Business Media, LLC, 2008.
14. Nergui, M., Acharya, U.S., Acharya, U.R., and Wenwei, Y. Reliable and robust transmission and storage techniques for medical images with patient information. Springer Netherlands, *J Med Sys*, 2009.
15. Announcing the Advanced Encryption Standard, Federal Information Processing Standards Publication 197, November 26, 2001. Available at http://csrc.nist.gov/publications/fips/fips197/fips-197.pdf.

16. Daemen, J. and Rijmen, V. AES Proposal Rijndael. Version 2, 1999.
17. Wicker, S.B. and Bhargava, V.K. *Reed-Solomon Codes and Their Applications*. IEEE Press, 1994.
18. Shannon, C.E. A mathematical theory of communications. *Bell System Technical Journal* **27**:379–423 (Part I), 623–656 (Part 2), 1948.
19. Moon, T.K. *Error Correction Coding: Mathematical Methods and Algorithms*. John Wiley & Sons, Inc., Hoboken, New Jersey, 2005.
20. Blahut, R.E. *Algebraic Codes for Data Transmission*. Cambridge University Press, 2002.
21. Pretzel, O. *Error Correcting Codes and Finite Fields*. Imperial College, London, Clarendon Press, Oxford, 1992.
22. Lin, S. and Costello, Jr., D.J. *Error Control Coding Fundamentals and Applications*. Prentice Hall, Englewood cliffs, NJ, 2004.
23. Wicker, S.B. *Error Control Coding for Digital Communication Systems*. Prentice-Hall, 1995.
24. McEliece, R.J. *The Theory of Information and Coding*. Second Edition, Cambridge University Press, 2002.
25. Berrou, G., Glavieuc, A., and Thitmajshima, P. Near Shannon limit error-correcting coding: turbo codes. In: *Proc. 1993, Int. Conf. Com.*, Geneva, Switzerland, pp. 1064–1070, 1993.
26. Benedetto, S., Divsalar, D., Montorsi, G., and Pollara, F. A soft-input soft-output maximum *a posteriori* (MAP) module to decode parallel and serial concatenated codes. TDA Progress Report 42–127, 20 pp., November 15, 1996.
27. Rekh, S., Rani, S.S., and Shanmugam, A. Optimal choice of interleaver for turbo codes. *Academic Open Internet Journal* **15**, July 2, 2005.

Chapter 12

Temperature Changes Inside the Human Eye During LTKP

Ooi, E.H. and Ng, E.Y.K.

12.1. Introduction to Laser-Thermokeratoplasty (LTKP)

We have developed a complete model of the human eye in the O_{rz} coordinate system to investigate the temperature changes inside the human eye during LTKP. Heat that is absorbed inside the eye during laser radiation is described using the Beer-Lambert law. Two types of lasers that are commonly used in LTKP, which are pulsed and continuous-wave lasers, are investigated. The model is numerically created using the time-stepping dual reciprocity boundary element method. The temperature achieved inside the cornea during pulsed laser radiation is found to be higher than continuous-wave laser radiation, albeit for a short duration. In addition to the analysis of temperature distribution inside the eye, a thermal damage assessment is carried out using the thermal damage model of Henriques and Moritz.

LTKP is a corneal refractive surgery intended to correct vision. LTKP is a subset of the thermokeratoplasty technique that involves heating the cornea via laser treatment to induce corneal shrinkages. When the cornea is subjected to heat, the collagen bonds inside the stroma break, leading to contractions (shrinkages) of the collagen fibrillae.[1] The shrinkages of the cornea cause the curvature to change, which ultimately affects the refractive power of the cornea. Based on the aforementioned behavior, ophthalmologists have been able to correlate the amount of corneal shrinkages due to heat, with parameters such as heating duration and laser energy to produce

School of Mechanical and Aerospace Engineering, Nanyang Technological University, 50 Nanyang Avenue, Singapore 639798.

the desired level of change in the corneal refractive power. Although LTKP is a less popular method than other laser surgery techniques, such as laser *in situ* keratomilieusis (LASIK), it is used in clinical practice, especially to treat presbyopia and astigmatism.

One of the main concerns involved in LTKP is the risk of causing irreversible thermal damage that may induce permanent scarring to the cornea. LTKP also suffers from poor predictability and repeatability, which often lead to unreliable clinical outcomes. A proper understanding of the thermal behavior of the cornea during LTKP is, therefore, essential in designing treatment procedures that may help to reduce the risk of thermal injury and to improve the predictability and repeatability of clinical results. Mathematical modeling plays an important role in the studies of corneal temperature changes during LTKP. The role of mathematics in determining the accuracy of the treatment is largely due to the hazardous nature of the treatment procedure, which makes experimental studies on human subjects difficult. Most experimental studies of LTKP have been performed *in vitro* using samples of animal cornea.[1,2]

One of the earliest mathematical models that was developed for investigating corneal temperature changes during LTKP was presented by Mainster in 1979.[3] Although various studies of the thermal interaction between lasers and cornea have been reported prior to the work of Mainster,[3] these investigations mainly focused on the thermal effects of lasers that were not associated with the treatment of LTKP.[4,5] In the investigation carried out by Mainster,[3] the cornea was modeled as a semi-infinite region where a one-dimensional (1D) heat transfer was assumed. By the early 1990s, numerical models that investigated the thermal profile of the cornea during LTKP began to emerge. Most of the models developed during this period considered only the cornea as the solution domain.[6,7] The cornea was modeled as a cylinder in the axisymmetrical coordinate system. Numerical solutions were largely obtained using the finite difference method.

Between the early 1990s and the early 2000s, the studies of LTKP were eagerly pursued by Brinkmann *et al.*[8–10] from both the experimental and numerical perspectives. The models, which comprised of the cornea and the anterior chamber, were developed in the axisymmetrical coordinate system and were assumed as a two-layered cylinder. The finite element method was used to obtain a numerical solution. One of the unique features of the models developed by Brinkmann *et al.*[8–10] was the inclusion of intra-corneal focusing of the laser beam, i.e. the distribution of laser energy within the cornea. Various types of lasers used in LTKP, such as Er: Glass, Ho: YAG, and CO_2 lasers, have been investigated, and, in most cases, the numerical predictions showed results comparable with those observed during *in vitro* experiments.

Since the early 2000s, numerical investigations of LTKP have been actively carried out by Manns *et al.* and Borja *et al.*[11–13] Unlike the models developed by Brinkmann *et al.*[8–10] the Manns *et al.* and Borja *et al.* model region consisted only of the cornea and was modeled as a single-layered tissue slab in the two-dimensional (2D) coordinate system. Analytical solutions that were expressed as summations of an infinite series were derived based on the Green's function approach. The transient temperature profile of the cornea was found to be biologically reasonable.[11–13] No comparisons with experimental data have been performed, however. In addition to the investigations on the transient temperature changes of the cornea, Manns *et al.* and Borja *et al.*[11–13] predicted the degree of thermal damage and the degree of corneal shrinkages inside the cornea during LTKP using the Arrhenius equation, the damage integral, and the second-order kinetic model of corneal shrinkage.

One common feature shared by most previous LTKP studies is the use of simplified models for developing the mathematical models. The choice of using the simplified model approach may be due to the need-to-maintain mathematical simplicity. This approach remains valid since the temperature changes inside the cornea during LTKP are localized to the area where the laser beam is applied. One of the limitations of the simplified model approach, however, is that the assumptions that have to be made when specifying the boundary conditions may not be physiologically correct.

In this study, a model of the human eye is developed to investigate the changes in the temperature distribution inside the human cornea during the treatment of LTKP. In addition to improving our understanding of the thermal behavior of the cornea when subjected to heat, this study is designed to examine the feasibility of using an anatomically more realistic eye model for the studies of ocular heat transfer during LTKP. Problems associated with the increase of computer memory due to the use of a more complex model are avoided by numerically solving the model using the boundary element method. Investigations are carried out for the treatments of LTKP using pulsed and continuous-wave lasers. Additionally, the degree of thermal damage induced onto the cornea during laser radiation is estimated based on the transient temperature profile of the cornea using the thermal damage model of Henriques and Moritz.[14]

12.2. Characteristics of LTKP

The premise behind the treatment of LTKP is the shrinkage experienced by the corneal collagen when subjecting the cornea to heat. According to Stringer and Parr,[15] the corneal collagen shrinks when the cornea is heated to a temperature between 55°C and 65°C. However, this observation was made without considering the factors involved in ensuring a successful treatment of LTKP. Subsequent investigations carried out by other researchers have revealed that the cornea has to be heated to a minimum temperature of 64°C in order to produce permanent corneal shrinkage.[16] Likewise, Brinkmann *et al.*[8–10] found that the deep stromal tissues must be heated to temperatures beyond the shrinkage threshold to guarantee long-term effects of corneal shrinkage.

It is necessary to maintain the corneal temperature below 100°C during LTKP to avoid corneal relaxation.[17] The relaxation causes the bonds between the collagen molecules inside the cornea to break, thus countering the intended corneal shrinkage. Another important issue during treatment of LTKP concerns the temperature at the corneal endothelium, which is highly sensitive toward heat.[18] According to Wirbelauer,[18] endothelial cell deaths occur when the temperature at the endothelium increases to 65°C. Therefore, an ideal treatment of LTKP is expected to produce the desired amount of permanent shrinkage while minimizing the amount of corneal relaxation and thermal damage inside the cornea.

In a typical clinical treatment of LTKP, eight equally distanced laser spots, arranged concentrically around the center of the corneal surface, are applied to the cornea. Each spot has a radial distance of 3–5 mm from the center of the corneal surface. In some cases, 16 spots are applied, where the additional eight spots are located on an outer ring at a radial distance of approximately 7 mm from the center. Although the threshold for corneal shrinkage has been found to be in the range of 55–65°C, it is common for the cornea to be heated beyond this threshold. The temperature that the cornea is eventually heated to depends on the type of lasers used. Two types of lasers, pulsed and continuous-wave, can be used for LTKP.

12.3. Pulsed Laser

Each laser pulse delivers a large amount of energy at a duration that is usually less than 0.25 s. This output is irregular.[19] One of the most commonly used pulsed lasers in the treatment of LTKP is the Ho: YAG laser, which emits a laser beam at a wavelength of 2.1 μm.[7,16] Other types of pulsed lasers that have been used in LTKP include the Er: Glass and the Tm: YAG laser. In a typical LTKP procedure using a pulsed laser, seven laser pulses, each with the duration of 200 μs, are applied to the corneal surface at a repetition rate of 5 Hz.[12]

12.4. Continuous-Wave Laser

Unlike pulse lasers, continuous-wave lasers produce steady and continuous output.[19] Radiation is usually carried out at a low-energy rate. Among the type of continuous-wave lasers that have been used in LTKP are laser diodes, CO_2 lasers, and $CoMgF_2$ lasers. Laser diodes are usually preferred over CO_2 and $CoMgF_2$ lasers due to their tunability. This tunability allows the laser beam to be emitted at different wavelengths, hence, making it a more versatile choice for LTKP. Commonly used wavelengths range from 1.85 to 1.87 μm.[9] LTKP treatment using continuous-wave lasers may be carried out using two approaches, the 10-s coagulation, where the cornea is heated continuously for 10 s at a laser power of 0.125 W, and the minute coagulation, where the cornea is heated continuously for 60 s are both carried out using a 0.10-W laser.[10]

12.5. Mathematical Model

12.5.1. *Model Description*

Our model of the human eye is developed based on dimensions found in literature.[20,21] Figure 12.1 displays the model developed with reference

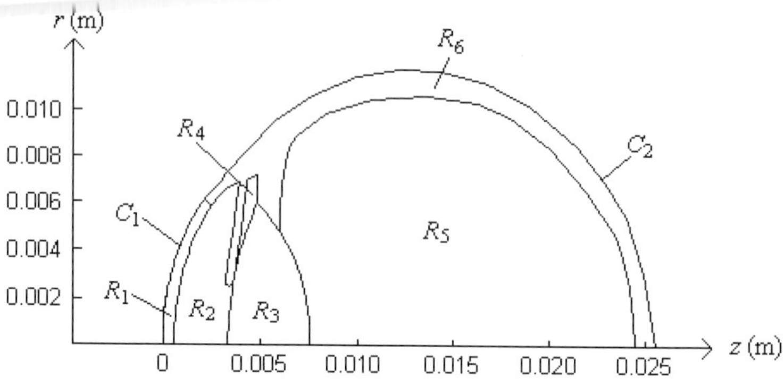

Fig. 12.1. Model of the human eye used in the study of LTKP.

to the axisymmetrical coordinate system, O_{rz}. The model comprises six regions, which are the cornea, the anterior chamber, the lens, the posterior chamber, the vitreous, and the sclera. These regions are denoted by R_1, R_2, R_3, R_4, R_5, and R_6, respectively.

The retina and choroid, which are relatively thin when compared to the sclera, have been modeled with the sclera as one homogeneous region.[22,23] Similarly, the iris and sclera are treated as one homogeneous region, since they have been found to exhibit the same thermal properties.[24] The exterior surfaces of the cornea and sclera are denoted by boundaries C_1 and C_2, respectively.

12.5.2. *Governing Equations*

Assuming that blood perfusion and metabolic heat generation inside the eye are negligible, and treating each ocular region as a solid, the transient temperature distribution inside the human eye during LTKP, with reference to the model illustrated in Fig. 10.1, may be described using

$$\rho_i c_i \frac{\partial}{\partial t}[T_i(r, z, t)] = \nabla(\kappa_i \nabla T_i(r, z, t)) + S_i(r, z, t)$$
$$\text{for } i = 1, 2, 3, 4, 5 \text{ and } 6, \quad \text{for } t > 0, \tag{12.1}$$

where ρ_i, c_i, and κ_i are the density, specific heat, and thermal conductivity of region R_i, respectively, T_i is the temperature distribution in R_i, t is time, and S_i is the heat generated inside region R_i due to the absorption of laser

Table 12.1. Thermal properties of the human eye.

Region	Thermal conductivity, κ ($Wm^{-1}K^{-1}$)	Density, ρ (kgm^{-3})	Specific heat, c ($Jkg^{-1}K^{-1}$)
Cornea, R_1	0.58[25]	1050[26]	4178[25]
Anterior chamber, R_2	0.58[27]	996[24]	3997[24]
Lens, R_3	0.40[28]	1050[26]	3000[28]
Posterior chamber, R_4	0.58[27]	996[24]	3997[25]
Vitreous, R_5	0.60[25]	1000[25]	4178[25]
Sclera, R_6	1.00[24]	1100[24]	3180[24]

energy. Each region inside the human eye is assumed to be homogeneous and thermally isotropic. Note that the temperature distribution is now a function of space and time.

Values of the thermal conductivity, density, and specific heat of each ocular region may be found in literature.[24–28] These values are tabulated in Table 12.1.

Heat that is generated inside the eye, S_i due to the absorption of laser energy may be described using the Beer-Lambert law.[19] For the typical range of laser wavelengths used in LTKP (10.85–2.1 μm), S_i may be mathematically expressed as:

$$S_i(r, z, t) = \begin{cases} \psi(t)\mu(1 - F)E(r, z)\exp(-\mu z), & \text{for } i = 1 \\ 0, & \text{for } i = 2, 3, 4, 5 \text{ and } 6 \end{cases}, \tag{12.2}$$

where μ is the corneal absorption coefficient, which is dependent on the wavelength of the laser (unit of m^{-1}), F is the Fresnel reflectance of the corneal surface, $E(r, z)$ is the incident irradiance at the center of the corneal surface, and $\psi(t)$ is given as:

$$\Psi(t) = \begin{cases} 1, & \text{if laser is off} \\ 0, & \text{if laser is on} \end{cases}, \tag{12.3}$$

which dictates the time that the laser is being applied onto the cornea. When the laser is off, $\psi(t)$ has a value of zero. Consequently, the value of

S_i becomes zero; implying that no heat generation takes place inside the human eye.

Equation (12.2) implies that the laser beam that is incident on the corneal surface generates heat only inside the cornea. This implication is largely attributed to the penetration depth of laser, cornea d_p, and thickness. Penetration depth generally describes the depth inside the ocular media where radiation from the laser approaches zero, as it decreases exponentially from the irradiated surface.[19] Mathematically, the penetration depth is given by:

$$d_p = \mu^{-1}. \tag{12.4}$$

The typical laser wavelengths used in the treatment of LTKP (as described in Sec. 12.2) ranges from 1.87 to 2.1 μm. These values correspond to those of the corneal absorption coefficient of 1900–2000 m^{-1}. Using Eq. (12.4), the range of d_p corresponding to the range of the corneal absorption coefficient is calculated as 0.48–0.50 mm. This calculation suggests that if a laser with a wavelength of 1.87 μm is applied to the corneal surface, the laser radiation approaches zero at a depth of 0.50 mm from the corneal surface. Similarly, if a laser with a wavelength of 2.1 μm is used, radiation at regions beyond the depth of 0.48 mm from the corneal surface becomes negligible. Since the thickness of the cornea in the present model is approximately 0.588 mm, no radiation is expected in the ocular regions located beyond the cornea.

The mathematical expression of the incident irradiance, $E(r, z)$, in Eq. (12.2) depends on the profile of the laser beam that arrives at the corneal surface. Ideally, the laser beam may be assumed to have a flat profile so that the distribution of irradiance is uniform. In most practical cases, the laser beam has a distribution that follows the profile of a Gaussian-type Equation,[19] such as shown in Fig. 12.2. For laser beams that are centered near the center of the corneal surface, $(r, z) = (0, 0)$, the function of $E(r, z)$ for a Gaussian beam profile is mathematically given as:

$$E(r, z) = E_o \exp\left(-\frac{2r^2}{w^2}\right), \tag{12.5}$$

where E_o denotes the peak irradiance and w denotes the radius of the laser beam where the laser irradiance decreases to exp(−1) times its maximum value.[19]

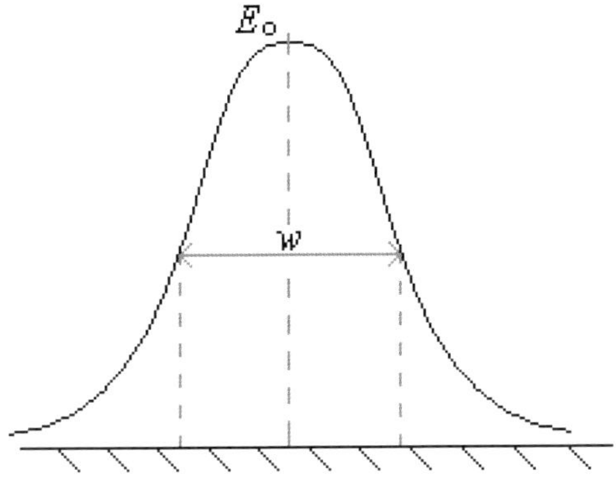

Fig. 12.2. Distribution of the Gaussian laser beam.

12.5.3. *Initial-Boundary Conditions*

Boundary conditions are specified on both the surfaces of the cornea, C_1, and on the sclera, C_2. Laser radiation during LTKP is assumed to produce no significant effect on the heat transfer between the environment and the corneal surface. Similarly, heat transfer between the blood flow and the eye is assumed not to be affected by the treatment of LTKP.

On the exterior surface of the cornea, C_1, heat is transferred to the environment via convection and radiation. Cooling is aided by the evaporation of tears from the tear film on top of the corneal surface. Mathematically, this condition is given as:

$$-\kappa_1 \frac{\partial T_1}{\partial n} = h_{\text{amb}}(T_1 - T_{\text{amb}}) + \varepsilon\sigma(T_1^4 - T_{\text{amb}}^4) + E_{\text{vap}}, \quad \text{on } C_1 \quad \text{and} \quad \text{for } t > 0, \tag{12.6}$$

where the first, second, and last term on the right-hand side refer to the heat loss due to convection, radiation, and tear evaporation, respectively, h_{amb} is the ambient convection coefficient, T_{amb} is the ambient temperature, ε is the corneal emissivity, σ is the Stefan-Boltzmann constant, E_{vap} is the heat loss due to tear evaporation, and $\partial T_1/\partial n$ is the rate of change of T_1 in the outward unit vector normal to the external corneal surface C_1.

On the exterior surface of the sclera, C_2, heat from blood flow across the sclera enters the eye and diffuses via conduction to the corneal surface.

Table 12.2. Values of the various control parameters.

Parameter	Value
Ambient temperature, T_{amb} (°C)	25
Ambient convection coefficient, h_{amb} ($Wm^{-2}K^{-1}$)	10
Blood temperature, T_{bl} (°C)	37
Blood convection coefficient, h_{bl} ($Wm^{-2}K^{-1}$)	65
Heat loss due to tears evaporation, E_{vap} (Wm^{-2})	40
Emissivity of the cornea surface, ε	0.97
Stefan-Boltzmann constant, σ ($Wm^{-2}K^{-4}$)	5.67×10^{-8}

Thus, we may write

$$-\kappa_6 \frac{\partial T_6}{\partial n} = h_{bl}(T_6 - T_{bl}), \quad \text{on } C_2 \quad \text{and} \quad \text{for } t > 0, \tag{12.7}$$

where h_{bl} is the blood convection coefficient, T_{bl} is the blood temperature, and $\partial T_6/\partial n$ denotes the rate of change of T_6 in the outward unit vector normal to the external corneal surface, C_2. Values of the parameters used in Eqs. (12.6) and (12.7) are the same as those employed by Ooi *et al.*[23] and are summarized in Table 12.2.

The variation of the temperature and heat flux at the interfaces between two contiguous ocular regions, I_{ij}, may be described using the continuity condition, which is mathematically given as:

$$T_i = T_j \quad \text{and} \quad \kappa_i \frac{\partial T_i}{\partial n} = \kappa_j \frac{\partial T_j}{\partial n}, \quad \text{on } I_{ij}, \tag{12.8}$$

where T_i and T_j are temperature of regions R_i and R_j, respectively, and $\partial T_i/\partial n$ and $\partial T_j/\partial n$ are the rate of change of T_i and T_j in the outward unit vector normal to the interface I_{ij}. Since the problem in the present study is a transient one, an initial condition is required to complete the formulation of the problem. For this purpose, the temperature distribution inside the normal unexposed human eye during a steady state is obtained by solving the following:

$$\nabla(\kappa_i \nabla T_i(r, z)) = 0, \quad \text{for } i = 1, 2, 3, 4, 5 \text{ and } 6. \tag{12.9}$$

The result is subjected to Eqs. (12.6) to (12.8), as an initial, specified condition.

12.6. Numerical Scheme

12.6.1. *Integro-Differential Equation*

To obtain a numerical solution of Eq. (12.1) subjected to Eqs. (12.6) through (12.9) using the boundary element method, the integro-differential representation of Eq. (12.1) has to be derived. Following the steps outlined by Brebbia *et al.*[29] the integro-differential representation of Eq. (12.1) may be written as:

$$\begin{aligned}
&\lambda(\xi_i, \eta_i)T_i(\xi_i, \eta_i, t)\\
&\quad = \int_{\Gamma_i} T_i(r, z, t)\frac{\partial}{\partial n}[\Phi_{\text{axis}}(r, z; \xi_i, \eta_i)] \cdot r \cdot ds(r, z)\\
&\qquad - \int_{\Gamma_i} \Phi_{\text{axis}}(r, z; \xi_i, \eta_i)\frac{\partial}{\partial n}[T_i(r, z, t)] \cdot r \cdot ds(r, z)\\
&\qquad + \iint_{R_i} \Phi_{\text{axis}}(r, z; \xi_i, \eta_i)\left[\frac{1}{\alpha_i}\frac{\partial}{\partial T}[T_i(r, z, t)] - \frac{1}{\kappa_i}S_i(r, z, t)\right] \cdot r \cdot dA(r, z)\\
&\qquad\qquad \text{for } (\xi_i, \eta_i) \in R_i \cup \Gamma_i \quad \text{and} \quad \text{for } i = 1, 2, 3, 4, 5, \text{ and } 6,
\end{aligned} \tag{12.10}$$

where $\alpha_i = \kappa_i/\rho_i c_i$ is the thermal diffusity of region R_i, Γ_i denotes the curve boundary of the region R_i, $ds(r, z)$ denotes the length of an infinitesimal portion of curve Γ_i, $dA(r, z)$ denotes the area of an infinitesimal portion of the region R_i, $\lambda(\xi_i, \eta_i) = 1$, if (ξ_i, η_i) lies inside the region R_i, $\lambda(\xi_i, \eta_i) = 0.5$ if (ξ_i, η_i) lies on a smooth part of Γ_i, and Φ_{axis} and $\partial\Phi_{\text{axis}}/\partial n$ are the axisymmetric fundamental solution of the Laplace equation and its normal derivatives, respectively, which are expressed as[29]:

$$\begin{aligned}
&\Phi_{\text{axis}}(r, z; \xi, \eta)\\
&\quad = -\frac{K(m(r, z; \xi, \eta))}{\pi\sqrt{a(r, z; \xi, \eta) + b(r; \xi)}}\frac{\partial}{\partial n}[\Phi_{\text{axis}}(r, z; \xi, \eta)]\\
&\quad = -\frac{1}{\pi\sqrt{a(r, z; \xi, \eta) + b(r; \xi)}}x\\
&\qquad \times \left\{\frac{n_r}{2r}\left[\frac{\xi^2 - r^2 + (\eta - z)^2}{a(r, z; \xi, \eta) - B(r; \xi)}E(m(r, z; \xi, n)) - K(m(r, z; \xi, \eta))\right]\right.\\
&\qquad \left. + n_z\frac{n - z}{a(r, z; \xi, \eta) - b(r; \xi)}E(m(r, z; \xi, \eta))\right\},
\end{aligned} \tag{12.11}$$

where n_r and n_z are the components of the outward unit normal vector on Γ_i in the r and z direction, respectively, K and E denote the complete elliptic integral of the first and second kind, respectively, as defined by Abramowitz and Stegun,[30] and m, a, and b are, respectively, given as:

$$m(r, z; \xi, \eta) = \frac{2b(r; \xi)}{a(r, z; \xi, \eta) + b(r; \xi)},$$

$$a(r, z; \xi, \eta) = \xi^2 + r^2 + (\eta - z)^2,$$

$$b(r; \xi) = 2r\xi.$$

Note that in the axisymmetric formulation, the z-axis does not form part of the curve boundary Γ_i.

The boundary element method is implemented by discretizing the boundary Γ_i into small straight-line segments, also known as boundary elements. The domain integral and the time-dependent variable in Eq. (12.10) are treated using the dual-reciprocity boundary element method and the time stepping scheme, respectively. For a detailed derivation, one may refer to the work carried out by Ooi *et al.*[31]

12.7. Results

Investigations on the transient temperature distribution inside the cornea during LTKP are carried out for both pulsed and continuous-wave lasers. Only lasers with a Gaussian beam profile are considered. The values of laser parameters used in the present study are selected based on the values used in a typical clinical LTKP treatment. Although a minimum of eight laser spots are usually applied to the corneal surface (see Sec. 12.2), in the present study, only a single laser spot, assumed to be applied to the center of the corneal surface, is considered. This assumption is necessary in order to maintain the axisymmetric feature of the human eye, which is of great computational advantage. With the laser spot placed at the center of the corneal surface, an increase in the temperature near the center of the corneal surface is expected.

To capture the large thermal variation over the small heated area around the center of the corneal surface accurately, the level of boundary discretization used has to be sufficiently fine. Likewise, the number of interior points

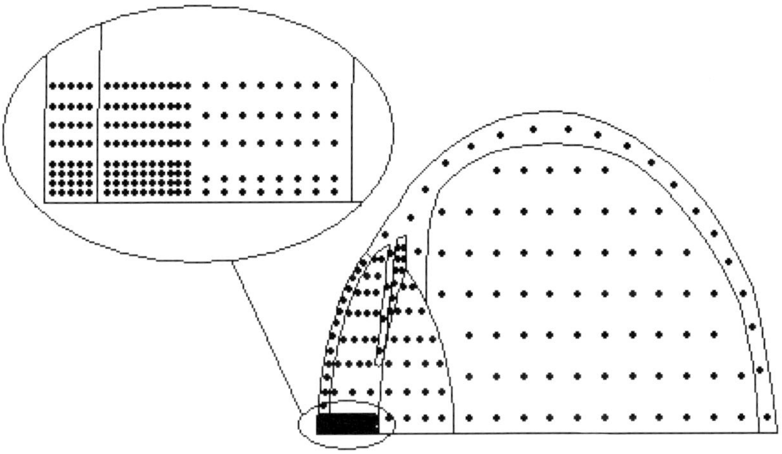

Fig. 12.3. Interior points selected inside each region of the human eye.

used to implement the dual reciprocity method has to be sufficiently large. In the present study, the human eye model is discretized into 253 boundary elements, with 72 of them located on the boundary of the cornea. A total of 316 interior points are selected inside the human eye model, with 53 and 142 of them placed inside the cornea and anterior chamber, respectively. The interior points selected inside each region of the human eye model are illustrated in Fig. 12.3.

12.7.1. *Pulsed Laser*

Table 12.3 summarizes the values of laser parameters of a typical LTKP treatment using pulsed laser. These values are obtained from Manns *et al.*[12] Energy per pulse describes the amount of laser energy that is delivered onto the corneal surface during each laser pulse. The laser absorption coefficient of the cornea is obtained by assuming that the cornea has optical properties similar to those in water. The value of peak irradiance, E_o, in Table 12.3 is obtained using the following expression,[11]

$$E_o = \frac{E_{pp}}{\pi w^2 t_p},$$

where E_{pp} is energy per pulse and t_p is pulsed duration.

Table 12.3. Typical laser parameters chosen for the pulsed laser.

Parameter	Value
Energy per pulse, E_{pp} (mJ)	30
Pulse duration, t_{p} (μs)	200
Pulse repetition rate (Hz)	5
Number of pulse	7
Wavelength (μm)	2.1
Laser beam radius, w (mm)	0.3
Laser absorption coefficient, μ (m^{-1})	2000
Peak irradiance, E_{o} (Wm^{-2})	5.31×10^8

Based on the values given in Table 12.3, the function $\psi(t)$ in Eq. (12.3) that describes the period when the laser is on and off may be expressed as:

$$\psi(t) = \begin{cases} 1, & \text{if } t \in J \\ 0, & \text{if } t \notin J \end{cases}, \tag{12.12}$$

where t is time taken (in seconds) and J is the time interval defined by:

$$J = \bigcup_{m=0}^{6} \{t : 0.2002\,\text{m} \leq t \leq 0.2002\,\text{m} + 0.0002\}. \tag{12.13}$$

In carrying out the time-stepping scheme, the time step, Δt is chosen to be

$$\Delta t = \begin{cases} 0.0002\text{ s}, & \text{if } t \in J \\ 0.05\text{ s}, & \text{if } t \notin J \end{cases}.$$

Figure 12.4 shows the transient temperature changes along the pupillary axis ($r = 0$) at various depths of the cornea during treatment of pulsed LTKP. The seven temperature peaks seen in Fig. 12.4 corresponds to the seven laser pulses that are applied to the corneal surface. Cooling is observed at intervals between laser pulses where heat that is absorbed inside the cornea is diffused into the environment via convection and radiation and into the other ocular regions inside the eye. Throughout the treatment of LTKP, the corneal surface experiences the largest increase in temperature. At the end of the seventh laser pulse, temperature as high as 111°C is reached; albeit for only a

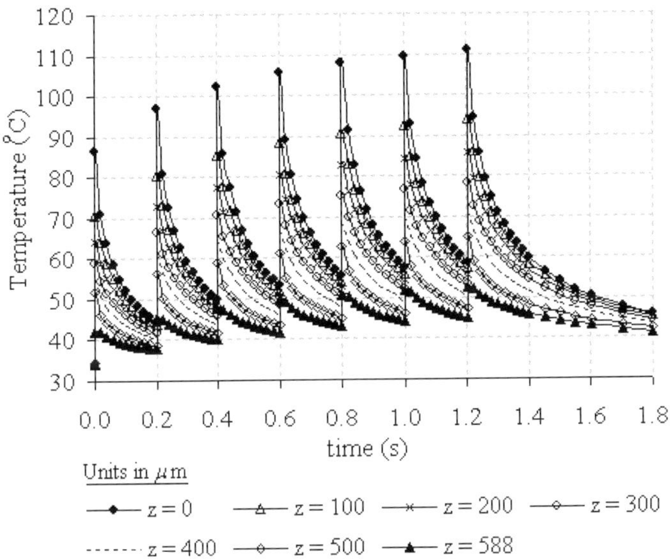

Fig. 12.4. Transient temperature changes along the pupillary axis at various depth inside the cornea during pulsed LTKP with a Gaussian beam profile.

short duration. At the corneal endothelium, i.e. at $z = 588\,\mu$m, temperature at the seventh laser pulse is found to be approximately 53°C. Temperature at the corneal stroma, which is represented by the curves between $z = 100\,\mu$m and $z = 500\,\mu$m, increases beyond the threshold for corneal shrinkages after the third laser pulse.

Figure 12.5 illustrates the temperature profile along the pupillary axis at various depths of the cornea during each laser pulse, i.e. for $t = 0.0002$, 0.2002, 0.4006, 0.6008, 0.801, 1.0012, and 1.2014 s. The vertical lines separate the z-axis into regions occupied by the epithelium, stroma, and endothelium, while the horizontal lines represent the lower threshold for corneal shrinkages ($T = 55$°C) and corneal relaxation ($T = 90$°C). A large part of the corneal stroma, where the corneal collagens are located, is found to have temperatures that are greater than the threshold for corneal shrinkages. The high temperature indicates corneal shrinkages. At the corneal epithelium, temperature increases beyond the relaxation threshold. However, this may not severely affect the overall shrinkage of the cornea, since the collagens that are responsible for the contraction and relaxation of the cornea are not found in the corneal epithelium. At the end of the seventh laser pulse, the

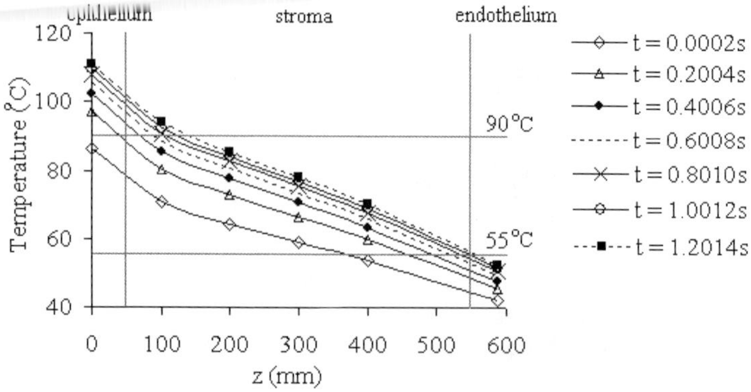

Fig. 12.5. The corneal temperature along the pupillary axis (against z at $r = 0$) at the end of each laser pulse.

maximum temperature achieved by the corneal endothelium is smaller than 65°C; implying an absence of endothelial cell deaths.

The spatial temperature profiles over a small cross-section of the cornea and the anterior chamber on the rz plane are shown in Fig. 12.6. The cross-section is selected in such a way that significant variation in the temperature can be observed. For a clearer visualization, the temperature plots in Fig. 12.6 are given with the mirror images about the z-axis.

There appears to be a large temperature gradient at approximately $z = 0.5$ mm. This may be attributed to the penetration depth of the laser inside the cornea, which according to Eq. (12.4) and based on the properties of the pulsed laser given in Table 12.3, is calculated to be 0.5 mm. Temperature increase at regions defined by $z < 0.5$ mm is, thus, caused by the absorption of laser energy, while temperature increase at $z > 0.5$ mm is due to the diffusion of heat from the regions at $z < 0.5$ mm.

12.7.2. *Continuous-Wave Laser*

The transient temperature changes inside the eye during the treatment of LTKP using a continuous-wave laser are examined in this section. Both the 10-s coagulation and the minute coagulation (see Sec. 12.2) are investigated. Parameters of the continuous-wave laser used in a typical treatment of LTKP are tabulated in Table 12.4, which are obtained from Brinkmann *et al.*[10] Wavelengths of the lasers used to treat LTKP using continuous-wave lasers

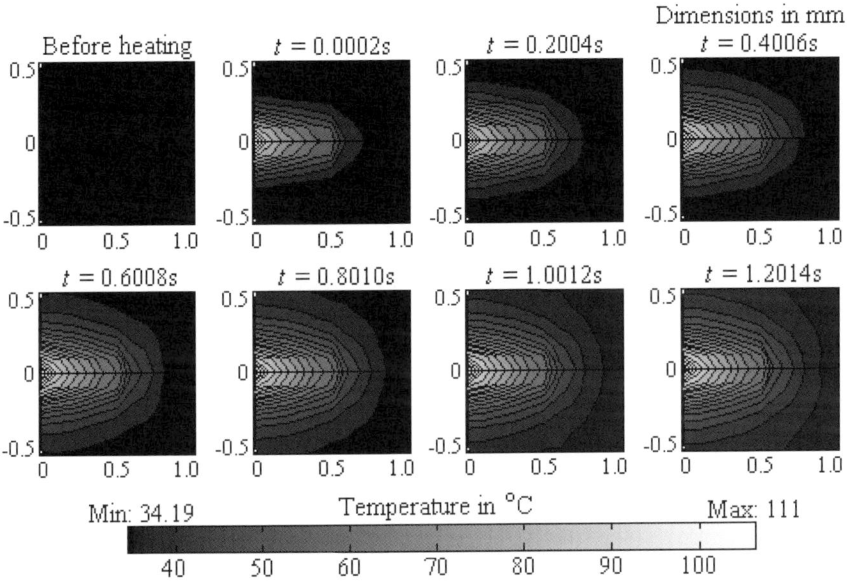

Fig. 12.6. Spatial temperature profiles over a selected cross-section of the eye subject to pulsed laser irradiation.

Table 12.4. Typical laser parameters chosen for the continuous-wave laser.

Parameter	10-s coagulation	Minute coagulation
Laser power, P (W)	0.125	0.10
Peak irradiance, E_o (Wm^{-2})	4.42×10^5	3.54×10^5
Heating duration, t(s)	10	60
Time step, Δt(s)	0.1	0.5
Wavelength (μm)	1.87	1.87
Laser absorption coefficient, μ(m^{-1})	1900	1900

may vary between 1.85 and 1.87 μm. However, in the present study, only the laser that is emitted at a wavelength of 1.87 μm is considered.[10] Values of peak irradiance, E_o, are obtained using the expression:

$$E_o = \frac{P}{\pi w^2},$$

where P is laser power. The expressions of $\psi(t)$ for the 10-s and minute coagulations are given by:

$$\Psi(t) = \begin{cases} 1, & \text{if } t \leq 10\,\text{s} \\ 0, & \text{if } t > 10\,\text{s} \end{cases} \tag{12.14}$$

and

$$\Psi(t) = \begin{cases} 1, & \text{if } t \leq 60\,\text{s} \\ 0, & \text{if } t > 60\,\text{s} \end{cases}, \tag{12.15}$$

respectively. When executing the time stepping scheme, a value of $\Delta t =$ 0.1 s is chosen for the 10-s coagulation while for the minute coagulation, a value of $\Delta t = 0.5$ is selected.

Figures 12.7 and 12.8 illustrate the transient temperature profiles along the pupillary axis ($r = 0$) at various depths of the cornea during the treatment of LTKP for the 10-s and minute coagulations, respectively. Both approaches show uniform increases in the corneal temperature, which are unlike the temperature profiles observed in Fig. 12.4 during pulsed laser radiation. At the end of the laser treatment, a sharp decrease in temperature is found because of the rapid dissipation of heat to the environment and adjacent regions inside the human eye.

The largest temperature reached inside the cornea during the 10-s and minute coagulations are 75.6°C and 70.1°C, respectively. The higher

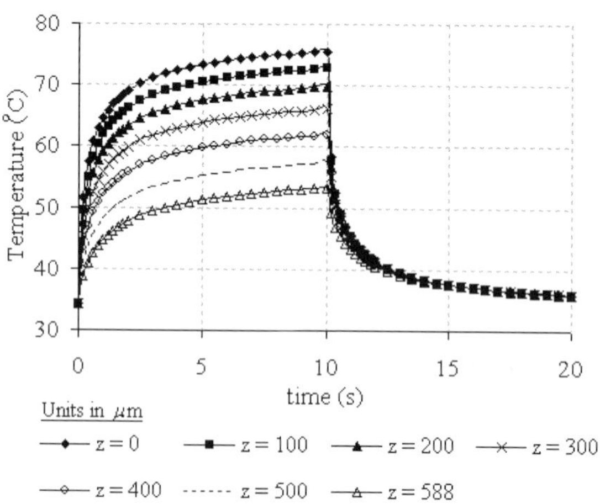

Fig. 12.7. Transient temperature profiles at various depths of the cornea in the 10-s coagulation.

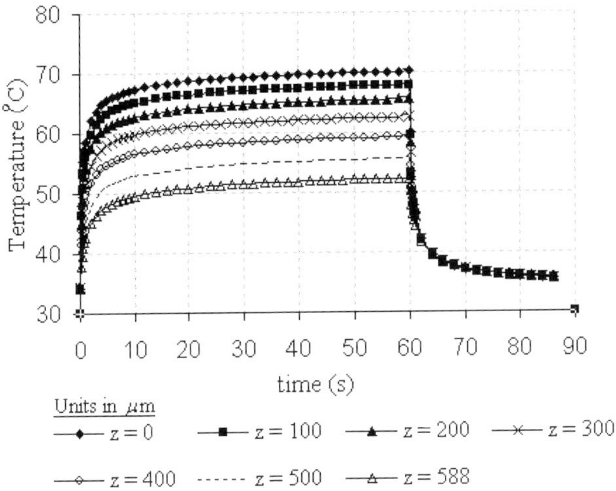

Fig. 12.8. Transient temperature profiles at various depths of the cornea in the minute coagulation.

temperature observed in the 10-s coagulation may be attributed to laser power, which is higher than the laser power used in minute coagulation (0.125 vs. 0.10 W). A large portion of the corneal stroma, defined by depths between 50 and 550 μm, are heated beyond the threshold for corneal shrinkages. This heating is true for both the 10-s and minute coagulations. In both coagulations, no corneal relaxation occurs, since the largest temperatures achieved are not greater than 90°C.

Figures 12.9 and 12.10 correspondingly show the temperature profiles along the pupillary axis ($r = 0$) of the cornea at various time levels for both the 10-s and minute coagulations. During the course of radiation, temperatures at the endothelium ($z > 550\,\mu$m) are found to be less than 55°C for both the 10-s and minute coagulations. These observations imply that no endothelial cell deaths occur during continuous-wave laser radiation.

The spatial temperature distribution over a small cross-section of the cornea and anterior chamber at $t = 2, 4, 6, 8$, and 10-s for the 10-s coagulation and at $t = 12, 24, 36, 48$, and 60 s for the minute coagulation are illustrated in Figs. 10.11 and 10.12, respectively. The temperature plots in Figs. 10.11 and 10.12 have been presented with the mirror images throughout the z-axis. The isotherms observed in Figs. 12.11 and 12.12 appear to

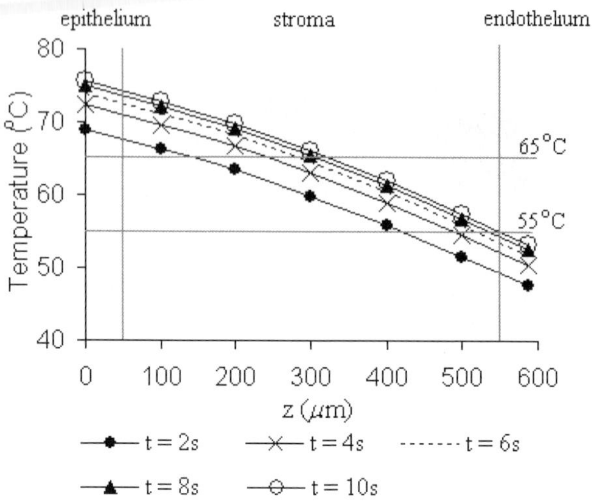

Fig. 12.9. Temperature distribution inside the cornea at various time levels in the 10-s coagulation.

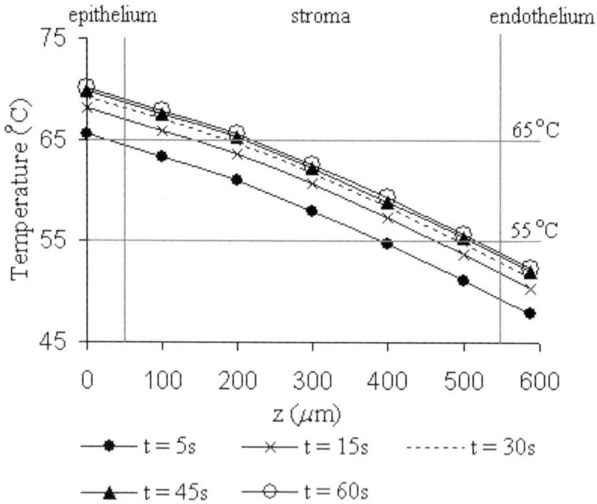

Fig. 12.10. Temperature distribution inside the cornea at various time levels in the minute coagulation.

be more uniformly distributed when compared to the isotherms observed during pulsed laser radiation. A large thermal gradient is observed at about $z = 0.52$ mm, which is approximately the point defined by the penetration depth of the continuous-wave laser.

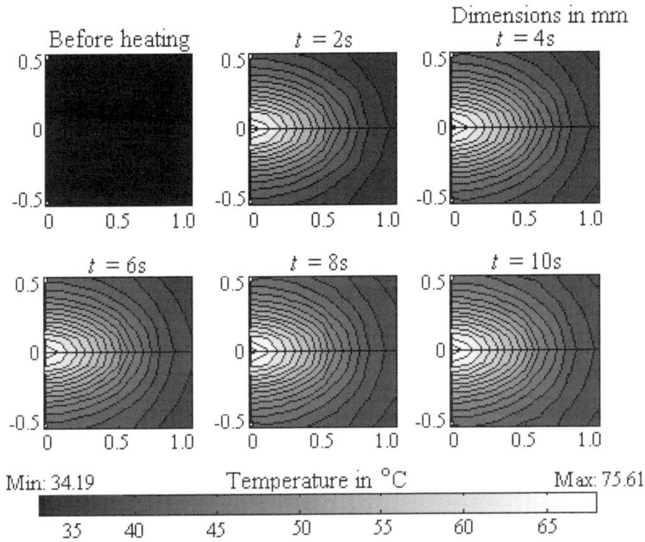

Fig. 12.11. Spatial temperature profiles over a selected cross-section of the eye subject to 10-s coagulation radiation.

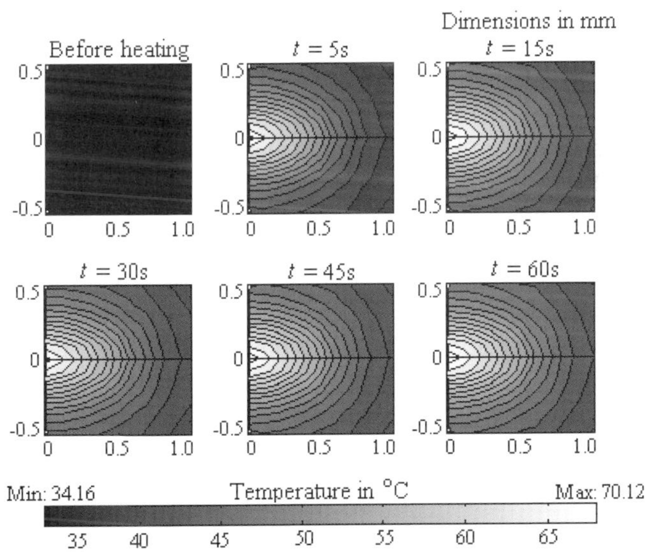

Fig. 12.12. Spatial temperature profiles over a selected cross-section of the eye subject to minute coagulation radiation.

12.7.3. Thermal Damage Assessment

One of the major concerns of LTKP, as we have pointed out in Sec. 12.1, is the risk of inducing thermal damage onto the cornea, since the temperature to which the cornea is heated can be very high.[9,10,12] The numerical results obtained in the present study have shown this rise in temperature to be as high as 111°C during pulsed laser treatment (see Figs. 12.4, 12.5, and 12.6).

Based on the magnitude of the temperature rise inside the cornea alone, the pulsed laser appears to produce a greater amount of thermal damage than the continuous-wave laser. However, the degree of thermal damage inflicted on human tissues is dependent not only on the magnitude of temperature, but also on the duration of heating.[14]

In this section, we carry out a thermal damage assessment to estimate the degree of thermal damage experienced by the cornea during treatment of LTKP. Investigations are carried out using the thermal damage model of Henriques and Moritz.[14]

In 1947, Henriques and Moritz[14] developed a mathematical model to predict the degree of thermal damage inside the human skin when the skin is burned. Henriques and Moritz assumed the thermal damage of the human skin was a biochemical process defined by the thermal denaturation of proteins, such that the process could be described in terms of an Arrhenius relationship.[32] Assuming that the same biochemical process takes place inside the cornea during LTKP, we may then express the degree of thermal damage inside the cornea, Ω at $r = 0$ using the Arrhenius integral, which is given by[14]:

$$\Omega(z,t)_{r=0} = \int_0^{t_f} A \exp\left(-\frac{\Delta E_a}{RT(0,z,t)}\right) dt, \tag{12.16}$$

where ΔE_a is the activation energy, R is the universal gas constant, t_f is the time thermal damage is evaluated, and A is an empirically valued coefficient given as:

$$\mathrm{A} = \frac{\mathrm{k}_b T(0,z,t)}{h} \exp\left(\frac{\Delta S}{R}\right), \tag{12.17}$$

where k_b is the Boltzmann constant, h is the Planck constant, and ΔS is the entropy of activation. The values of the parameters used for calculating Eqs. (12.16) and (12.17) are obtained from Kampmeier *et al.*[33] and Incropera and DeWitt[34] and are summarized in Table 12.5.

Table 12.5. Values of the parameters used in estimating the damage integral.

Parameter	Value
[a]Activation energy, E_a ($Jmol^{-1}$)	106×10^3
[a]Entropy of activation, ΔS($Jmol^{-1}K^{-1}$)	39
[b]Universal gas constant, R ($Jmol^{-1}K^{-1}$)	8.314
[b]Boltzmann constant, k_b ($kgm^2s^{-1}K^{-1}$)	1.38×10^{-23}
[b]Planck constant, h (kgm^2s^{-1})	6.63×10^{-34}

[a]Values obtained from Ref. [33].
[b]Values obtained from Ref. [34].

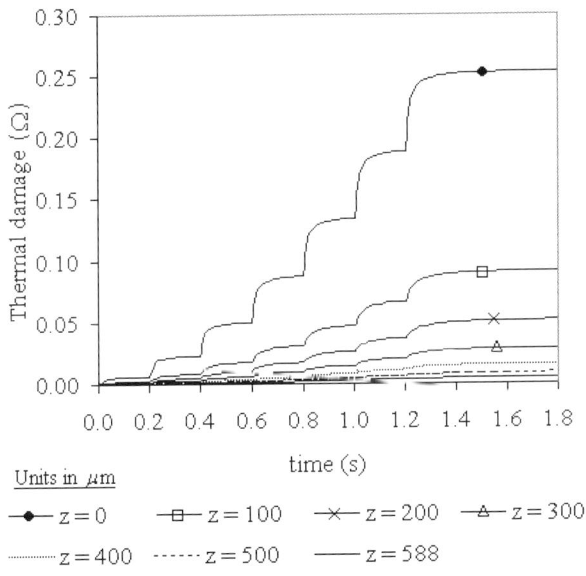

Fig. 12.13. Thermal damage inside the cornea during pulsed laser radiation.

Using the values tabulated in Table 12.5, the degree of thermal damage at various depths inside the cornea are estimated by evaluating Eq. (12.16) numerically using the trapezoidal rule. Figure 12.13 shows the thermal damage profile of the cornea during pulsed laser radiation. The largest thermal damage is observed at the surface of the cornea where the increase in temperature is at a maximum. The "step" profile observed in Fig. 12.13 is due to the irregular radiation of the pulsed laser. A large increase in thermal damage occurs only when there is an increase in the corneal temperature, i.e. during the application of each laser pulse.

In order to assess the severity of the damage illustrated in Fig. 12.13, we assume that the thermal damage characterization for skin is applicable in the case of the cornea. Therefore, a value of $\Omega < 0.53$ describes reversible thermal damage; $0.53 \leq \Omega < 1$ defines the range of irreversible thermal damage, and $\Omega \geq 1$ describes complete tissue necrosis.[14] From Fig. 12.13, it is found that pulsed laser radiation produces no irreversible thermal damage, since the largest value of Ω is less than the threshold of 0.53.

The thermal damage induced onto the cornea during continuous-wave laser radiation is illustrated in Figs. 12.14 and 12.15 for the 10-s and minute coagulations, respectively. Both figures show similar profiles. The sharp increase in thermal damage observed in the 10-s and minute coagulations plots are attributed to the continuous heating of the cornea during continuous-wave laser radiation. Once laser radiation is terminated, no major changes in the thermal damage are observed. The thermal damage induced onto the cornea during the 10-s coagulation appears smaller than the

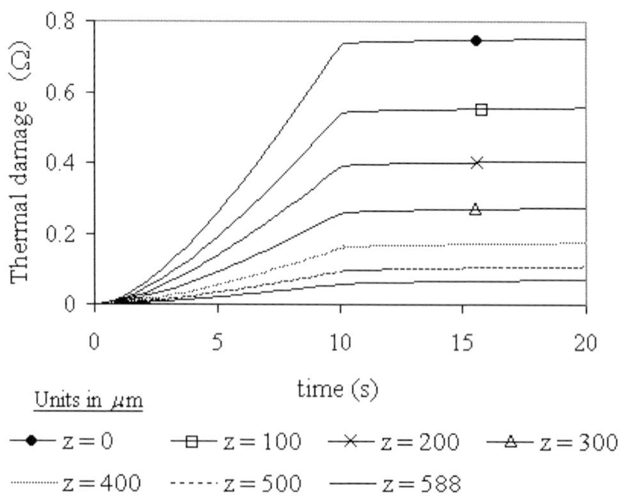

Fig. 12.14. Thermal damage inside the cornea during 10-s coagulation of continuous-wave laser radiation.

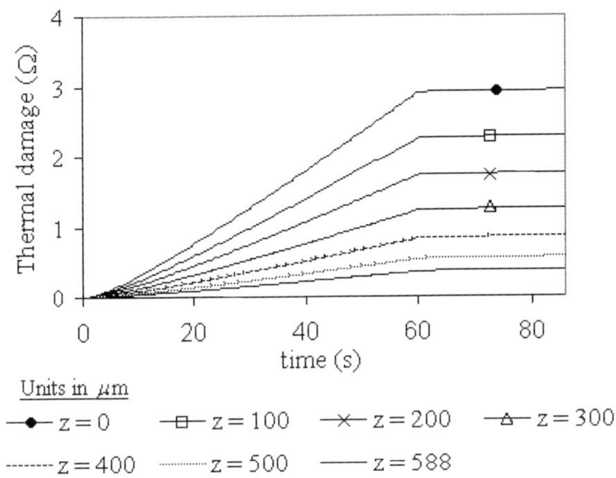

Fig. 12.15. Thermal damage inside the cornea during minute coagulation of continuous-wave laser radiation.

minute coagulation. This smaller size is surprising given that the increase in corneal temperature during the 10-s coagulation is higher than during the minute coagulation. Similarly, larger thermal damage is predicted for the pulsed laser radiation than for the continuous-wave laser radiation, since temperature increases during pulsed laser treatment are greater than during the continuous-wave laser.

From Fig. 12.14, it is established that the 10-s coagulation produces irreversible thermal damage inside the cornea. The irreversible thermal damage occurs only at depths up to 200 μm from the corneal surface, however. In the minute coagulation, complete necrosis is found at depths of $0 \leq z \leq 300\ \mu$m.

12.8. Discussion

The numerical results obtained from the present study are validated with the experimental and numerical data obtained that can be found in literature provided by other researchers. Generally, the temperature profiles of the cornea during pulsed laser radiation shown in Figs. 12.4 and 12.6 agree with those predicted by Manns *et al.*[12] Comparisons of the transient variations of the maximum (at $z = 0$) and minimum temperatures ($z = 588\ \mu$m) in Fig. 12.4 with the experimental results presented by Papaioannou *et al.*[2]

also show good qualitative agreement. The corneal temperatures measured by Papaioannou *et al.*[2] however, are smaller than the predictions obtained from the present study.

In continuous-wave laser radiation, the temperature distributions in Figs. 12.11 and 12.12 show isotherms that agree with the predictions obtained by Brinkmann *et al.*[10] The magnitudes of temperature, however, are different. In the present study, the largest corneal temperatures during the 10-s and minute coagulations are found to be 75.6 and 70.1°C, respectively, while Brinkmann *et al.*[10] predict values of 110 and 90°C, respectively. Discrepancies may be due to the absence of intra-corneal focusing in the present study, which was taken into account by Brinkmann *et al.*[10] The use of the simplified model approach may also be a contributing factor.

One of the differences between the model in the present study and those found in the literature is the geometry of the cornea. In earlier studies, the assumption that the cornea can be modeled as a cylinder or a tissue slab with finite thickness has been highly idealized. In the present study, the model is anatomically and physiologically more complete and realistic where various components of the human eye have been taken into consideration. The use of a complete model provides a more accurate anatomical representation of the human cornea. At the same time, boundary conditions can be specified based on the physical observation of the actual human eye. In maintaining the axisymmetrical feature of the human eye model, we have assumed that only one laser spot is applied to the center of the corneal surface. While this single spot does not represent the actual treatment of LTKP, the assumption is not expected to introduce significant errors to our analysis, since the corneal temperature changes are localized to the area where heating takes place. The thermal response of the cornea in each laser spot is not affected by the temperature changes in the other parts of the cornea where the other laser spots are applied.

From the numerical results presented in Sec. 12.5, the pulsed laser has been found to cause a greater increase in the corneal temperature than the continuous-wave laser causes. In particular, the temperature at the corneal epithelium has been predicted to be as high as 111°C, which may imply a total destruction of the epithelial layer. Based on the corneal temperature alone, one may be inclined to deduce that the pulsed laser produces greater thermal damage on the cornea than the continuous-wave laser. On the contrary, the assessment of thermal damage carried out using the Arrhenius integral appears to suggest otherwise. In Sec. 12.6, the amount of thermal damage suffered by the cornea during continuous-wave laser radiation is estimated to be greater than the pulsed laser radiation. This observation may be partially due to the duration of heating.

From Eq. (12.16), we find that the degree of thermal damage inside the cornea depends on both the temperature and the duration of heating. In the treatment of LTKP using pulsed laser, the duration when heating actually takes place is only a few milliseconds. Consequently, while the cornea is subjected to intense heat during each laser pulse, the period when heating actually takes place is short. On the other hand, the continuous exposure of the cornea toward heat makes it more susceptible to thermal damage. This explanation is further supported by the results presented in Figs. 12.14 and 12.15, where the minute coagulation that has a prolonged heating duration, causes a more severe thermal damage than the 10-s coagulation despite the smaller increase in corneal temperature during the minute coagulation. Results from the thermal damage assessment suggest that the duration of heating plays a more important role than the magnitude of temperature increase in determining the degree of thermal damage inside the cornea during LTKP.

According to Jean *et al.*[16] continuous-wave lasers are better suited to treat LTKP than pulsed lasers are. This suitability has been attributed to the smaller corneal temperature, which helps to avoid overheating and corneal relaxation. The ability of continuous-wave lasers to produce various coagulation depths (due to the tunability of the laser wavelength) and the consumption of less energy loss than pulsed laser also makes the continuous-wave laser a preferred choice. If the temperature profile of the cornea is used for determining the type of laser that is best suited for LTKP, the results obtained from the present study would then support the conclusion drawn by Jean *et al.*[16] However, if one considers the degree of thermal damage to be a more important criterion, the results from the thermal damage assessment would suggest that the pulsed laser is a better choice, since the thermal damage induced onto the cornea is less severe.

It may be premature, in this study, to draw a conclusion on the type of laser that is best suited for the treatment LTKP, since the degree of corneal shrinkages has not been taken into consideration. Additionally, the accuracy of the thermal damage estimated in the present study depends on how well the Arrhenius damage integral models the thermal damage inside the human cornea. While the thermal damage of the human skin may be accurately described,[32] the same may not be said of the cornea. In the present study, the severity of the thermal damage inside the cornea has been categorized based on the classification for the human skin. More studies have to be carried out in the future to determine if such a classification can accurately represent the severity of the thermal damage inside the cornea.

The model in the present study can be further improved by taking into account the effects of intracorneal focusing, such as in the work carried out by Brinkmann and *et al.*[8–10] Some issues that we have presented in this study, which regard the eye model, should be addressed in future studies. During LTKP, shrinkages alter the dimension and shape of the cornea. This aspect has not been considered in the present study. Inclusion of this feature, which involves a moving boundary problem, may produce results that are different from those presented in Sec. 12.5.

The absorption coefficient of the cornea has been taken to be the same as the absorption coefficient of water, which we assume to be constant. Previous studies, however, have shown this assumption to be misleading.[35] According to Ith *et al.*[35] the absorption coefficient of water is subjected to dynamic changes during laser–tissue interaction. Some studies have found the accuracy of simulated results to be severely affected by the assumption of constant optical properties.[36,37] The variation of the thermal conductivity of the cornea with temperature has also been reported. According to Bhattacharya and Mahajan,[38] the thermal conductivity of the cornea increases linearly with temperature. This linear increase in temperature has not been considered in the present study, since the relationship was obtained using sheep cornea and is only valid for heating temperatures up to 55°C. More research has to be carried out in future studies to examine the effects of these factors on the thermal response of the cornea during LTKP.

12.9. Concluding Remarks

A model of the human eye has been successfully developed for simulating temperature changes inside the cornea during LTKP. Its accuracy is demonstrated by the good qualitative agreement between the numerical predictions of our model and those obtained from the literature. The use of the boundary element method helps to alleviate the high computational cost of solving a complete eye model by allowing discretization to be carried out only at the boundaries. In the axisymmetric formulation, the boundary is given by a set of curves, which are easily discretized into straight-line elements.

Numerical results show that temperature increases during pulsed laser radiation are greater than during continuous-wave laser radiation. The duration that the corneal temperature actually exceeds 100°C is small, however. Intuitively, one may consider the thermal damage induced onto the cornea during pulsed laser radiation to be greater than continuous-wave laser radiation, due to the higher increase in temperature. Our assessment of thermal damage has shown otherwise. It appears that the duration of heating plays a more dominant role than temperature rise in determining the amount of thermal damage induced on the cornea during LTKP.

References

1. Brinkmann, R., Micka, D., Strunge, C., Nowack, C., and El-Hifnawi, E. Laser thermokeratoplasty: a comparative *in vitro* study with an Er:Glass and a Cr:Tm:Ho:YAG laser. In: Parel, J.M. (Eds.), *Ophthalmic Technologies II, Proc SPIE* **1644**: 294–300, 1992.
2. Papaioannou, T., Maguen, E., and Grundfest, W.S. Spatiotemporal temperature profiling of corneal surface during (LTK). In: Manns, F., Söderberg, P.G., and Ho, A. (Eds.), *Ophthalmic Technologies XII, Proc SPIE* **4611**:110–114, 2002.
3. Mainster, M.A. Ophthalmic applications of infrared lasers-thermal considerations. *Invest Ophthalmol Vis Sci* **18**:414–420, 1979.
4. Peppers, N.A., Vassiliadis, A., Dedrick, K.G., Chang, H., Peabody, R.R., Rose, H., and Zweng, H.C. Corneal damage thresholds for CO_2 laser radiation. *Appl Optics* **8**:377–381, 1969.
5. Mainster, M.A., White, T.J., and Tips, J.H. Corneal thermal response to the CO_2 laser. *Appl Optics* **9**:665–667, 1970.
6. Valderrma, G.L., Fredin, L.G., Berry, M.J., Dempsey, M.J., and Harpole, G.E. Temperature distributions in laser irradiated tissues. In: Jacque, S.L. (Eds.), *Laser-Tissue Interaction II, Proc SPIE* **1427**:200–213, 1991.
7. Zhou, Z., Ren, Q., Simon, G., and Parel, J.M. Thermal modeling of laser photo thermo keratoplasty (LTKP). In: Parel, J.M. (Eds.), *Ophthalmic Technologies II, Proc SPIE* **1644**:61–71, 1992.
8. Brinkmann, R., Koop, N., Dröge, G., Grotehusmann, U., Huber, A., and Birngruber, R. Investigations on laser thermokeratoplasty. In: Melamed, S.T. (Eds.), *Laser Applications in Ophthalmology, Proc SPIE* **2079**:120–130, 1994.

9. Brinkmann, R., Geerling, G., Kampmeier, J., Koop, N., Radt, B., and Birngruber, R. Laser thermokeratoplasty: analysis of in vitro results and refractive changes achieved in a first clinical study. In: Laffitte, F., Hoenigsmann, H., Fercher, A.F., Birngruber, R., Hibst, R., Fante, M.D., and Krasner, N. (Eds.), *Medical Applications of Lasers in Dermatology, Ophthalmology, Dentistry, and Endoscopy*, *Proc SPIE* **3192**:180–186, 1997.
10. Brinkmann, R., Radt, B., Flamm, C., Kampmeier, J., Koop, N., and Birngruber, R. Influence of temperature and time on thermally induced forces in corneal collagen and the effect on laser thermokeratoplasty. *J Cataract Refract Surg* **26**:744–754, 2000.
11. Manns, F., Borja, D., and Parel, J.M. Calculation of corneal temperature and shrinkage during laser thermokeratoplasty (LTK). In: Manns, F., Söderberg, P.G., and Ho, A. (Eds.), *Ophthalmic Technologies XII*, *Proc SPIE* **4611**:101–109, 2002.
12. Manns, F., Borja, D., and Parel, J.M. Semianalytical thermal model for subablative laser heating of homogeneous nonperfused biological tissue: application to laser thermokeratoplasty. *J Biomedical Optics* **8**:288–297, 2003.
13. Borja, D., Manns, F., Lee W., and Parel, J.M. Kinetics of corneal thermal shrinkage. In: Manns, F., Söderberg, P., and Ho, A. (Eds.), *Ophthalmic Technologies XIV*, *Proc SPIE* **5314**:79–87, 2004.
14. Henriques, F.C. and Moritz, A.R. Studies of thermal injury, I. The conduction of heat to and through skin and the temperatures attained therein. A theoretical and experimental investigation. *Am J Pathol* **23**:531–549, 1947.
15. Stringer, H. and Parr, J. Shrinkage temperature of eye collagen. *Nature* **204**:1307, 1964.
16. Jean, B., Bende, T., Oltrup, T., Matallana, M., and Walker, R. Depth adjusted thermal keratoplasty using a cw diode laser and a new focusing handpiece. In: Parel, J.M., Joos, K.M., and Rol, P.O. (Eds.), *Ophthalmic Technologies VI*, *Proc SPIE* **2673**:61–67, 1996.
17. Spoerl, E., Schmalfuss, K., Genth, U., Seiler, T., and Huebscher, H.J. Thermomechanical behaviour of the cornea. *Invest Ophthalmol Vis Sci* **36**:39, 1995.
18. Wirbelauer, C., Geerling, G., Koop, N., Brinkmann, R., Tüngler, A., Birngruber, R., and Laqua, H. Acute endothelial cell changes after laser thermal keratoplasty with a cw-IR laser diode. *Invest Ophthalmol Vis Sci* **38**:541, 1997.
19. Henderson, R. and Schulmeister, K. *Laser Safety*, Taylor and Francis Group, New York, 2004.
20. Charles, M.W. and Brown, N. Dimensions of the human eye relevant to radiation protection. *Phys Med Biol* **20**:202–218, 1975.
21. Fontana, S.T. and Brubaker, R.F. Volume and dimensions of the anterior chamber of the normal aging human eye. *Arch Ophthalmol* **98**:1803–1808, 1980.
22. Ng, E.Y.K. and Ooi, E.H. FEM simulation of the eye structure with bioheat analysis. *Comput Methods Programs Biomed* **82**:268–276, 2006.
23. Ooi, E.H., Ang, W.T., and Ng, E.Y.K. Bioheat transfer in the human eye: a boundary element approach. *Eng Anal Bound Elem* **31**:494–500, 2007.
24. Cicekli, U. Computational model for heat transfer in the human eye using the finite element method. M.Sc. Thesis, Department of Civil and Environmental Engineering, Louisiana State University, 2003.
25. Scott, J.A. A finite element model of heat transport in the human eye. *Phys Med Biol* **33**:227–241, 1988.

26. Neelakantaswamy, P.S. and Ramakrishnan, K.P. Microwave-induced hazardous nonlinear thermoelastic vibrations of the ocular lens in the human eye. *J Biomech* **12**:205–210, 1979.
27. Emery, A.F., Kramar, P.O., Guy, A.W., and Lin, J.C. Microwave induced temperature rises in rabbit eyes in cataract research. *J Heat Transf* **97**:123–128, 1975.
28. Lagendijk, J.J.W. A mathematical nodel to calculate temperature distributions in human and rabbit eyes during hyperthermic treatment. *Phys Med Biol* **27**:1301–1311, 1982.
29. Brebbia, C.A., Telles, J.C.F., and Wrobel, L.C. *Boundary Element Techniques: Theory and Applications in Engineering*, Spinger-Verlag, Heidelberg, 1984.
30. Abramowitz, M. and Stegun, I.A. *A Handbook of Mathematical Functions with Formulas, Graphs, and Mathematical Tables*, Dover Publications, New York, 1972.
31. Ooi, E.H., Ang, W.T., and Ng, E.Y.K. A boundary element model of the human eye undergoing laser-thermokeratoplasty, *Comput Biol Med* **38**:727–737, 2008.
32. Diller, K.R. Analysis of skin burns. In: Shitzer, A. and Eberhart, R.C. (Eds.), *Heat Transfer in Medicine and Biology: Analysis and Applications*, Volume 2, Plenum Press, New York, pp. 85–137, 1995.
33. Kampmeier, J., Radt, B., Birngruber, R., and Brinkmann, R. Thermal and biomechanical parameters of porcine cornea. *Cornea* **19**:355–363, 2003.
34. Incropera, F. and DeWitt, D.P. *Fundamentals of Heat and Mass Transfer*, Fifth Edition, John Wiley and Sons, New Jersey, 2002.
35. Ith, M. Frenz, M., and Weber, H.P. Scattering and thermal lensing of 2.12 μm laser radiation in biological tissue. *Appl Optics* **40**:2216–2223, 2001.
36. Walsh, J.T. and Cummings, J.P. The effect of dynamic changes in the water absorption coefficient on mid-infrared laser ablation. *Lasers Surg Med* **15**:295–305, 1994.
37. Jansen, E.D., Leeuwen, T.G., Motamedi, M., Borst, C., and Welch, A.J. Temperature-dependence of the absorption coefficient of water for mid-infrared laser radiation. *Lasers Surg Med* **14**:258–268, 1994.
38. Bhattacharya, A. and Mahajan, R.L. Temperature dependence of thermal conductivity of biological tissues. *Physiol Meas* **24**:769–783, 2003.

Chapter 13

Optical Eye Modeling and Applications

Ying-Ling Chen*, Bo Tan*,†, Lei Shi*,
J.W.L. Lewis‡ and Ming Wang§

13.1. Introduction to Optical Eye Modeling

In recent years, ocular optics has advanced dramatically in three areas. First, the developing interest in the study of disease progression in addition to the emmetropization in the growing eyes has resulted in the demand for more accurate and noninvasive measurement methods. Second, advances in optical detector and analysis technologies have resulted in instruments with increased accuracy for the measurement of essential parameters of the physiological optical system. Third, dramatic improvements in computational capability have enabled more complex and accurate optical computations and simulations of the eye. The high precision, ophthalmic patient data (including ocular element biometry), topographies of the corneal surfaces, and ocular wavefront aberrations (WFAs) can be incorporated into the computational construction of a personal-tailored, optically functional, and analytical eye model. The results are eye models that render the exact clinical measure and optical performance of the eye. Customized eye modeling may assist in the planning of ocular surgeries, such as LASIK, PRK, CK, and Intacs, and in the design of personalized spectacles and contact or

*University of Tennessee Space Institute, Tullahoma, TN 37388.
†University of California at Berkeley, School of Optometry, CA 9472.
‡E-Vision Technologies, Inc. Tullahoma, TN 37388.
§Wang Vision Institute, Nashville, TN 37203.

intraocular lenses. Another promising modeling direction is the population-based eye modeling that can be established using traditional eye-modeling techniques with the statistical information and correlations of the ocular parameters of the targeted population. Simulating ocular device measurements including eye models could be useful, for not only rapid prototyping and sensitivity evaluations of ocular instruments, but also for the teaching and training of medical personnel. Computational personalized eye modeling techniques also allow the integrated information to be quantitatively evaluated for disease development studies without the repetitive examination of human subjects. The revelations that the ocular optics may be clinically altered have sped the renovation of physiological optics from an observational to a quantitative, and possibly even a predictive, science.

The optical elements of an anatomically accurate eye model include the cornea, the anterior chamber, the iris (pupil stop), the posterior chamber, the crystalline lens, the vitreous chamber, and the retina. Schematic eye models in the early 20th century used spherical ocular elements and constant indexes of refraction.[1] Later in the 1980s, aspherical ocular elements and complex models of the eye lens were incorporated to represent better than average ocular monochromatic and chromatic imaging properties of the eyes. These general eye models were important to evaluate optical performance, to investigate ocular properties, and to design new ophthalmic corrections, including spectacles and contact lenses. Among the current published generic eye models, the wide-angle Navarro model[2] was built by incorporating published anatomical values of conic constant into the Gullstrand-Le Grand spherical surfaces[3] and by updating the values of the anterior radius and refractive index of the cornea. This model has been shown to produce on-axis image quality, as well as off-axis aberrations that are in agreement with mean human measurements, and, therefore, this model will be used in this chapter as the initial base emmetropic model for demonstration purposes.

13.1.1. *Ocular Measurements for Optical Eye Modeling*

The numerical measurement of the ocular biometry and optical parameters are essential for mathematical eye modeling. The ophthalmic measurements that are useful for modeling are briefly reviewed below.

13.1.1.1. *Curvature, dimension, thickness, or distance parameters of ocular elements*

Because of the difficulty and intrusiveness of reaching the internal ocular tissues *in vivo*, the noninvasive reflection signals of light or sound waves from the interfacial surfaces of different ocular layers are typically used to provide information about the ocular dimensions and locations. These techniques include the ultrasoundgraphy A- and B-scans, the pachymeter, which measures corneal thickness, partial coherence interferometry (PCI) techniques, such as the IOL mater, OCT devices including Cirrus™ HD-OCT, the RTVue-100 Fourier-Domain OCT, the Visante OCT, and the Spectral OCT/SLO, MRI devices, spectacular microscopy, and keratometer.

13.1.1.2. *Three-dimensional (3D) corneal topography*

Corneal topography, also known as photokeratoscopy or videokeratography, is a noninvasive technique for imaging the surface curvature of the cornea. Since the cornea is typically responsible for about 70% of the eye's refractive power, its surface condition, especially the anterior interface, is of critical importance in both determining the quality of vision and in eye modeling work. The topographic techniques include the Placido disk imaging, such as Humphrey® Atlas™ corneal topography systems and TMS-4a Topographer, the slit-scanning system of Orbscan, and the Scheimpflug imaging of Pentacam. The Orbscan and the Pentacam devices provide both anterior and posterior corneal elevation maps as well as mapping the thickness of cornea, and, therefore, are especially valuable for the eye modeling.

13.1.1.3. *Crystalline lens parameters*

The crystalline lens can be measured *in vitro* or *in vivo*. Ultrasound, OCT, and Purkinje images are the most popular *in vivo* methods. Though the first two methods can measure curvatures and thicknesses, they have the disadvantages of touching the cornea and being slow in measurement. Purkinje images are the sequence of reflections from the refracting surfaces of the eye elements. PI, PII, PIII, and PIV indicate the refraction images formed at the boundary of air/cornea (or tear), posterior cornea/aqueous, aqueous/anterior lens, and posterior lens/vitreous, respectively. The Purkinje images method can measure curvatures while the eye is viewing a target and can, therefore, monitor accommodation variations.

13.1.1.4. *Refractive index*

Refractive indexes of ocular elements are not often measured in clinical environments. However, they do not vary significantly across individuals. The human lens has a gradient refractive index. The Liou and Brennan eye model[4] uses a gradient index to mimic the measured data. For simplified computation, an "equivalent" constant index is often used instead of the actual gradient index. The Navarro 1985 model used 1.42 for relaxed eye, and this model describes the index as a function of accommodation. The Atchison 2008 study used Purkinje imagery to describe a lens equivalent refractive index as a function of age and applied the parameter in the four-surface eye model.[5] Unless age or accommodation factors are encountered, most eye-modeling work simply adapts the constant values of refractive index.

13.1.1.5. *Wavefront aberration*

In recent years, WFA has become a popular method for evaluating monochromatic performance of the human eye and is sometimes described as "the fingerprint of the individual eye." In physical optics, the wavefront represents the surface locus of points that have the same optical path or phase from the light source. The 2D function of WFA describes the point-to-point departure from perfect wavefront, normally a spherical surface, in the exit pupil. In ocular optics, WFA becomes an important addition to the conventional refractive error measurement. The second-order WFA corresponds to the refractive error, and high-order WFA is responsible for reduced visual acuity (VA) in spite of an optimal correction from spherical or cylindrical refractive lens. VISX WaveScan Wavefront System and the Allegro Analyzer are examples of commercial devices of the Shack-Hartmann and the Tscherning types of aberrometer, respectively. The clinically measured WFA of an eye can be used for customized eye modeling.

13.1.2. *Eye Modeling Using Contemporary Optical Design Software*

Most of current generic eye modeling research utilizes optical design software, such as ZEMAX, Code V, and OSLO, for both the construction of models and the subsequent applications in optical engineering. These programs assist the design of optical systems by providing optical optimization and analysis that is based on ray tracing technology. The optical parameters of an optical system or an eye model are entered in a spreadsheet format. In this contribution, ZEMAX will be used as an example of the modeling tool. Using parameters of Navarro eye model, Table 13.1 shows the form of the lens data editor in ZEMAX. From top to bottom in this table, the rows describe the light source (OBJ, object), the two surfaces of the cornea (surfaces 1 and 2), the pupil of the iris (STO; aperture stop), the two crystalline lens surfaces (surfaces 4 and 5), and the imaging surface of retina (surface IMA). From left to right, the first column "Surf: Type" allows the user to select many surface types from ZEMAX. The most commonly used optical surface is an aspherical surface named "Standard Surface." The required two parameters of a "Standard Surface" are radius of the curvature (R) and conic constant (Q). ZEMAX treats planes as a special case of the sphere (i.e. a sphere with infinite radius of curvature). The surface is centered on the "current" optical axis, with the vertex located at the "current" Z-axis position unless otherwise specified. The "sag" or z-value of the standard

Table 13.1. Parameters of Navarro eye model in ZEMAX lens editor.

Surf:	Type	Radius	Thickness	Glass	Semi-Diameter	Conic
OBJ	Standard	Infinity	Infinity		0.00	0.00
1*	Standard	7.72	0.55	CORNEA	5.00	−0.26
2*	Standard	6.50	3.05	AQUROUS	5.00	0.00
STO*	Standard	Infinity	0.00	AQUROUS	3.00	0.00
4*	Standard	10.20	4.00	LENS	5.00	−3.1316
5*	Standard	−6.00	16.32	VITREOUS	5.00	−1.00
IMA*	Standard	−12.00	—	VITREOUS	12.00	0.00

surface is given by:

$$z = \frac{cr^2}{1 + \sqrt{1 - 1(1 + Q)c^2r^2}},$$

where c is the curvature (the reciprocal of the radius R), r is the radial coordinate in the "lens unit," and Q is the conic constant. The default "lens unit" is millimeters. The radius of the surface vertex curvature is entered in the second column, "Radius," in mm. The conic constant, Q, is assigned at the sixth column. If the conic constant is less than -1, then it describes a hyperbolic surface. If it is -1, then it describes parabolas. If it is between -1 and 0, then it describes ellipses. If it is 0, then it defines spheres, and, if it is greater than 0, then it depicts oblate ellipsoids. As shown in Fig. 13.1, the colored lines illustrate the anterior corneal surfaces for different conic constants with the same corneal radius of curvature, $R = 7.72$ mm. The small influence of conic constant at the periphery of the cornea is not visible to human eye. Though the human corneal surface extends roughly 5.5 mm in radius, the most effective visual zone falls inside the center 2 mm of radius due to the

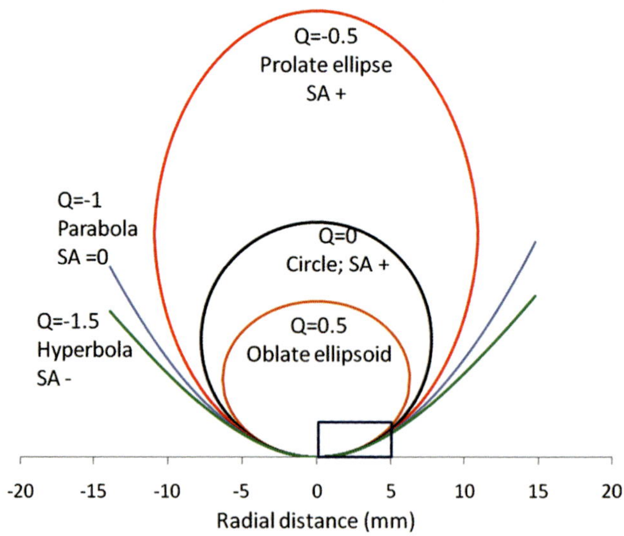

Fig. 13.1. Standard surfaces that are described with a curvature of radius = 7.72 mm and various conic constants.

limitation of the pupil stop. Although the conic constant does not seem to cause much variation inside the 2 mm visual zone, it produces significant spherical aberration (SA) and affects the imaging quality appreciably.

The third column in the table, "Thickness," expresses the distance from the vertex of the present surface to the vertex of the next surface in millimeters. The fourth column, "Glass," represents the refractive index data of the material between the current surface and the next surface. For each "glass" name that is specified in the table, the corresponding refractive index parameters must be prepared and entered in one of the currently loaded glass catalogs in the ZEMAX program. If the optical computation considers multiple wavelengths, the data should include dispersion information over the spectral range. The fifth column "Semi-Diameter," (diameter/2) describes the aperture size of each surface. Columns after the sixth describe the decentering of the apex and the tilting parameters of the surface. Since all the surfaces in Navarro model are centered and symmetric to the optical axis as well as most optical system, they are not shown in Table 13.1.

After the data are entered in the lens data editor, the analysis tools of ZEMAX can be used to illustrate the result. Figure 13.2 shows a typical 3D layout of an eye model in ZEMAX. With an eye model constructed in ZEMAX, light rays can be traced from the object space (OBJ) sequentially through the system to the image plane (IMA), i.e. the retina, according to the Snell Law. Optical analysis, including point-spread function (PSF), WFA, and spot diagram (SPD), are available in ZEMAX for examining the optical

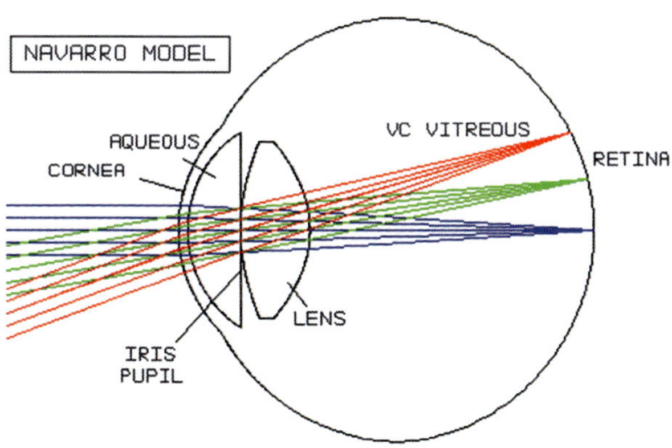

Fig. 13.2. A 3D layout of Navarro eye model in ZEMAX.

performance. If the optical performance does not meet the required target or purpose, "Optical optimization" can be used to approach the target. With specified merit functions, ZEMAX uses a numerical algorithm to perform the optical optimization iteration until the specified target criteria are met. The validation of an eye model is determined by the closeness of the optical performance of the model to the target eye. In general, analysis is performed on the aberrations, and the final model is examined using the SPD, the point-spread function, and the modulation transfer function.

A systematic general eye modeling procedure can be found on the ZEMAX Web site.[6] The ZEMAX modeling of a more complex Liou and Brennan model[4] that uses a gradient refractive index lens can also be found on the Web site. In addition, a forward, a backward, and a nonsequential eye model module can be downloaded.[7] The sequential forward and backward models allow ray tracing to be performed in one direction, while the nonsequential model treats the eye model as a single optical element. Light rays are allowed to diffract/reflect multiple times on the same surfaces until they exit the element or are absorbed.

13.1.3. *Optical Optimization and Merit Function*

For more specific or customized eye modeling, ocular parameters must be mathematically tailored in order to describe better the properties of the target eye. "Optical optimization" is the iteration algorithm that takes the initial optical design and changes the values of the assigned parameters (variables) in steps to approach the specified targets. The starting layout should have a suitable number of optical surfaces of appropriate types, since optimization can change only the values of the selected parameters, but not the number or types of surfaces. Optimization can be accomplished in three steps.

1. Construct a reasonable initial layout so that rays can be traced from the object plane to the image surface.
2. Specify free variables to be "optimized" and the corresponding tolerances to prevent unrealistic results or convergence to local minima.
3. Define the merit functions that describe the ultimate goals at the end of iteration.

Many previously published eye models can be used as the initial model. The parameters of a selected base model should be entered first. The variables of iteration, which are required for the optimization algorithm to progress, are specified next. Since optics are very precise (even distances of micrometers can make a big difference), we need to determine the values of all our variables at each step of the optimization carefully. The selection of variables is important for optimization. The less rigid parameters are assigned with more tolerance. In the eye modeling, variables are allocated on different ocular components at different modeling stages. After the variables are selected, suitable metrics are used as the indicators of progress of optical optimization. These metrics are defined as the merit functions. A merit function is a numerical representation of how closely the optimization result meets a specified set of goals. Usually, different merit functions will lead to different optimization results. Then, the final values of merit functions after optimization are indicators to evaluate the success of eye modeling. Therefore, the optimization and selection of merit functions are the most important process in the eye modeling procedure.

The optimization feature provided by the contemporary optical software programs is powerful. ZEMAX optimization uses an actively damped least squares or an orthogonal descent algorithm. The algorithms are capable of optimizing a merit function that is composed of weighted target values. ZEMAX has several default merit functions. For the majority of applications, the optical optimization is performed to achieve optimal imaging quality. In the other word, the default merit functions include attempts to minimize optical aberration or obtain the smallest focus spot (or PSF).

In eye modeling applications, the goals of optimizations are to produce a realistic human eye with personal clinically measured or validated ocular measurements. These specific merit functions are assigned using the Merit Function Editor in ZAMAX. If the clinical measured WFA map is available for an eye, the personalized eye modeling will aim to reproduce the exactly measured wavefront on the modeled eye. Since the clinical measured WFA data is typically expressed in Zernike polynomial coefficients, the merit function at the final optimization produces a wavefront of the exact series of Zernike coefficients. The ZEMAX operand, ZERN, which designates the intended set of Zernike coefficients of the target wavefront, will be used for this purpose. However, when the wavefront data are not obtained from the patient, the most common clinical eye examinee record, the sphero-cylindrical refraction prescription and the VA, will be the targets of optimizations.

ZEMAX default optimization is applied to improve the performance of wide-ranging optical systems. Generally, the goal is to produce the best imaging quality for the final optical system. The default merit functions are designed to approach either the minimum focus size or the spot radius in the SPD (i.e. the geometric optics approach) or the minimum aberration or root mean square WFA (RMS WFA, i.e. the wave optics approach). In eye modeling, RMS is typically used instead of peak-to-valley (PTV) optimization because of the weakness of PTV. The PTV approach considers only two points, the highest and the lowest, and ignores all points that lie between. Important issues, such as roughness are ignored, while a very small high or low point may be exaggerated beyond their significance. RMS greatly improves the PTV method since it takes into account areas on the optic that may vary when compared to the optic's general surface characteristics.

The numerical value of the merit function is physically significant when using RMS as the optimization type. If the merit function is RMS-Wavefront-Centroid, then the numerical value of the merit function is the RMS wavefront error in the unit of waves (λ). If the merit function is RMS-Spot Radius-Chief, then a value of 0.145 means the RMS spot radius is 0.145 lens units. If the lens units were millimeters, the RMS spot radius will correspond to a focus radius of 145 micrometers RMS. If more than one field or wavelength is defined, then the merit function numerical value is the weighted average of the RMS values for the various fields and wavelengths.

Note that optimization using the RMS spot radius merit function will, in general, yield an optimum design different from the RMS wavefront merit function. The reason for this difference in design is that ray aberrations are proportional to the derivative of the wave aberrations. Therefore, it is unreasonable to expect that the minimum of one aberration corresponds to the minimum of the other aberration. A general rule is to use wavefront error if the system is close to diffraction limited (for instance, a PTV wavefront error of less than two waves). Otherwise, use the spot radius.

In the eye modeling work, a typical focus size is larger than 2 micron for green light (555 nm). This size is derived by $1.22 * \lambda * f/d$, where f is the equivalent focal length of the eye (17 mm), and d is the pupil diameter (~6 mm). Therefore, usually, SPD is used first to run optimization. Then at the end of the optimization, as the system approaches close to the diffraction limit, the optimization is changed to WFA, and a "fine-tune" optimization is performed to get the desired system performance.

The default numbers of Rings and Arms in the ZEMAX optimization are 3 and 6. However, an eye model with a clinically measured corneal topography does not present a well-behaved rotational symmetry as those of the typical industrial optical elements. To prevent falling into local minimum, higher numbers of arms and rings are generally used in the optimization. Both damped least squares and orthogonal descent algorithms are usually performed and compared for the same reason.

13.2. Personalized and Population-Based Eye Modeling

The general eye models that were described were based on average ocular parameters obtained from young emmetropic adults. These models are generally rotationally symmetric, centered, and aligned surfaces, whereas the real eyes show degrees of irregularities with no well-defined optical or symmetry axes. In other words, both monochromatic aberration and transverse chromatic aberration are known to vary widely across subjects. This section considers more specific eye modeling.

13.2.1. *Customized Eye Modeling*

Customized eye models are based on the clinical measurement of the individual human eye. In the recent years, high precision ophthalmic patient data have become available to characterize anterior and posterior surfaces of the cornea, ocular WFAs, and ocular element biometry accurately. These measurements can be incorporated into the construction of an optically functional and analytical, personal-tailored, eye model. Figure 13.3 shows the Pentacam data of layers of corneal topography that can be used for modeling. In 2004, Navarro published the first construction of 19 personalized eye models of healthy eyes with the optical design software, CodeV.[8] The concept of personalized eye modeling is displayed in Fig. 13.4. The patient data may include modeling parameters, such as refractive index and ocular geometric data, and information that evaluates the optical performance, such as total WFA, refractive error measurement, and VA. It is

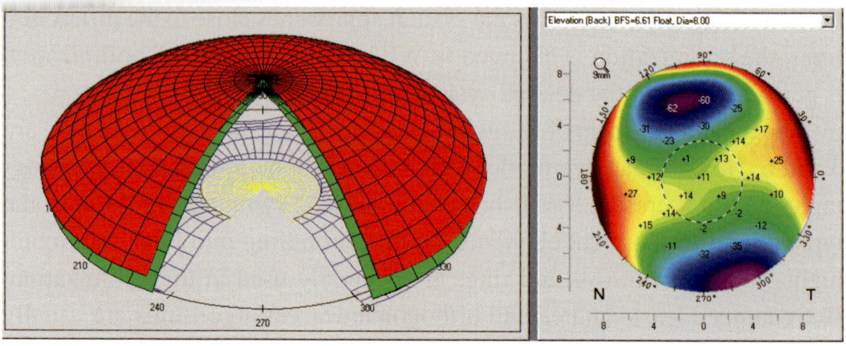

Fig. 13.3. Pentacam data of corneal surfaces for eye modeling.

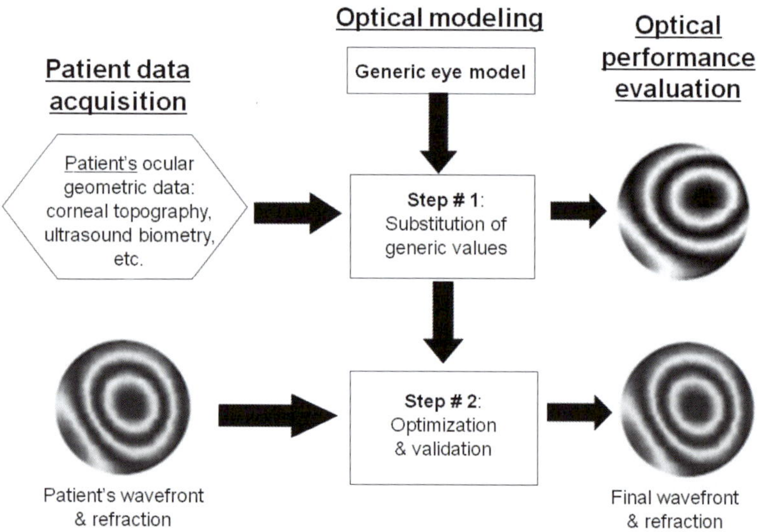

Fig. 13.4. Schematic diagram of the method of personalized eye modeling.

almost guaranteed that a generic eye model is needed as the base model that provides a draft functioning model with a complete set of mean optical parameters. For an adult eye, we have used the emmetropic Navarro model of 1985 that uses constant index refraction for the lens rather than the more computationally burdensome gradient index. The unaccommodated state model was based on the Gullstrand-LeGrand model with modifications in the asphericities (conic constants) of the refractive surfaces as well as the radius of the anterior corneal surface and the dispersions of the refractive indexes to fit the longitudinal chromatic aberration. This model varies continuously with accommodation and reproduces the overall average optical performance (aberrations, polychromatic MTF and PSF) of the eye both on-axis and off-axis remarkably well.

As shown in Fig. 13.4, after constructing the base model, the first step in optical modeling is the substitution of the optical and geometric parameters with the clinically measured data. Most research groups that perform eye modeling assume that the refractive indices are constants among individuals and are equal to those of the generic model used. However, with the published research results of the age dependence of refractive indices, it is feasible to adapt age-corrected refractive indices, as well as many other age dependent parameters. The clinically obtained geometric data of the patient's eye will remain invariable parameters throughout the modeling process. One potentially important aspect is the dependence of the result on the initial generic model used, because the final personalized model will depend on those parameters of the initial model that are not changed (iterated) during the optimization process.

The replacement of the thicknesses data of ocular elements obtained from the ultrasound biometry is straightforward. The topography substitution requires description. The corneal topographic data exported from the clinical instrument require mathematical interpolation and extrapolation to obtain a sufficient area, namely a 10-mm diameter, of an anterior corneal map. The topography map(s) should be presented in elevation instead of power. The corneal map obtained in the clinic often contains missing data points due to the interference of eyelids and eyelashes. C^{++} can be used to create a ZEMAX-readable grid sag from this corneal map. Before the insertion of the user-defined surface, the grid sag surface should be selected in the anterior corneal surface in the lens-data-editor of the draft model. Misalignment between separate sets of clinical data should be prevented during the measurements and minimized in the modeling.

After the step 1, an initial personalized eye model is formed. The second step is the validation of this model through the optical optimization to approach the expected optical performance of the patient vision. As mentioned earlier, the most commonly acquired clinical eye data that indicates vision quality are the refractive errors and the VA. The VA measurement gives an estimated minimum requirement for the focus size of retinal blur (i.e. the size of the PSF). Refractive error is the second-order aberration that corresponds to the three variables that are related to sphero-cylindrical refraction prescription (S, C, X).

13.2.1.1. *Optimization to the refractive error*

When a specified refractive error of a model eye is requested, unless the ocular axial length is known for the individual eye, the optimization may be performed in two steps. Because the axial length of the human eye is the dominant factor of the spherical refractive error, the first step of optimization may require assigning the axial length as a free variable in the iteration. Anterior chamber depth (ACD) and lens thickness (LT) are generally age related and less dependent on refractive error. ACD and LT may be modified from the base model according to the age if these parameters are not obtained from the clinic. The free variable that is used to achieve the desired spherical equivalent refraction is normally the vitreous chamber depth (VCD). The merit function at this stage is the demand for an exact spherical equivalent. An ideal Gaussian thin lens of the desired refractive correction will be entered in front of the cornea of the model. For example, to achieve a myopic (hyperopic) eye of 5D, a −5D (+5) thin lens will be added. With the correction lens in place, we expect that the typical optical optimization such as the use of default merit function in ZEMAX will bring the final model to the optimized image quality. The object (light source) is set at infinity. The point source at infinity will be best focused on the retinal surface after iteration. In the second optimization step, the cylindrical refractive error will be achieved. The correcting Gaussian thin lens that is anterior to the corneal surface must be replaced with the exact refractive error prescription of all three parameters, including astigmatism. Therefore, at least three free variables need to be assigned for the optimization. For the cylindrical refractive error, both crystalline lens and corneal are the possible sources. Axial length plays no role here. If the corneal topography was assigned from the patient's clinical record, then the corneal surface parameters are fixed. The free variable can only be placed at lens parameters such as surface shape, de-center, or tilt. Another simple approach to simulate or approximate reality is to add a thin lens with three variables near the crystalline lens position. Axial length remains constant during this stage of optimization. After the iteration, the three variables are optimized to produce the sharpest retinal image from a point source at infinity.

One important issue that deserves much attention in the refractive error optimization is the pupil diameter. With the presence of aberration, the refractive error always varies with the pupil size. A typical example of such variation is the existence of (positive and negative) spherical aberration (SA) that induces (myopic and hyperopic) refractive error in the periphery. As the pupil size increases (which occurs in darkened environments and night vision), the eye tends to be shifted myopically (or hyperopically). For this reason, when performing the optimization in either step for refractive error, a reasonable pupil size of 3–4 mm, which is about the condition for typical reading or eye examinations should be assigned before the iteration. When the pupil diameter is set as 3.0 mm, the entrance pupil will be approximately 3.33 mm due to the magnification from the anterior chamber and cornea. In clinical language, the pupil size means the "pupil appearance," which is the entrance pupil in the optical system, not the physical pupil size. After the two-step optimization is performed, the correcting Gaussian thin lens in front of the cornea should be removed. An eye model with appropriate refractive error is then obtained.

13.2.1.2. *Optimization to the wavefront measurement*

In the recent years, the WFA map is available in clinics. WA provides not only the second-order aberration, but also the aberration information to the sixth or seventh-order. If the WFA measurement of the eye is available, this clinical vision assessment could be used for the final targeted merit functions. Variables will be assigned in the initial personalized model for iteration to obtain the optical quality target values given by the merit function. If the WFA map is obtained from the patient eye, it will be assigned as the merit function in the final optimization. A backward eye model will be used and the wavelength (typically infrared) of the wavefront aberrometer should be assigned. If the wavefront data are available, the second optimization for cylindrical refractive error is not required. The optimization for axial length is still sensible if it is not a known condition. At the beginning of the optimization, the type of anterior lens surface should be selected as the Biconic Zernike and the Zernike coefficients will be assigned as free variables for iteration. In this step, the merit function will be changed from SRX to Zernike coefficients. The optimization target is to approach the measured WFA, which is reported in the Zernike polynomial format. The merit function will require the optimization operand "ZERN." The parameters may be set as Term = 1, 2. . . in the order of Zernike coefficients in ZEMAX, Wave = 1 (only one wavelength used in each of the calculations), Samp = 2 (pupil sampling = 64 $*$ 64), field = 1 (only one field set in our calculations), and Type = 1 (Zernike standard coefficient), and Zernike coefficients of the clinical WFA will be input at the column of the "target" values, and the weight of each coefficient will be set equally. Figure 13.5 shows the comparison of the measured and the final wavefront after optimization in a test run. The result reveals a successful determination of a personalized eye model that has the same anterior corneal map and WFA as the individual subject.

13.2.1.3. *Tolerance analysis*

The method described above has intrinsic uncertainties in the sense that the whole lens geometry is initially unknown. The geometry is then adjusted to fit the clinically measured optical performance with the sole constraint that the result is as close as possible to the initial base model. This uncertainty is an optical design problem, in which it is essential to perform a tolerance

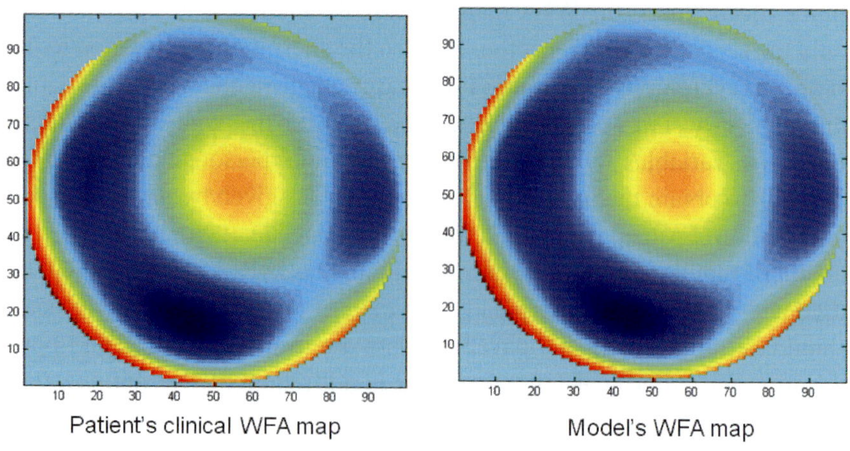

Fig. 13.5. Comparison of measured and reproduced WFA of an eye and eye model.

analysis to determine how critical the optimized values of the variables are for the prediction of the total wave aberration of the eye. Navarro has developed the following procedure for this particular optical design problem: once the optimization algorithm finds the minimum of the merit function, 1D plots of this merit function are obtained in the vicinity of the optimal or minimal value versus each independent variable (such as surface curvature, conic constant, decentrations, and tip/tilt angle in Navarro's study). Here, we perform a similar procedure. Once the optimization algorithm finds the minimum of the merit function, we obtain plots of this merit function against the varied RMS for each order of Zernike coefficients of anterior lens around the optimal (minimum) value. From these plots, we can see which order is the most significant and dominant (the smallest tolerance).

13.2.2. *Population-Based Eye Modeling*

13.2.2.1. *Accommodative eye modeling*

Most schematic eyes represent emmetropic and relaxed adult eyes. Under accommodation demand for near vision, the ciliary muscles holding the crystalline lens tighten, thereby causing the lens to become more rounded. The thickness and curvatures on both surfaces of lens increase. In the history of eye modeling, a few accommodative models exist. The Gullstrand No. 1 model[1] is constructed at about 10.9 diopter accommodation. The Gullstrand-Emsley[9] and Le Grand models[10] present full schematic eyes that are in accommodated forms of 8.6 and 7.1 diopters, respectively. The Navarro model eye is "adaptive" in the sense that it offers the variability of lens parameters (the thickness, the radius of curvatures, and conic constants on both surfaces) and the anterior chamber and vitreous depth. In this popular eye model, these ocular parameters are given as functions of accommodation level. Atchinson summarized the four accommodative eye models in Appendix A3.[11]

As shown in previous studies,[12–15] the lens biometry is fundamentally independent of eye refractive error. However, in addition to the variation under accommodation, lens shape and refractive index are also significantly related to age. With an increase in age during adulthood, the lens becomes thicker and more curved in its relaxed state, and its refractive index distribution changes. Therefore, modeling the accommodative eye should be performed with specification of a limited age range.

13.2.2.2. *Ametropic eye modeling*

In 2006, Atchison published the optical models for myopic eyes[11] according to the analysis of statistical relevance obtained from the subjects and studies primarily of his research group. Table 13.2 compares the refraction dependence of the Atchison myopic eye model and the emmetropic Navarro eye model.[2] Not included in the table is the information regarding the decenters of the pupil and the lens, the tilt of the lens, and the fovea location that are also used in the models. These parameters contribute to ocular

Table 13.2. Comparison of parameters between Atchison myopia model[11] and Navarro emmetropic model.[2]

Ocular parameter	Model	Emmetropic condition (K = 0)	Refractive error (K) dependence
Anterior corneal radius of curvature	Atchison	7.77 mm	+0.022 mm/diopter
	Navarro	7.72 mm	
Asphericity of anterior cornea	Atchison	−0.15	Not significant
	Navarro	−0.26	
Central corneal thickness	Atchison	0.55 mm	Not significant
	Navarro	0.55 mm	
Index of refraction of cornea	Atchison	1.3975, 1.3807, 1.37405, 1.3668, for	
	Navarro	$\lambda = 365, 486.1, 656.3, 1014$ nm	
Posterior corneal radius	Atchison	6.40 mm	Not significant
	Navarro	6.50 mm	
Asphericity of posterior cornea	Atchison	−0.275	Not significant
	Navarro	0	
Anterior chamber depth	Atchison	3.15 mm	Not significant
	Navarro	3.05 mm	
Index of refraction of aqueous humor	Atchison	1.3593, 1.3422, 1.3354, 1.3278, for	
	Navarro	$\lambda = 365, 486.1, 656.3, 1014$ nm	

(Continued)

Table 13.2. (*Continued*)

Ocular parameter	Model	Emmetropic condition (K = 0)	Refractive error (K) dependence
Anterior lens radius	Atchison	11.48 mm	Not significant
	Navarro	10.20 mm	
Anterior lens asphericity	Atchison	−5	Not significant
	Navarro	−3.1316	
Lens thickness	Atchison	3.6 mm	Not significant
	Navarro	4.0 mm	
Refractive index of crystalline lens	Atchison	Gradient index	
	Navarro	1.4492, 1.4263, 1.4175, 1.4097, for $\lambda = 365, 486.1, 656.3, 1014$ nm	
Posterior lens radius	Atchison	−5.9 mm	Not significant
	Navarro	−6.0 mm	
Posterior lens asphericity	Atchison	−2	Omit; not significant
	Navarro	−1	
Vitreous chamber depth	Atchison	16.28 mm	−0.299 mm/diopter
	Navarro	16.32 mm	
Refractive index of vitreous humor	Atchison	1.3565, 1.3407, 1.3341, 1.3273, for	
	Navarro	$\alpha = 365, 486.1, 656.3, 1014$ nm.	
Radius of retina curvature	Atchison	−12.91 mm (x); −12.72 mm(y)	−0.094 mm/diopter (x); +0.004 mm/diopter (y)
	Navarro	−12 mm	

asymmetry and introduce astigmatism, coma, and irregular aberrations, and, therefore, have important effect upon optical performance. The significance of using these parameters depends on the types of applications.

13.2.2.3. *Modeling with consideration of ocular growth and aging*

Age is an important factor in ocular biometry. The ocular dimension increases significantly during the first year of life. During infancy, the ocular refraction develops from mild hyperopia to emmetropia. From the age of one year to adulthood, ocular dimensions continue to grow to their final values but at a much reduced rate. After adulthood, the outer dimension and the shape of the eye ball are invariable, while the lens shape and positions continue to change. For infants, the anterior lens surface is much steeper. With growth, it becomes flatter, until maturation, and afterwards, the anterior lens surface will become steeper again. Similarly, the posterior lens surface becomes steeper after adulthood. The thickness of the lens continues to increase at a varying rate through life. Infants have shorter VCD. During the growth period, the VCD increases; it then decreases in older age.

Although many studies have investigated the correlation between the ocular biometric parameters and age, no age-dependent eye model has been published. The age-dependent eye modeling can be performed in three age groups: infants (newborn to 12 months old), children (approximately 1–16 years old), and adults (16 years old and over). As in the accommodative and the ametropic eye modeling, the validation of the models is required via optical optimization and proper selections of the free variables necessary to achieve the targeted ocular optics.

13.2.2.4. *Modeling for disease development*

Eye modeling can be used to study ocular disease and its development. One such example is the keratoconus (KC) eye modeling based on the statistical description of the disease biometry conditions. KC is a degenerative non-inflammatory disorder of the eye for which structural changes occur within the cornea and result in the thinning of the cornea and a change to a more conical shape than its normal gradual surface curvature. KC can cause a substantial distortion of vision in multiple images, streaking, and sensitivity to light. One purpose of KC modeling is to study and understand the influence of the properties of the KC cone(s) on the optical performance of human eyes. With the general KC eye models, the effects and visual impacts of different parameters of the KC cone, such as the cone location, volume, and shape, were investigated. The research results of this subject have been published in Ref. 16.

13.2.3. *Validation of Eye Models*

Patient vision performance is usually used to confirm the modeling success. The refractive errors and the WFA are normally guaranteed in the modeling process using optical optimization. Further validation with VA and street vision in the day or night conditions can also be examined after the eye modeling is complete.

13.2.3.1. *Point spread function and modulation transfer function*

The size and shape of the PSF, which is the image of a point source, provide an indication of VA. In general, the real image on the retina can be calculated by the spatial convolution of the PSF with the object in the object space. Knowing the PSF of one eye model, we can estimate the subject VA by the degree of concentration of the PSF. The dimension and the profile of the PSF can be compared with the clinical VA report to evaluate the degree of success of the model. The PSF can be directly obtained in the ZAMAX analysis. Notice that the PSF depends on the object distance and the field angle that are assigned in the lens editor. For a 20/20 visual acuity, the PSF should be comparable to a spot size of $\sim 5\ \mu$m on the retina. A similar way of this measure is the modulation transfer function that represents spatial resolution in frequency domain. A well resolved 20/20 Snellen VA corresponds to a frequency of ~ 100 cycle/mm.

13.2.3.2. *Letter chart simulation*

The PSF can be used to infer how well the subject can see a point source, but using the parameter alone is neither straightforward nor reliable to describe the subjective vision alone. Before calculating the PSF, we must first set a pupil sampling number of ray tracing (i.e. the size of the grid of rays to trace to perform the computation). Higher sampling densities yield more accurate results at the cost of longer computation times. Second, the PSF depends on the location of the point source in the field of view. When the subject looks at an object, especially a large object, calculation of PSF over a large field angle range is required for accurateness. Third and most importantly, the PSF is a 2D function, which is difficult to directly quantify or directly show correspondence to the single parameter VA. A single index that is derived from PSF, such as the FWHM or STR, does not directly correspond to VA, especially when the PSF profile is far from a Gaussian or Lambertian type of symmetric shape. For these reasons, the

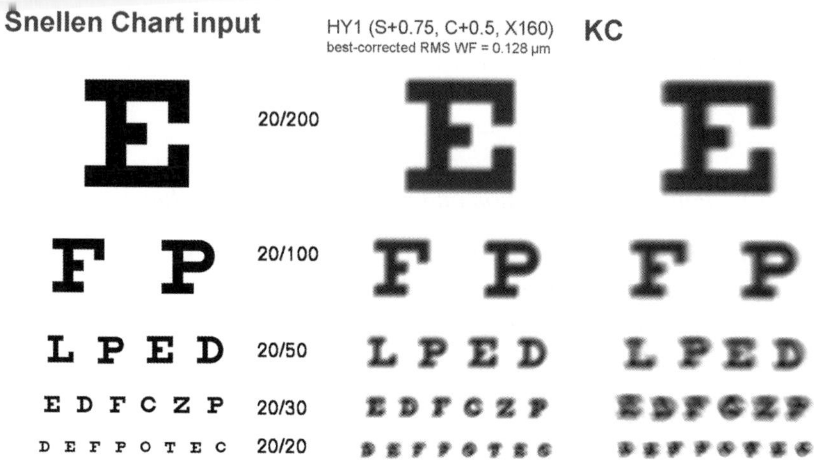

Fig. 13.6. Input letters (left) and the best-corrected Snellen chart vision of an astigmatism patient (middle) and a KC patient (right).

best way to validate the personalized eye models is to simulate the subject's vision of an extended object, using, for example, a Snellen letter chart.

ZEMAX geometric image analysis (GIA), rather than PSF convolution, is used to provide such vision simulation. GIA is based strictly upon geometrical ray tracing. It can be used to model extended (light) sources, analyze useful resolution, represent the appearance of imaged objects, and provide intuition as to image rotation. A perfect alphabetic letter E, for example, is assigned as the object image (or the light source) at the desired distance. Figure 13.6 shows the input Snellen letter chart at 20 feet and the resulting retinal images of a hyperopic eye and a KC eye using the customized eye models. Each letter in the letter chart is simulated individually with its corresponding resolution. A letter *E* at 20/20 line corresponds to 0.873 cm at 6 m and ideally forms a reversed image of $\sim 24\,\mu$m on retina.

13.2.3.3. *Night/day vision simulation*

The human vision is different in daylight than in a nighttime environment because light, or the lack, thereof causes a change in pupil size. In the darkness, the pupil is naturally dilated to include more light signals. In nature, the human visual procedure and correction are designed and adapted for daytime vision. However, since human activities extend long after nightfall, the performance of night vision becomes important and requires more concern. Night vision problems have been the reason for most complaints from LASIK patients. Fortunately, eye modeling can be used to predict a patient's night vision. The object image, such as the street view, can be entered in ZEMAX at the assigned object distance. The retinal image of the street view can then be obtained by running the image analysis procedure through the desired eye model. Pupil size and accommodation level must be assigned adequately. The day and night vision predictions using ZEMAX BMP Image–Analysis can serve as one way to validate eye modeling.

13.3. Other Modeling Considerations

In the eye modeling work described previously, many ocular conditions are not encountered. Some of these missing conditions will be addressed here.

13.3.1. *Stiles Crawford Effect (SCE)*

SCE is an optical feature of the eye that is caused by the wave-guide property of the cone photoreceptors of the eye. A photoreceptor acts like an optic fiber on the retina; it captures light that hits it at a narrow angle from its normal. As a result, the rays of light passing through the periphery of the pupil are more oblique to the cone. The acceptance angle of a cone is narrow, approximately 5°; rods have larger acceptance angles. SCE reduces the disadvantageous effects of aberration and the light scatter on the retina at photopic levels. A number of mathematical functions have been used to describe the SCE, the most popular of which is a Gaussian distribution as first used by Stiles and Crawford in 1937.[17] This function is usually an excellent fit to experimental data out to 3 mm from the peak of the function, and has the additional virtue of simplicity. This SCE function $L_e(r)$ is described as $L_e(r) = \exp(-\beta r^2)$, where r is the distance in the pupil from the peak of the function. The function is normalized to have a value of 1 at the peak. The Stiles-Crawford co-efficient β describes the steepness of the function, and is assumed to reflect the directionality (variation in alignment) of the photoreceptor population being tested. It may not have the same value for measurements in eyes affected by retinal pathology. Measured β coefficients for the large-scale study are given by Applegate and Lakshminarayanan.[18] Combining the data across many studies gives a mean value of 0.12. The SCE can be included easily in optical modeling of the eye as an apodization effect, which means that it can be treated as an optical filter of variable density attenuation placed at the pupil. The apodization filter can be simply added into our model by entering a user-defined surface and loading a "dll-file" to define the surface that represents the equation of SCE.

13.3.1.1. *Multiple reflection and scattering of the retina*

In some applications of eye modeling, especially the simulation of ophthalmic measurement, the reflection and scattering properties of the retina need to be addressed. Because the human retina is not self-luminous, an external light source is necessary to make the retina visible, and the light reflected from the retina will bring the information behind the cornea to the instrument. Thus, the multiple reflection and scattering properties of the retina should be addressed. Although most layers in the retina are virtually transparent, there are small refractive index variations between cells. Such a deviation from homogeneity will give rise to scattering and reflections. A proper model of any ophthalmic or optometric instrument measurement that uses the double pass reflection must consider the uniqueness and complexity of this retinal reflection. Even if the simplifications of any model are used, the implications of such simplifications should be understood. Up to this point, only sequential ray tracing is described for eye modeling. Rays are always traced from the object surface to the assigned surface numbers in a strict sequential order. Each ray "hits" each surface once and only once in this predetermined sequence. The sequential model is straightforward, numerically fast, and extremely useful and complete for many important cases. However, when multiple scattering/reflection is involved, a nonsequential ray tracing may be required. Nonsequential means the rays are traced in the physical order that they encounter the various objects or surfaces, and not necessarily in the order the objects are listed in the software user interface. Note that rays in a nonsequential trace may hit the same object repeatedly, and entirely miss other objects. Generally, the order in which objects are hit by rays depends upon the object geometry and the angle and position of the input ray. Certain types of analysis, such as stray or scattered light effects, are only practical in a completely nonsequential environment. In the ophthalmic simulation, the inclusion of the back scattering from retina and the use of the nonsequential system may well be the only solution in ZEMAX to mimic the multiple reflection and scattering from the retina.

13.3.1.2. *Other retinal properties*

The spectral properties of retina can be easily handled by assigning the wavelength and weight at the beginning of each of the double path simulations. A birefringent retinal surface can be modeled using the birefringent in/out surface types as described in "Birefringent In and Birefringent Out" in ZEMAX. As to the position-dependent reflectance, a retinal component with schematic layers, using the nonsequential component in ZEMAX and setting the scattering type and parameters of each layer, can be constructed. However, using this multiple retinal layer model to make any simulation could significantly increase the computation time, and the result may not be appreciably different from the calculation based on single-layer retina model.

13.3.1.3. *Tear film influence and tear film breakup*

The tear film has not been included or discussed in any acknowledged schematic eye models. Although the typical tear film is very thin (3–40 μm) compared to the corneal thickness (greater than 500 μm), vision image quality as well as the ophthalmic measurements can be influenced significantly by the tear film condition. The tear film quality is determined by its structure, composition, and thickness. The tear film possesses a free surface, which is secreted by the lacrimal gland, is lost via evaporation and drainage at the lacrimal ducts at the nasal side, and will eventually breakup in the absence of blinking.

The tear film breakup time (TBUT) is normally greater than 10 second for healthy tears. Pathologies of this film, or its production, are typically responsible for dry eye syndrome that possesses many features that are encompassed by the fluid mechanical and associated solute transport processes of the tear film. Studies in this are far from complete. Recently, the studies of tear film models.[19–21] These studies show that the optical parameters of the tear film that can affect the accuracy and completeness of optical eye modeling include tear film thickness, post-blink tear undulation, tear breakup pattern, eyelid-produced bumps and ridges, bubbles, and rough pre-contact lens tear surfaces. These tear film characteristics in spatial and temporal domains can be included in the schematic eye models. The predictive modeling and simulation could yield insightful information regarding the dry eye vision and promote the diagnostics technology for the disease.

13.3.1.4. *Optical opacity*

The only published cataract eye model is proposed by Donnelly.[22] In this study, Scheimpflug cameras characterized the anterior segment and backscatter from cataract. He discussed how to measure and model intraocular light scatter with SH wavefront-sensing data.[22] The key to simulate the cataract based on our personalized eye models is to simulate the surface and volumetric scattered light in the eye, both of which contribute to contrast reduction of the image at the retina. The scattering theory to be applied is Rayleigh-Mie scattering, since cataracts are volumes composed of small scattering particles.[22,23] The scattering of cataracts can be simulated with a bi-directional scatter distribution function (BSDF) in a nonsequential component in ZEMAX.

13.4. Examples of Ophthalmic Simulations

Eye models have applications in scientific research and industrial engineering. Patient vision simulation is useful for medical education and for patient consultation. Ophthalmic measurement simulations are advantageous for even more applications. As demonstrations, the simulations of retinoscopy and the photorefraction (PR) measurement are described in this section.

13.4.1. *Simulation of Retinoscopy Measurements with Eye Models*

Retinoscopy was introduced more than 100 years ago and is still practiced today to yield important clinical results. A traditional "spot retinoscope" projects a spotlight onto a patient's eye at a distance of 0.5–1.0 meter, while a contemporary "streak retinoscope" projects a straight-filament image. The size of the spot or streak projection is adjustable by moving a condenser lens that is located above the light source (or filament) in the handle of the device. The retinal reflex is observed by the examiner through a peephole on the scope. When moving the streak projection across the patient's pupil, the reflex of a myopic or hyperopic eye appears to move with or against the projection motion. The moving speed and direction of reflex depend on the position of the condenser lens. Subsequently, the examiner uses a phoropter or manually places trial lenses over the examinee's eye to "neutralize" the reflex movement.

When the refraction is neutralized, the pupil will suddenly appear bright, as the light projection aligns to the center of pupil, and will turn dark if the projection slightly misaligns toward either side. No movement should be seen under the neutralization condition. The compensation lens indicates the required defocus correction. Retinoscopic measurement is objective and, therefore, especially useful in prescribing corrective lenses for patients who are unable to undergo a subjective refraction test that requires a response and judgment from the examinee. Retinoscopic measurement is also used to evaluate the accommodative ability of the eye and to detect latent hyperopia. Although the retinoscope optical structure is simple, the thorough analysis is not easy due to frequent utilization of low-cost, imprecise optical elements. As a consequence of the absence of detailed analysis, medical textbooks illustrate retinoscopy with over-simplified portraits. Further, ambiguous observations occur when the ocular aberrations are significant and when overlap occurs in the light path of the multiple aperture stops, including eye pupils. These limitations and difficulties have perhaps discouraged the quantitative of retinoscopy in clinical practice.

With eye modeling, retinoscopic measurements can be predicted through simulation. The optical layout of the simulation is shown in Fig. 13.7. Using a program such as ZEMAX, light rays from the filament source of retinoscope are traced through the scope and enter the targeted eye model. In the return path, the retinal image at the end of the first ray tracing serves as the light-source object, and the light rays are traced through the peephole of the scope and imaged on the retinal plane of the observer. The eye model of the retinoscopist can be a perfect thin lens with an image

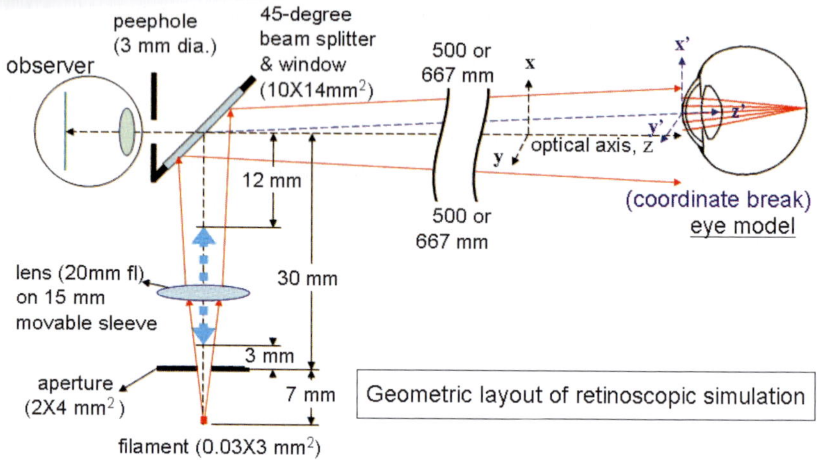

Fig. 13.7. Optical layout in the retinoscopy simulation.

plane. As in the real condition, four effective apertures were involved in this retinoscopic simulation. These apertures were the small apertures in front of the filament, the window on the beam splitter (along both paths), the pupil of the eye (along both paths), and the peephole of the observation. A coordinate break in the ZEMAX program is required to simulate the movement of the streak projection, as it moves across the examinee's pupil. Ray aiming is necessary to ensure that all of the vignetting or cut-off effects are encountered when using coordinate breaks.

Figure 13.8 demonstrates some example results of the retinoscopy simulations with eye models. On the left of the figure is the famous scissors reflex of a KC eye, as the streak retinoscope projection moves along the 45-degree meridian. Typically, 100 million rays are traced in each of these double-path simulations to produce one reflex image. A KC patient's topography and manifest refractive prescription are used to construct a personal eye model. Similar simulations can be performed to predict the observation when the eye is focused on a different direction and when the eye pupils are dilated to different sizes. Likewise, the retinoscope simulation can be assigned to move along any meridian with any sleeve position of the scope. In Ref. 24, both plane- and concave-mirror practices of retinoscopy are simulated. The typical ammetropia reflex movements of with- and against-motion, and the

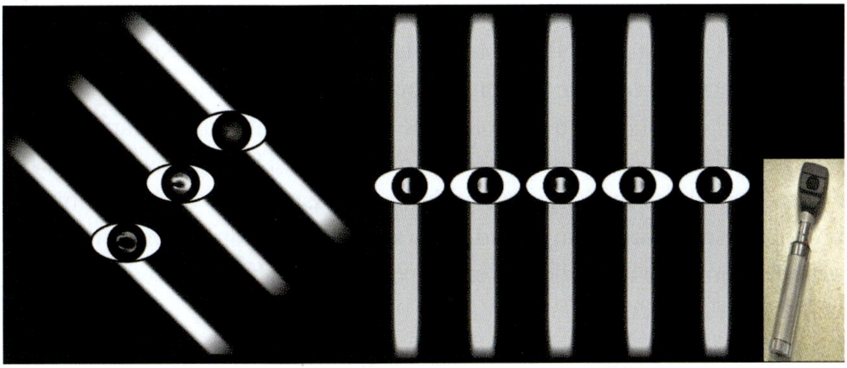

(a) (b)

Fig. 13.8. Simulations of retinoscopic observations on (a) a 38-year-old Caucasian female patient's KC eye and (b) an eye with high degree of SA resulting from a surgical procedure. Scissors reflex and hourglass reflex are illustrated, respectively.

so-called "anomalous with-motion" of the high myopia condition are produced. This type of simulation can be applied to medical training, without the need for using human subjects.

A second retinoscopic example is the hourglass reflex that is shown in Fig. 13.8(b). In the recent years, case observations have suggested that inadvertently induced SAs from surgical procedures, such as the Schachar's sclera band procedure and the use of intraocular lenses produce "pseudoaccommodation" in presbyopia patient vision. One such case was reported by Dr. Guyton in Johns Hopkins.[25,26] Dr. Guyton had the opportunity to examine a patient after surgery for presbyopia. The patient had relied on reading glasses to see objects that were nearby and had undergone Schachar's scleral band procedures for presbyopia. Two to three months later, she went without glasses entirely, with 20/20 uncorrected VA for viewing both objects in the distance and objects nearby. However, in the dynamic retinoscopic reflex examination, Guyton observed the static hourglass shape of reflex, rather than the streak reflex, for both near and distant visions, thereby demonstrating the absence of actual accommodation. He suspected that the hourglass reflex indicated a condition of a high-degree of SA, which provides an effect of long focal depth, or the so-called "pseudo-accommodatrion." This hourglass reflex observation can be reproduced with the retinoscopy simulation using general eye models.

In the simulation of Fig. 13.8(b), the Navarro eye model is used as the base model, and a ZERNIKE phase plate is utilized on either the surfaces of the anterior cornea or the lens to produce an aberration of the desired magnitude. The vision simulation of the letter chart at near and far distances describes the effect of pseudo-accommodatrion. The simulation with the eye model allows investigation on how various degrees of aberration can produce visible clinical signs and allow scientists to examine the consequent patient vision due to the degree and type of ocular aberration without using real human subjects.

13.4.2. *Simulation of PR*

With some similarity to retinoscopy, stationary PR was introduced during the 1970s and 1980s. Kaakinen,[27] Howland,[28] and Bobier and Braddick[29] replaced the observer with a camera. The stationary PR is classed into two types: the coaxial PR (CPR) and the eccentric PR (EPR). CPR has the light source positioned in front and at the center of the camera lens.[29,30] The retinal reflex coming from the eye to the camera is defocused into a blur pattern, which changes with the eye's defocus magnitude. CPR bases the judgment of the refractive error on the extent of the defocused retinal reflex. In contrast, EPR places a light source eccentric from the camera lens aperture, and the camera is focused on the examined eye to form a sharp pupil image of reflex. Like the spot retinoscope, the EPR measurement is objective, and the light source and camera are placed away from the examinee. EPR calculates the state of refractive error from the size of the bright crescent that appears in the focused pupil image. EPR is currently the most used PR method to screen for binocular refractive errors in children and to detect accommodation in lab animals.

The commercial PR instruments for pediatric vision screening include the iScreen Photoscreener,[32] the MTI Photoscreener,[33] and the PediaVision Vision Screener[34] (previously Power Refractor II[35]). Shown in the upper portion of Fig. 13.9 is a photograph using the iScreen Photoscreener. Contact lenses were worn on the two eyes to produce refractive errors of +4 and −4 diopters. For comparison, the simulated pupil reflex images using model eyes of various degrees of refractive errors are illustrated at the bottom of Fig. 13.9.[36] The Navarro eye models are used with additional thin lenses at

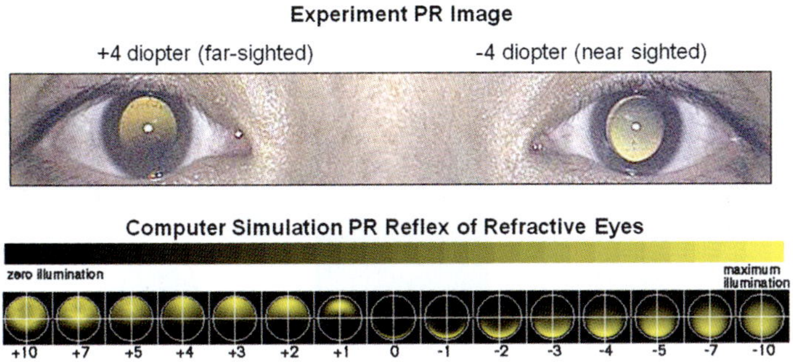

Fig. 13.9. PR image of refractive eyes acquired with aid of contact lenses (upper) and simulation images of $+10$ to -10 D refractive eye models (lower).

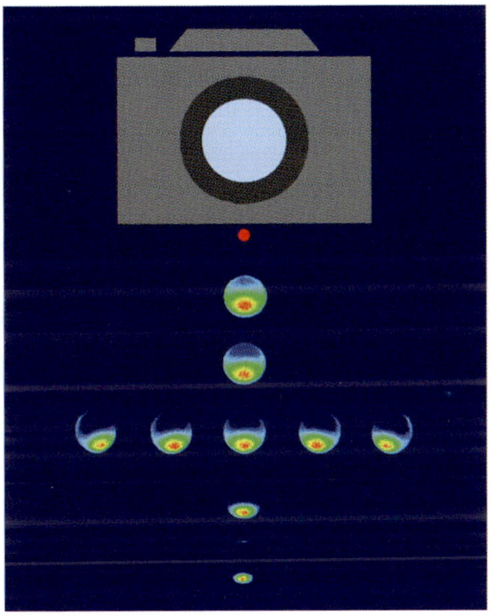

Fig. 13.10. Gazing angle effect on the EPR measurement of the same eye using eye model.

the corneal plane. Simulation of the measurement permits the investigation of various human factors and environmental conditions. For example, the simulation result in Fig. 13.10 predicts the tilting and size changes of the reflex due to the alignment of ocular orientation or gazing angle. The effects of pupil size, light source location and dimension, and the wavelength of light source can be evaluated through the computation with eye models. Further demonstrated in Figs. 13.11 and 13.12 are the investigation of the sensitivity in detecting high-order ocular aberrations with a new PR set-up.[37]

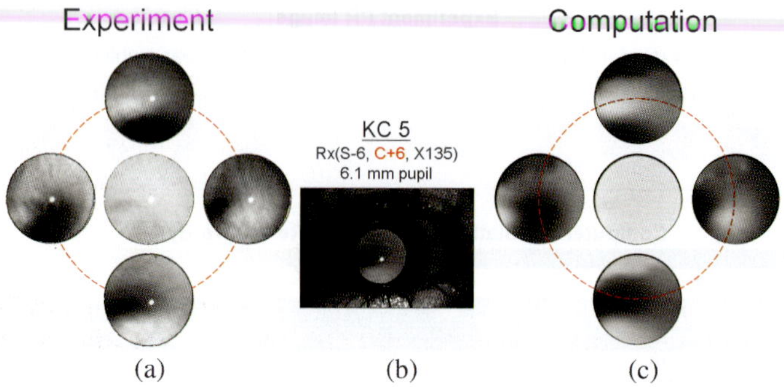

Fig. 13.11. (a) Five measured infrared PR KC pupil reflex images (pupil only), (b) the lowest of the five original reflex images, unprocessed digital photograph, and (c) simulated images using the personalized model of the patient.

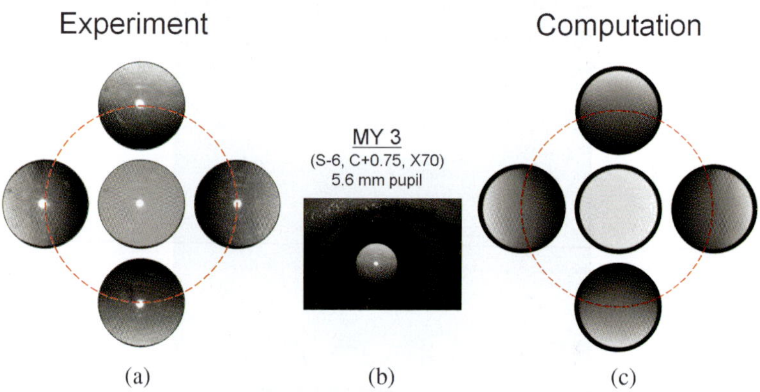

Fig. 13.12. (a) 5 measured infrared PR myopic pupil reflex images (pupil only). (b) The lower one of the 5 original, unprocessed digital photograph. (c) Simulated images using the personalized model of the patient.

An infrared camera with five NIR LEDs that are placed on center, above, below, and on both sides of the camera aperture is assembled. Photographs of one CPR and four EPR images are acquired in sub-second. One raw PR photograph image of a KC eye with a high-order aberration is illustrated in Fig. 13.11(b). The pupil reflexes are identified and cropped from the five raw images of the KC eye with computer image analysis program and shown in Fig. 13.11(a). Figure 13.11(c) shows the corresponding simulation images using the personalized eye model that is constructed with the patient's topography and WFA. The simulation result clearly predicts the measurement. Figure 13.12 illustrates the same investigation with a myopic eye. The prediction and experiment results shows similar feature as the eye with little high-order aberration shown in Fig. 13.9. These results show the potential applications that may result from developing new ophthalmic devices and technology.

13.5. Conclusion

In this contribution, we have indicated the broad spectrum of potential applications of the accurate computational modeling of human ocular optics. The application areas are diverse, and the ability to provide reliable predictions offers significant advantages and benefits.

General and population based eye modeling are significant for gaining knowledge of visual optics and studying the development of eye diseases. The personalized eye modeling, on the other hand, provides promising features in assisting ocular surgery and in designing customized spectacle, contact, or intraocular lens. Both types of modeling could be applied to predict visual changes under specified environmental or physical conditions. Furthermore, the computer simulations of ophthalmic measurements that utilize eye-modeling techniques offer a comprehensible tool for medical training.

Our selection of a specific commercial optics code is offered only as an example of the capabilities that are available today. The detailed procedures and selections employed for this code that are described in this work are indicative of the options and parameters necessary to obtain quantitative solutions. Finally, careful verification of computational techniques is essential to ensure the reliability and accuracy of the predictions.

The authors would like to acknowledge the research support of the Center for Laser Applications of the University of Tennessee Space Institute and the support for eye modeling and KC research of the National Eye Institute grants R21 EY18385 and R21 EY018935.

References

1. Gullstrand, A. The optical system of the eye, Appendix 11.3. In: Helmholtz, H. Von, *Physiological Optics*, Third Edition, 1:350–358, 1909.

2. Escudero-Sanz, I. and Navarro, R. Off-axis aberrations of a wide-angle schematic eye model. *JOSA A* **16**:1881–1891, 1999.
3. Le Grand, Y. Optique physiologique. *Tome I, La dioptrique de l'oeil et sa correction* (Masson, Paris). Revised edition translated to English: Le Grand, Y. and El Hage, S.G. (1980). *Physiological Optics*, Springer-Verlag, Berlin, 1956.
4. Liou, H.-L. and Brennan, N.A. Anatomically accurate, finite model eye for optical modeling, *J Opt Soc Am A* **14**:1684–1695, 1997.
5. Atchison, D.A., Markwell, E.L., Kasthurirangan, S., Pope, J.M., Smith, G., and Swann, P.G. Age-related changes in optical and biometric characteristics of emmetropic eyes. *J Vis* **8**:1–20, 2008.
6. Tocci, M. How to model the human eye in ZEMAX. *ZeMax Knowledge Base*. Available at http://www.zemax.com/kb/articles/186/1/How-to-Model-the-Human-Eye-in-ZEMAX/Page1.html, 2007.
7. Watkins, R. ZEMAX models of the human eye. *ZeMax Knowledge Base*. Available at http://www.zemax.com/kb/articles/193/1/ZEMAX-Models-of-the-Human-Eye/Page1.html, 2007.
8. Navarro, R., González, L., and Hernández, J.L. On the prediction of optical aberrations by personalized eye models. *II Physiological Optics Topical Meeting of the European Optical Society*, Granada, Spain, 2004.
9. Emsley, H.H. *Visual Optics*, London, Butterworth, 1952.
10. Le Grand, Y. and El Hage, S.G. *Physiological Optics*, Springer Series in Optical Sciences, Springer, Berlin, 1980.
11. Atchison, D.A. and Smith, G. *Optics of the Human Eye*, Elsevier Health, 2000.
12. Goss, D.A., Van Veen, H.G., Rainey, B.B., and Feng, B. Ocular components measured by keratometry, phakometry, and ultrasonography in emmetropic and myopic optometry students. *Optom Vis Sci* **74**:489–495, 1997.
13. Dubbelman, M., Sicam, V., and Van der Heijde, G.L. The shape of the anterior and posterior surface of the aging human cornea. *Vision Res* **46**:993–1001, 2006.
14. Atchison, D.A. Optical models for human myopic eyes. *Vision Res* **46**:2236–2250, 2006.
15. Stenstrom, S. Investigation of the variation and the correlation of the optical elements of human eyes. Part V–Chapter III (Woolf, D. Trans.), *Am J Optom Arch Am Acad Optom* **25**:438–449, 1948.
16. Tan, B., Baker, K., Chen, Y.-L., Lewis, J.W., Shi, L., Swartz, T., and Wang, M. How keratoconus influences optical performance of the eye. *J Vis* **8**:1–10, 2008.
17. Stiles, W.S. and Crawford, B.H. The effect of a glaring light source on extrafoveal vision. *Proc R Soc B* **122**:255–280, 1937.
18. Applegate, R.A. and Lakshminarayanan, V. Parametric representation of Stiles-Crawford functions: normal variation of peak location and directionality. *J Opt Soc Am A* **10**:1611–1623, 1993.
19. Barbero, S. Refractive power of a multilayer rotationally symmetric model of the human cornea and tear film. *J Opt Soc Am A Opt Image Sci Vis* **23**:1578–1585, 2006.
20. Heryudono, A., Braun, R.J., Driscoll, T.A., Maki, K.L., Cook, L.P., and King-Smith, P.E. Single-equation models for the tear film in a blink cycle: realistic lid motion. *Math Med Biol* **24**:347–377, 2007.

21. Wang, J., Fonn, D., Simpson, T.L., and Jones, L. Precorneal and pre- and postlens tear film thickness measured indirectly with optical coherence tomography. *Invest Ophthalmol Vis Sci* **44**:2524–2528, 2003.
22. Donnelly, W.J. 3rd, Pesudovs, K., Marsack, J.D., Sarver, E.J., and Applegate, R.A. Quantifying scatter in Shack-Hartmann images to evaluate nuclear cataract. *J Refract Surg* **20**:S515–S522, 2004.
23. Gilliland, K.O., Freel, C.D., Johnsen, S., Fowler, W.C., and Costello, M.J. Distribution, spherical structure and predicted Mie scattering of multilamellar bodies in human age-related nuclear cataracts. *Exp Eye Res* **79**:563–576, 2004.
24. Tan, B., Chen, Y.-L., Baker, K., Lewis, J.W., Swartz, T., Jiang, Y., and Wang, M. Simulation of realistic retinoscopic measurement. *Optics Express* **15**:2753–2761, 2007.
25. Guyton, D.L. Accommodative amplitude measurements after surgery for presbyopia. In: Tsubota, K., Wachler, B.S.B., and Azar, D.T. *Hyperopia and Presbyopia*, Informa Health Care, 2002.
26. Hunter, J.J. and Campbell, M.C.W. Potential effect on the retinoscopic reflex of sclera expansion surgery for presbyopia, *Optom Vis Sci* **83**:649–656, 2006.
27. Kaakinen, K. A simple method for screening of children with strabismus, anisometropia or ametropia by simultaneous photography of the corneal and the fundus reflexes. *Acta Ophthalmol (Copenh)* **57**:161–171, 1979.
28. Howland, H.C. Optics of photoretinoscopy: results from ray tracing. *Am J Optom Physiol Opt* **62**:621–625, 1985.
29. Bobier, W.R. and Braddick, O.J. Eccentric photorefraction: optical analysis and empirical measures. *Am J Optom Physiol Opt* **62**:614–620, 1985.
30. Howland, H.C. and Howland, B. Photorefraction: a technique for study of refractive state at a distance. *J Opt Soc Am* **64**:240–249, 1974.
31. Howland, H.C., Braddick, O., Atkinson, J., and Howland, B. Optics of photorefraction: orthogonal and isotropic methods. *J Opt Soc Am* **73**:1701–1708, 1983.
32. iScreen, Inc., Memphis, TN.
33. Medical Technology, Inc., Cedar Falls, Iowa.
34. PediaVision, LLC.
35. Plusoptix, Nuremburg, Germany.
36. Chen, Y.L., Tan, B., and Lewis, J.W. Simulation of eccentric photorefraction images. *Optics Express* **11**:1628–1642, 2003.
37. Chen, Y.L., Tan, B., Baker, K., Lewis, J.W., Swartz, T., and Wang, M. Simulation of keratoconus observation in photorefraction. *Optics Express* **12**:11477–11485, 2006.

Chapter 14

Automating the Diagnosis, Stratification, and Management of Diabetic Retinopathy Using Content-Based Image Retrieval in an Ocular Telehealth Network

Thomas P. Karnowski*, Luca Giancardo*, Deniz Aykac*, Kenneth W. Tobin*, Yaqin Li†, Seema Garg‡, Michael D. Abramoff§, Matthew T. Tennant¶, and Edward Chaum¶

Diabetic retinopathy (DR) is the leading cause of blindness among working aged adults in the industrialized world. Diabetes currently affects 246 million people worldwide and is anticipated to affect as many as 380 million people by 2025.[1] Every year an additional seven million people develop diabetes worldwide. Currently, 80% of people with diabetes live in low- and middle-income countries. Diabetics in many of these countries, including India (40.9 million population) and China (39.8 million population), have limited access to health care resources. The largest increases in diabetes prevalence are taking place in these and other developing countries. Based upon the epidemiology of diabetes, the number of patients that will

*Oak Ridge National Laboratory, Oak Ridge, TN 37831.
†Department of Electrical and Computer Engineering, University of Memphis, Memphis, TN 38152.
‡School of Medicine, University of North Carolina, Chapel Hill, NC.
§Department of Ophthalmology & Visual Sciences, University of Iowa Health Care.
¶The University of Tennessee Health Sciences Center, Hamilton Eye Institute, 930 Madison Avenue, Suite 731, Memphis, TN 38163.

need to be screened for DR will soon exceed one million patients per day, worldwide.

The seminal DR studies and the early treatment of DR studies have clearly demonstrated the efficacy of laser treatment in reducing the risk of severe and moderate vision loss from proliferative DR and macular edema, respectively. The Wisconsin epidemiology study of DR showed that 90% of Type 1 diabetics (insulin-dependent, onset <30 years of age) and up to 84% of Type 2 diabetics (onset after 30 years of age, +/− insulin dependence) have some degree of DR after 10–15 years.[2,3] Despite the known association of diabetes with vision-threatening DR, less than half of the diabetic patients in technologically advanced health care systems such as the United States are screened for retinopathy in any given year.[4] Remarkably, one-quarter of Type 1 diabetics and one-third of Type 2 diabetics have never had an eye examination.[5]

In its November 2005 report "Prevention of Blindness from Diabetes Mellitus," the World Health Organization (WHO) noted that in many developing countries, there are too few medical and diagnostic resources to provide even basic eye care to the general population, much less specialized care for advanced DR. However, the use of photographic systems coupled with expert interpretation could increase the ability of primary care providers (PCPs) to detect and manage early DR. The WHO identified key goals of integrating photographic diagnostic systems into DR management: (1) assessing system performance relative to the current gold standard of reading centers, (2) validating success in providing access, and (3) demonstrating health outcomes benefits over the current methods.

Commercially available high-quality mydriatic and nonmydriatic cameras generate high-quality digital data sets that have the potential to enhance our ability to screen large and at-risk populations for DR effectively to identify those in need of treatment, and thus, significantly reduced disease morbidity. Novel computer-based image analysis and diagnostic methods that can leverage this data have the potential to improve the sensitivity and specificity of remote retinal diagnosis in large patient populations, and to monitor therapeutic responses to treatment. The current paradigm applied to the remote assessment of retinal disease consists of "store and forward" digital retinal photography with subsequent image analysis by physicians or certified technicians at commercial and academic reading centers. The reading center method is an established and validated method, with high sensitivity and specificity for retinal disease detection and quantification in clinical trials.[6–8] This method is labor intensive, which means that it is more costly and slower than fully automated systems.

To achieve the goals of large population-based DR detection and management, automated methods must be developed and implemented. Current and even future public health care delivery models based upon manual analysis of retinal images simply cannot accommodate the need to screen over one million patients every day for DR. A new paradigm is required. An example paradigm is the office EKG, in which a computer algorithm generates a diagnostic reading at the point of care, in real time. The application of computer-based image analysis to aid in the diagnosis of DR has the potential to facilitate disease detection and management by increasing throughput, reducing costs, and potentially automating diagnostic capabilities within integrated telemedical network delivery systems.

Current image analysis algorithms can detect the anatomic features of the retina and retinal lesions using color and monochromatic retinal images.[9,10] A common approach is to locate of important anatomic structures in the eye, normalize images to accommodate illumination and contrast,[11] segment the vascular structures,[12,13] and exploit the geometric relationship that exists between the vasculature, optic nerve (ON), and macula in the retina.[14,15] Other approaches to image analysis in the retina include location regression, pixel classification, and graph search algorithms,[16] and techniques, such as dynamic contours, and model-based approaches[17,18] to name a few.

Content-based image retrieval (CBIR) is the process of retrieving related images from very large database collections, based on their pictorial content.[19,20] Pictorial content is defined by a set of intrinsic features extracted from an image that describe the attributes that the human brain uses to comprehend the image, such as the color, texture, shape, and regional structure of the image or specific objects within it. The pictorial content feature list becomes the index for storage and retrieval, and facilitates searches of large image datasets based upon the specific visual characteristics of a query image. An unknown query image is submitted to the CBIR library and its anatomic structures and lesions found, followed by feature extraction and analysis. The features become an index for searching in the database and are used to locate a population of similar images contained in the library, which can then be compared statistically to the query image. Low-level analyses employ feature description models and higher-level analyses use perceptual organization and spatial relationships to extract clinically relevant semantic information. The method also has the potential to utilize associated nonimage data (e.g. the clinical data such as disease history, previous treatments, and visual acuity), which may enhance the diagnostic capabilities. The database library representing the image population is constructed and an indexing tree is generated to facilitate rapid searching. The (Bayesian) probabilistic framework of CBIR permits us to make statistically relevant predictions regarding the presence, manifestations, and severity of common retinal diseases from digital images in an automated and deterministic manner.[21] The CBIR method provides a unique ability to describe "off-normal" occurrences in images based on deviation from normal retinal anatomy.

Our objective in this chapter is to describe our research of the use of CBIR with biomedical image databases in the clinical setting of a regional telemedical network designed for the remote diagnosis of DR in underserved patient populations using high-throughput methods to meet the growing need for disease assessment and management. We first describe a network infrastructure for the automated diagnosis of DR that provides a method for low-cost, real-time diagnosis and patient referral in the primary care environment. Our telemedical network is designed for remote diagnostics and, thus, the design of the underlying network infrastructure emphasizes high-speed data transmission for real-time image analysis, secure data encryption, ease of installation, and transmission of patient-sensitive information to meet federal regulations.

We then describe the capabilities of CBIR to classify, predict, diagnose, and otherwise learn from the informational content encapsulated in historical image repositories. We will demonstrate these principles using specific examples from our biomedical applications to describe anatomic segmentation, statistical feature generation and indexing, efficient retrieval architectures, and predictive results in the evolving diagnostic retinal image search and analysis (RISA) network. The benefits of automation will reduce the cost of health care management and speed up health care delivery, critical elements for managing the health of rapidly expanding at-risk populations in the coming decades.

This network and method addresses the needs of underserved populations and represents a novel asset to broad-based screening of DR and potentially other diseases of the retina. The ultimate aim of our work is to develop, validate, and implement a fully automated system for the detection and diagnosis of DR in real time from digital images in statistically significant and quantifiable terms. In our web-based network, retinal images from diabetic patients are transmitted from fundus cameras in DICOM format using secure protocols to a diagnostic server. The retinal images are then graded to stratify disease level, recommend a management plan, and generate a report accessible to the client. In the current stage, with the automatic diagnostic engine still under development, the system runs in a semi-automated fashion and disease stratification is carried out by an ophthalmologist. Eventually, the system will provide a fully automated real-time diagnosis of DR and other retinal diseases.

14.1. Network Infrastructure

The ubiquitous connectivity of the Internet provides an effective way of exchanging medical information and delivering services over a large geographic region, and the widespread availability of network bandwidth provides sufficiently high throughput for real-time data transmission. The advancement of technology makes remote delivery of consultative health care feasible and cost-effective. Telemedical network care has been applied in a variety of medical areas including home monitoring, emergency care, psychiatry, and health care in penal systems, to name just a few. However, most telemedicine systems are more focused on data storage and retrieval, such as teleconferences and televideo transmissions.[22] They have not tapped into the potential of computing power and image analysis for data processing. The ever-increasing computing capabilities of electronic devices open a new horizon for computer-aided and mobile decision-making and automated data management.

Ocular telehealth using digital images has proven to be an accurate and reliable method of diagnosing DR, which demonstrates that telehealth is a valid solution for the delivery of ophthalmic care. Our goal is to build a telemedical network for the automated diagnosis and management of DR in the primary care space. To achieve the goal of automating the detection and diagnosis of diabetic eye disease in real time from digital images taken in a primary care setting, a novel network infrastructure is established.

In our ocular telehealth network, the participating primary care clinics are connected to a diagnostic computer server. The web-based telemedical network adds portability and significant flexibility in health care delivery, providing access to expert diagnosis and high throughput potential to meet the growing need for disease assessment and management in rapidly expanding at-risk and underserved patient populations.

14.1.1. *System Requirements*

To implement an automated diagnosis system, it is necessary to design the underlying network infrastructure to meet the following requirements:

- Secure data transmission between the clinics and the server,
- Encrypted patient-sensitive information to meet the HIPAA-compliance requirements,
- Robust data transmission to guarantee that there is no loss of data and to address issues of unstable network connectivity,
- Data reliability and integrity, and
- High-speed data transmission for high throughput image analysis.

14.1.2. *Network Architecture Design*

In the network design for ocular telehealth, a client-server model approach is implemented for the data transmission between the client and the diagnostic server (Fig. 14.1).[23] After the fundus images of both eyes are captured by a commercial retina camera, they are exported and submitted to the server for an automated quality assessment (QA) metric to assure adequacy for the purpose of diagnosis. The QA results are communicated to the end user in real time. Inside the server, the automatic diagnostic service retrieves the images and patient metadata, perform an image-processing task to identify the anatomical structure location, extract features, detect lesions, and assign a diagnosis with an appropriate management plan to generate a report accessible to the referring physicians. An ophthalmologist then reviews the images and the diagnostic report, and provides validation before the

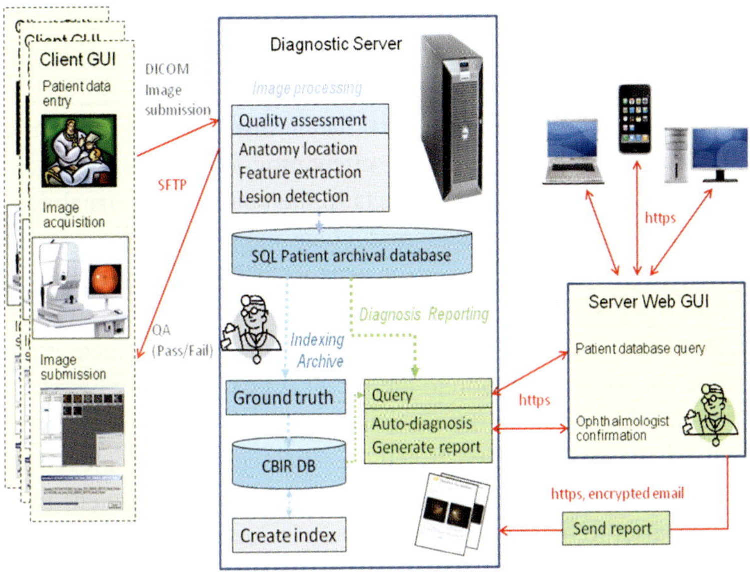

Fig. 14.1. Design for an HIPAA-complaint ocular telehealth network.

confirmed report is returned to the end user. An ophthalmologist can also override the computer-generated auto-diagnosis by manually retrieving the patient information and assigning an alternative diagnosis through the web interface to generate a confirmed report and management plan.

To provide secure data communications between the clinics and the diagnostic server over the Internet, and to meet HIPAA compliance requirements, all data transmission between the client and the server is performed using a cryptographic protocol. To transfer image files to the server and communicate the QA result (pass or fail) to the camera, a secure file transfer protocol (SFTP) is used. To access the report generated on the server, the referring physician can either login on a secure Web site using HTTPS (HTTP over a secure socket layer), or download an encrypted PDF report sent as an e-mail attachment.

To ensure the security of patient-sensitive information, data encryption is performed in the reviewing and confirmation process. When the review is complete, the ophthalmologist digitally signs and encrypts the report using an X.509 certificate, and then sends the confirmed report with digital signature and encryption to the referring physician in the participating teleophthalmology clinic. PDF encryption with an issued certificate guarantees data integrity and security, while providing password-protected access of the report to physicians as an e-mail attachment.

To fulfill the requirement of robust data transmission in the conditions of unstable network connectivity, the client program on the camera closely monitors the network status and reacts to the connectivity instability in a real-time manner. In the multi-thread client-monitoring program, one of the threads is devoted to scanning the network status constantly and reporting the network status to the user every second. Once connectivity issues occur, the client program reacts promptly. If the clinic is not transmitting any data when network connection to the server is lost, the client program first issues alert information to the user, then attempts to reestablish network connections for self-recovery. If self-recovery cannot be achieved within a preset time limit and the network problem persists, the client program prompts the user to fix the problem. If there is data transmission when network problems occur, the client program retransmits the exported images after the network status becomes normal. To guarantee that there is no loss of data in the client-server communication, a persistent acknowledgment is required for committed storage in the bidirectional data transmission.

Reliability and data integrity are critical in designing the ocular telehealth system. Reliability requires the completeness, timeliness, and accuracy of the data; integrity refers to the validity and consistency of the data. Reliability and data integrity can be compromised in a number of ways, for example, human errors when data is entered, errors that occur during data transmission, viruses, hardware malfunctions, and natural disasters. To minimize these threats, the following methods are incorporated in the system design: regular backup to independent physical media in the university server farm, control of data access via security mechanisms, design of user interfaces that prevent the input of invalid data, and bookkeeping of the critical status of every procedure or GUI-involved data manipulation to provide an audit trail of data flow in the system.

14.1.3. *Workflow*

The workflow of the ocular telehealth network is illustrated in Fig. 14.2. The modules in the detailed workflow diagram can be categorized into three

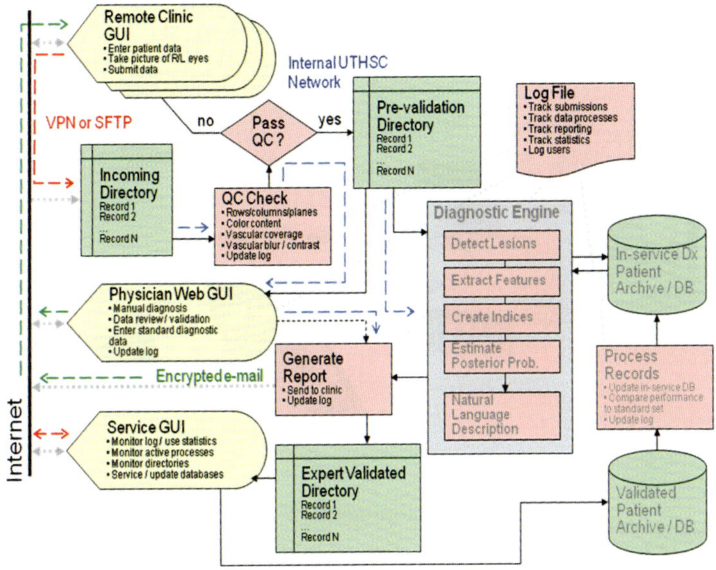

Fig. 14.2. Workflow schematic design for an HIPAA-complaint ocular telehealth network.

groups: graphical user interface (GUI) modules (shown in yellow blocks in the diagram), data storage (shown in green), and procedural modules (shown in pink). The GUI modules include three parts: the application GUI for the remote clinics, the web GUI for the authorized physicians, and the service GUI for monitoring and bookkeeping. A detailed discussion of the GUI design will be provided in the next section. Data storage can be either in file format inside directories or in a database format. The procedural modules include quality check and thresholding, a diagnostic engine running to perform disease stratification for incoming data and to build a CBIR library based on archived data, report generation, and bookkeeping functions such as log and process records.

Three mechanisms for data communications are involved in the network flow. The communication between the remote clinics and the server is through VPN or SFTP connections; the generated report is communicated back to the referring physicians in an encrypted e-mail, and the communications between the data storage modules and the procedural modules are completed through internal UHTSC network.

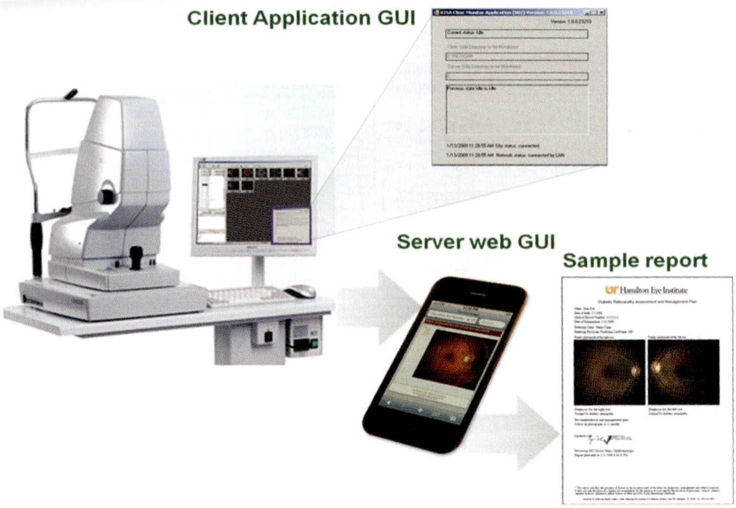

Fig. 14.3. Interfaces between the end-user clinic, the server, and referring physician.

14.1.4. *GUI Design*

The interface design in the ocular health network is composed of three parts: the client application GUI for the remote clinics, the physician web GUI, and the service GUI, as shown in the workflow chart in Fig. 14.3. In our web-based network, retinal images from diabetic patients are encrypted and transmitted from the fundus cameras at the primary care clinics to the diagnostic server using DICOM protocols. The retinal images are then graded to stratify disease level, recommend a management plan, and generate a report accessible to authorized users.

The client application GUI provides an interface for the regional primary care clinics for monitoring the data transmission and network status. The clinic acquires fundus images of the retina using VisuCam Pro NM (nonmydriatic) camera (Carl Zeiss Meditec). The client application GUI closely monitors the image exportation on the camera. New images are immediately encrypted, exported, and transmitted to the dedicated diagnostic server, along with patient metadata entries from the clinics. At the same time, the client application GUI constantly scans the network status, and reacts in real-time to prompt interactive information when connectivity issues occur.

After the images are submitted to the server, image processing and recognition procedures are triggered on the server to perform automated diagnosis. However, those procedures are transparent to the users. The procedures involved can be briefly described as the following: the submitted images are first subject to a QA to ensure adequacy for further analysis. Images that pass QA are put through the following steps: anatomical structure is analyzed, features are extracted, lesions are segmented, and a diagnosis is assigned according to the posterior probability, $P(w_i|\mathbf{v})$, of each defined disease state w_i using a content-based image retrieval (CBIR) method that evaluates the retrieval response and stratifies the disease state as described below.

The server web GUI provides an interface for authorized access to the diagnostic result. For an ophthalmologist, the diagnostic result, together with the images and patient metadata are reviewed through the web GUI to confirm digital signature, encrypt sensitive information, and report to the referring physician. For a referring physician, the web GUI provides secure access to the diagnostic result, along with patient information. The server web GUI is accessed through fixed or mobile electronic devices if the GUI has browser access to the web.

The encrypted PDF report (Fig. 14.4) contains patient metadata, the fundus image from both eyes, the diagnosis and recommendation plan, encrypted and electronically signed by a confirming ophthalmologist. The report can be accessed by an authorized user through the web GUI, or as a password-protected e-mail attachment for the referring physician.

14.1.5. *Performance Evaluation of the Network*

The combination of rapid data transmission and high throughput is one of the key features of the telemedical network. To validate the performance of the designed network infrastructure, the average time from image acquisition to report generation is applied as a metric for performance evaluation. The following system response time is measured by observing 60 sessions of data transmission between a fundus camera in an active clinic and the diagnostic server located in the university server farm. In our implementation of the ocular telehealth diagnostic network, we provide the ophthalmologist with the ability to assign a diagnosis manually before the automatic diagnosis engine is fully incorporated into the system. In the current experiment, we

UT Hamilton Eye Institute

Diabetic Retinopathy Assessment and Management Plan

Name: Dani c
Date of birth: 1/1/2000
Medical Record Number: 12345
Date of Examination: 9/5/2008

Referring Clinic: Hamilton Eye Institute, UTHSC Ophthalmology Clinic
Referring Physician: Yaqin Li, MD

Fundus photograph of the right eye | Fundus photograph of the left eye

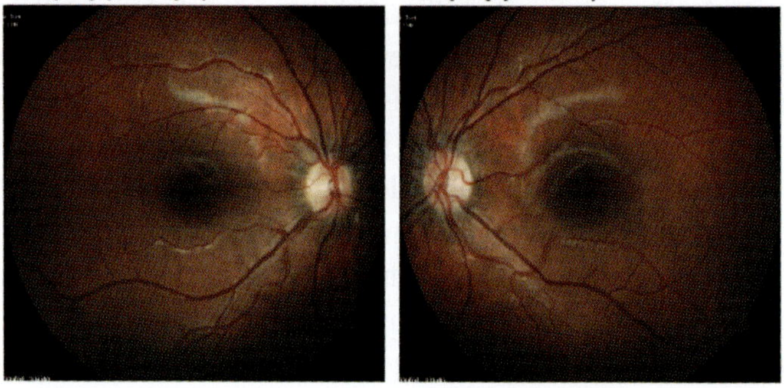

Diagnosis for the right eye:
Normal/No diabetic retinopathy

Diagnosis for the left eye:
Normal/No diabetic retinopathy

Recommendation and management plan:
Follow up photographs in 12 months.

Signature valid

Digitally signed by Edward Chaum MD PhD
Date: 2008.12.10 12:56:37 -06:00
Reason: Physician review complete.

Reviewing MD: Edward Chaum
Report generated on 12/10/2008 12:56:37 PM

Fig. 14.4. Sample encrypted patient report with electronic signature and certificate validation.

estimate that the automatic diagnosis can be assigned within 30 seconds. The average period (and variance) for each processing step is shown in Table 14.1. We estimate that the average response time from image acquisition to the generation of a diagnostic and management report for the end user is less than two minutes. The network structure provides sufficiently high throughput for image analysis and diagnosis of diabetic eye disease.

14.2. Image Analysis

The use of computer-based image analysis in ophthalmology has been a subject of research for some time. The use of telemedicine employing

Table 14.1. Average response time from image acquisition to generation of a diagnostic report in a fully automated system.

Process	Response time (seconds)
Export image	10.60 ± 0.61
Encryption and transmission of images	29.78 ± 1.57
Validate metadata and image data consistency	8.41 ± 0.82
Image quality assurance check	1.72 ± 0.45
Feedback to the camera	4.18 ± 0.62
Load into database	1.20 ± 0.38
Estimated automatic diagnosis	20.00 ± 0.10
Generate report	11.98 ± 1.18
Total	87.88 ± 2.37

human experts to diagnose DR from digital images is beginning to grow. Established retinal reading centers such as the Joslin Vision Network (Boston, MA) have shown that digital photography is an excellent tool for identifying DR when performed by experienced and certified readers.[24,25] In addition, a web-based telemedicine system for DR with open-source software has been demonstrated and other concepts have been explored.[26,27]

There has also been considerable research in the development of automated screening for DR. An excellent overview of this area is provided in Ref. [10], which summarizes image segmentation as a two-step process. The first step is summarized as identifying the regular or expected physiological features of the retina. The second step is concerned with identifying the pathology of the retina through the detection of abnormal phenomena (lesions). The final step uses an intelligent system to analyze and diagnose the segmentation results. Recent reports on this subject include Ref. [28], which describes information fusion and the use of a computer-aided diagnosis (CAD) system for the detection of normal and abnormal fundus images.

In our system, image processing and analysis falls into a sequence of steps, which are detailed as follows and shown in Fig. 14.5. First, quality estimation is performed on the image at the clinical site to ensure adequate image quality. Vessel segmentation is conducted in conjunction with this operation and is saved for further analysis. The image and examination

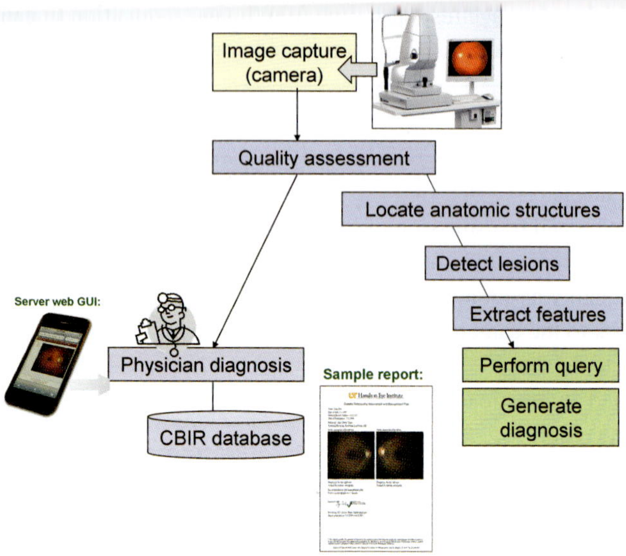

Fig. 14.5. Information flow in the telemedical network. The left path is manually driven and the right path is processed automatically.

information is saved in a database as described earlier for use in physician diagnosis, typically using a network-interfaced PC or smart phone. In a parallel track, the processing flow is automated for computer diagnosis of the retinal images. The vessel segmentation is used to estimate the location of the ON in each image. A complementary analysis is performed using an alternate ON detection method, and the results are compared to present a confidence level on the ON detection. The location of the macula is estimated using a nonlinear parabolic fit to the vessel tree and statistics on retina physiology. Lesions are detected next: our current work focuses on microaneurysms and exudates, as they are the main indicators of DR. The final estimated lesion population is then measured using numerical descriptors, which are then projected to a custom, semantic-driven index space. This "directed index" space allows the rapid grouping of similar images and, thus, permits our system to ultimately search through hundreds of thousands of examples using CBIR. The retrieved candidate images are then brute-force searched using distance metrics measured on the individual feature vectors. Confidence measurements (which will be described) are attached to the retrieval based on the population of the image database. In a complete system, computer diagnosis can replace an ophthalmologist, when some level of physician oversight is built-in for statistical sampling, or when confidence levels do not exceed an acceptable level. In our research system, the computer-generated reports are analyzed and compared with the actual physician diagnosis for statistical comparison and system improvement.

14.2.1. *Vascular Tree Segmentation*

Segmentation refers to the process of labeling an image to distinguish between the background and foreground of objects of interest. In vascular tree segmentation, our goal is to identify the vessels in the retinal image as distinct from the background or nonvessel pixels. There has been much research on this area in the literature dating back several years with newer work published fairly frequently; see for example Refs. [29–33]. Much of the interest in vessel segmentation stems from the argument that some ocular diseases, such as retinal vein occlusions and hypertensive retinopathy can be diagnosed through the analysis of the vascular tree (Fig. 14.6). The vessels can also serve to provide landmarks for the registration of multiple images of the same retina from different fields of view or from different time instances. Finally, they may also be used to locate key physiological elements in the retina, such as the ON and fovea. A valuable comparison of retina-vessel segmentation methods was performed in Ref. [34] in which medical experts hand-segmented a publicly available database. Their results were then compared to automatic segmentation methods. The comparison method was the agreement between pixel classification (as part of the vessel or the background) for one manual segmentation method and the automated

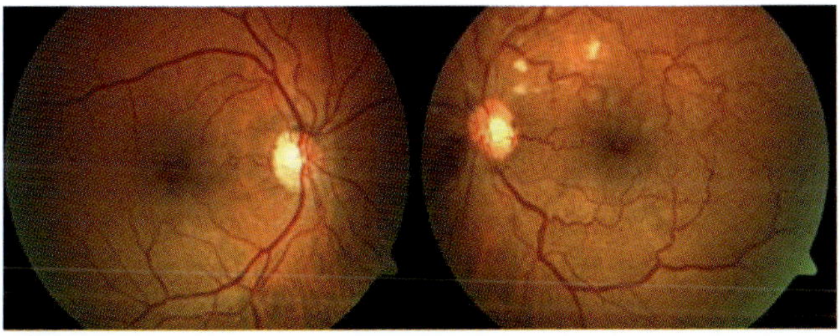

Fig. 14.6. Example retinal images showing an abnormal superior vascular tree of the left eye from a hemi-retinal vein occlusion.

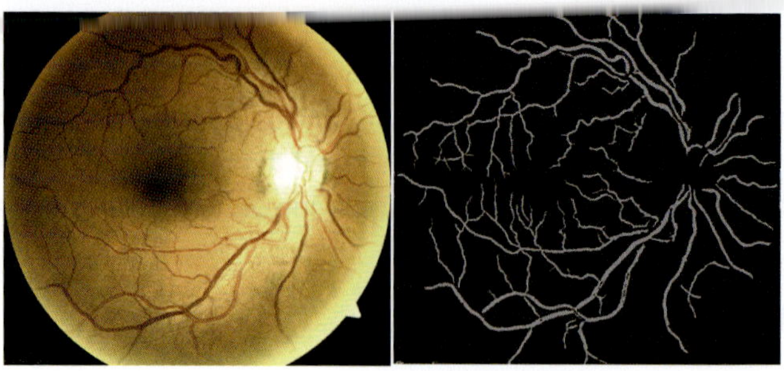

Fig. 14.7. Example of retinal vessel segmentation.

method of choice. The study found the method of Ref. [35] virtually identical to that of the second observer. However, several methods scored almost as well. In our work, we have implemented one of the studied methods.[36,37] In their method, linear elements are identified and enhanced by the use of morphological reconstruction techniques[38] followed by filtering and noise suppression techniques. For our current application, we are less interested in using the vascular tree to identify disease states and more interested in its measure of image quality and the localization of the ON and fovea. Our implementation in the telemedical network is fast, with a mean time estimate of less than one second running on a 1.66-GHz Intel-based PC server. An example of the image segmentation is shown in Fig. 14.7.

14.2.2. *Quality Assessment*

QA of fundus images is another topic of interest in the literature, and various methods have been proposed.[39–42] In our application, we found that most methods were either computationally too expensive or did not perform as well as desired. Thus, the QA method we use is ideal for a telemedicine network, for which it was created, due to its speed. The method allows quick feedback to the camera operator and requires no more computational power than is available on a regular personal computer commonly used in a fundus camera platform. We note, however, that while the method can be run on such a platform, in our current implementation, the quality check is run on the server for development purposes. We summarize the method here and note that more detail is available in Ref. [43]. An overview of the method is shown in Fig. 14.8. First, as discussed above, a segmentation of the vascular tree is

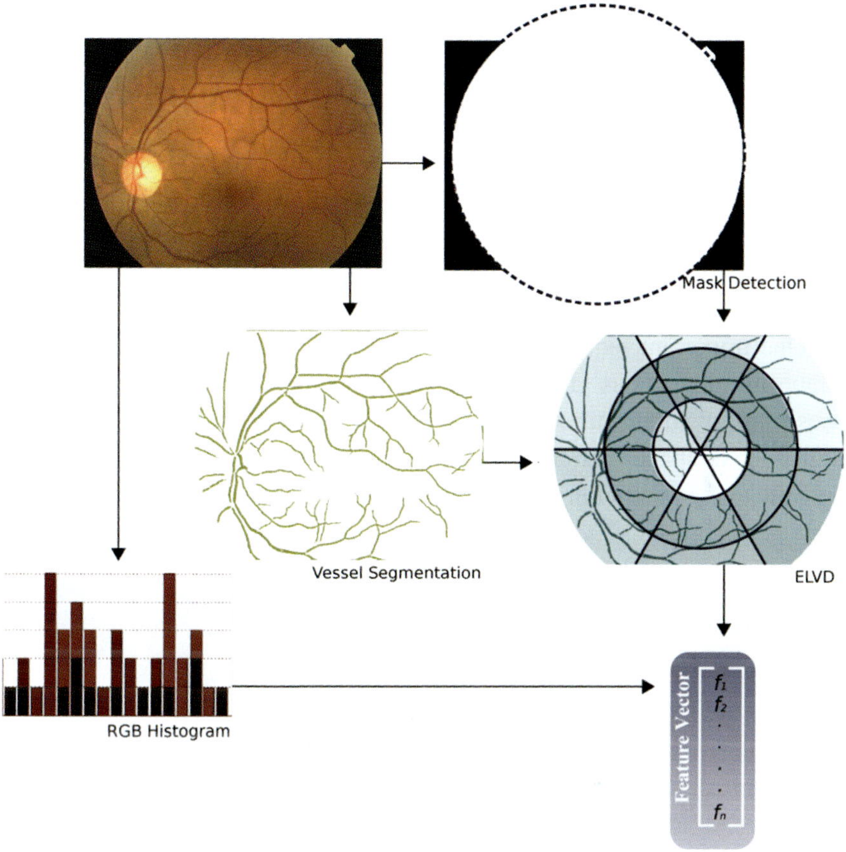

Fig. 14.8. Image QA processing.[44]

conducted. The image of the vascular tree is divided into annular and wedge-shaped regions to estimate accurately the coverage of the vascular tree. Poor-quality images will not achieve adequate coverage in some regions of the vessel coverage and, thus, will allow a pattern recognition engine to identify the image as an outlier (and, thus, of insufficient quality). The color of the fundus is also measured by using a histogram of the RGB values. These measurements are combined to build a complete feature vector that describes these aspects of the image, which our engineering efforts have indicated are key to quality. During development, a training set of images were manually labeled as good quality and poor quality, then a pattern recognition engine was trained to distinguish the two classes based on the measured numeric values. After creating the classifier, an additional set of unseen images was classified as a testing set, and a reviewing ophthalmologist (E. Chaum) sets a threshold. Note that this initial threshold is set to provide images of sufficient quality to be diagnosed by a human; generally, we note that this quality level may not be sufficient for automated diagnosis, and we show that improved results can be obtained when submitted images are of better quality images. An example of images of different quality from our telemedical network is shown in Fig. 14.9.

One important requirement in our implementation is standard image-capturing methods as given by the fundus camera. Our imaging protocol uses single-field, macula-centered images so that the vessel tree always has a distinctive shape that changes only slightly with the retinal physiology and due to the positioning of the macula or eye movement during imaging.

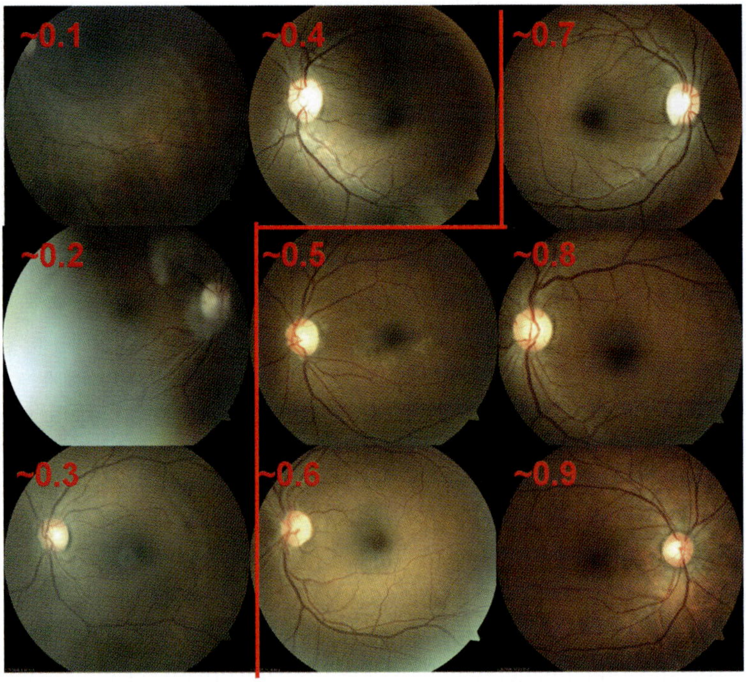

Fig. 14.9. Examples of acquired images of different qualities. A threshold of 0.4 was established by empirical methods.

In addition, the images submitted to the telemedical network are already labeled as right or left eye, and our method by virtue of its training examples is sensitive to this fact. Thus, we have two QA instances running, one for the left eye and the other for the right eye. Another issue is the field of view of the fundus photography; our original intention was to work with a 45-degree field of view, and, in some of our installations, this is the case, but others have begun to utilize 30-degree field of view, because it is easier to get consistent images in that setting. This setting is easily customizable by multiple classifiers, as well, and it is very simple to identify these different conditions based on the metadata of the images submitted.

We also note that the telemedical network allows the operator to have a "best-we-can-do" setting (BWCD). In this "mode," if an image fails a quality check, it may still be submitted for physician review. This option allows the system to (1) allow for potential failures in the quality estimation process, since the threshold chosen is not foolproof, (2) allow for the possibility that additional fundus images cannot be obtained, either because the patient has left or is otherwise unavailable, and (3) allow for possible intrinsic difficulty in obtaining fundus images from patients. In these cases, the image quality may still be sufficient for manual review. We save the vasculature tree segmentation, which can then be used in the ON detection method without having to be recomputed. Results from the initial network are shown in Fig. 14.10 where the quality levels of the images collected during the first eight months of operations for our first clinic are shown. We see from these results that the QA module is working to improve the quality of images submitted.

14.2.3. *ON Detection*

ON detection is a mature field, and there are many different ON detection methods that have been applied to a wide variety of data sets (see, for example, Refs. [15, 35, 45, and 46], which reference a publicly available STARE database). In our work, we have studied and implemented two methods. The first method relies on extracted features of the vessel segmentation.[47] Some methods emphasize that vessel segmentation can be a "weak link" in the estimate of the ON location. However, in our implementation, this problem is not an issue, because images with poor vessel segmentation will

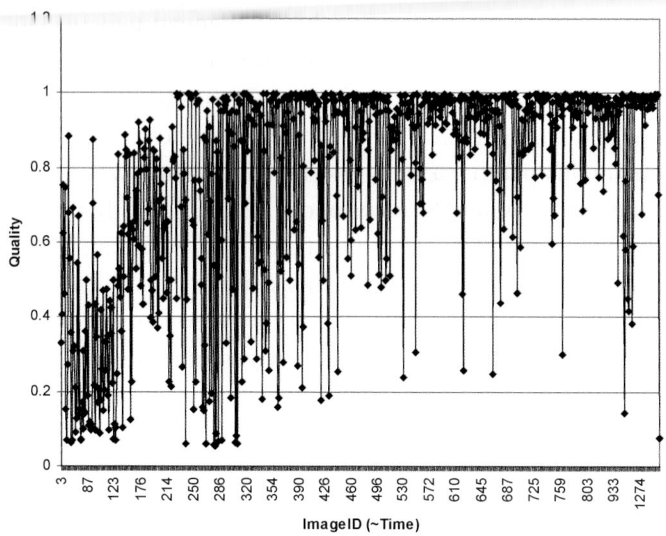

Fig. 14.10. Quality levels over an eight-month period. The overall quality of the image capture has improved after a roughly 2–3 month initial learning phase.

result in a failure in the QA and bypass automatic analysis. We summarize the processing portion of this method here.

The segmented vasculature tree of the image is utilized, and a set of four features are generated at each pixel by a neighborhood operation. The first feature is the brightness of the region. The remaining three features are measured on the vascular tree. The thickness of the vessels is measured by taking vessels in an area of interest and performing a thinning operation, which reduces all vessels to a single-pixel thickness. The difference between the thinned result and the original segmentation is measured perpendicular to the vessel direction and averaged within the neighborhood of the target pixel to estimate the thickness of the tree in that region. The second feature is the orientation of the vessels, which is estimated by measuring the vessel directions, scaled so that vertical vessels are emphasized over horizontal vessels. Finally, a measure of the density of the vasculature tree is measured in each neighborhood as well by simply summing up the number of estimated vessel pixels in this region. These operations are performed on a training set of images with hand-labeled ON centers. Feature values within the ON radius are chosen as examples of "ON," and all others are chosen as examples of "non-ON." These values are then used to estimate the parameters of a 4D Gaussian distribution describing the ON regions and the non-ON regions. We also use the hand-segmented training set to estimate the ON center probability density function (PDF). The use of this model is justified, because, as mentioned earlier, our imaging protocol is macula-centered images. The Gaussian parameters and the PDF are used to compute a likelihood ratio function using maximum *a posteriori* (MAP) estimation. Results on a pair of data sets are shown in Ref. [47], but these are more difficult data sets than seen in a true telemedical network. In our implementation, the results of preliminary testing are more accurate. These results are shown in Fig. 14.11.

Because one of our goals in our implementation is a level of physician oversight, we have also studied a complementary method of ON detection that does not rely on vascular tree segmentation. This method, which we will identify as the PCA-LDA method, is based on the model-based method of Ref. [18], which uses principal component analysis (PCA) to capture the information content of a training set of ON images. In their work, "candidate regions" of a retina image are projected to PCA space and then used to reconstruct the regions. The pixel with the smallest residual error is chosen as the ON location. In Ref. [48], we extended the method to include labeled

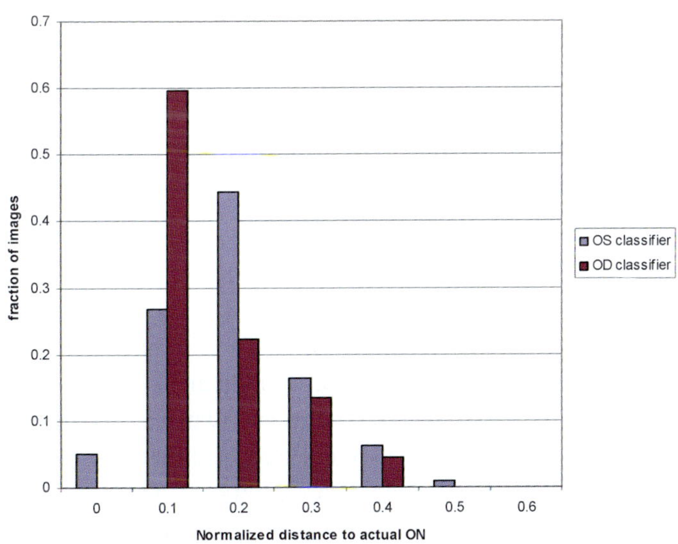

Fig. 14.11. Histogram of ON location estimates for a test set from the network. A normalized distance of 0.5 or smaller indicates the ON location estimate.

information about ON locations using linear discriminant analysis (LDA). The performance of the PCA-LDA method was shown to be superior to that of PCA alone, but we also investigated ways to combine this method with our vessel-segmentation-based method. In our first improvement, we trained the vessel segmentation method and estimated the performance on the training set to identify false positives where the vessel-based method had issues. These data were then used as part of the "non-ON" population in the LDA training phase. We refer to this as "directing" the PCA-LDA toward the images that give the vessel-based method trouble. This addition is an intuitive and straightforward modification to the method and essentially comes "for free" in that we simply choose different examples of the non-ON areas to train the LDA classifier; our main goal is to prevent the two methods from picking the same wrong location. The new classifiers were then tested on an unseen testing set and some improvement in performance was found Ref. [49].

We use the two complementary ON location methods to estimate the accuracy of the measurement. Our strategy was to measure the distance between the two estimates, and when this measurement exceeded a threshold (confidence level), we rejected the measurement. In practice, this would mean that we would refer the image directly to the oversight physician. In Ref. [50], we reviewed how these two complementary methods (one focusing on the vasculature tree and the other focusing on the ON appearance) could be used in tandem to produce a confidence measurement. We showed empirically, on two data sets, how limiting the number of automatically screened images could improve accuracy by simply thresholding the distance between the estimates of the methods.

14.2.4. *Macula Localization*

Once the ON is located, the macula is found using the method described in Ref. [47]. This method again uses the vascular tree segmentation and attempts to fit a parabola to the vascular tree as shown in Fig. 14.12. The vascular tree segmentation is "cleaned" slightly by focusing only on the core trunk values by removing branches that are smaller in thickness, which essentially makes the structure look much more like a parabola. The structure is then thinned so that the vessels are represented by a very small

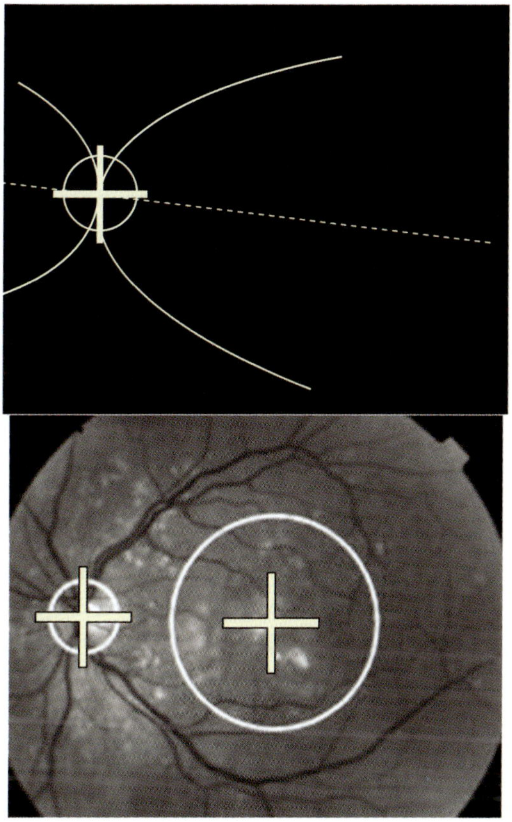

Fig. 14.12. Example of localizing the macula using parabolic model fit to the vessel structure.

number of pixels. Then, these pixel coordinates are taken and fit to a parabola using the ON estimate and the Marquardt-Levenberg nonlinear least-squares algorithm.[51] This method is similar to that of Ref. [15], but, in our case, we only seek to fit the orientation and curvature of the vessel structure, making the problem simpler. Once the orientation is found, we estimate the fovea position by using the mean of the ON–macula distances from a training set of images.

14.2.5. *Lesion Segmentation*

The two typical lesions that first appear in a patient affected by DR are microaneurysms and hemorrhages. While our system is intended to diagnose other retinal diseases, including age-related macular degeneration (AMD), the main driver is DR, and consequently, our initial focus is on these types of lesions. Microaneurysms are focal dilatations of retinal capillaries from 10 to 100 microns in diameter that appears as small red dots in a fundus image. Technically, they are not retinal lesions. However, we tend to group them in this category, because they are a phenomena associated with disease (as opposed to physiology) that we would like to detect. Hemorrhages are due to the leakage of blood from the wall of a damaged capillary or microaneurysm. These lesions are the most challenging to detect, but are the ones that allow the earliest disease detection. Exudates are yellowish in appearance and are sharp and bright structures caused by fluid leakage. They often have a circinate pattern of appearance regarding the entire population.[52]

There have been many approaches to lesion segmentation attempted in the literature. This work has been summarized in Refs. [10,53]. Some more recent approaches of interest for microaneurysms, exudates, and drusen (which appear similar to exudates but are more indicative of diseases such as AMD) include, but are not limited to Refs. [54–61]. An ongoing project of special note is the Retinopathy Online Challenge, which uses a publicly available database and evaluation method for algorithm comparison.[62]

In our work, we have developed a new algorithm for the segmentation of microaneurysms.[63] The algorithm is based on two steps, a background removal process that employs wavelets and the actual microaneurysms detection. The microaneurysm detection is performed with a new method, the Radon Cliff operator. Making use of the Radon transform, the operator is able to detect single noisy Gaussian-like circular structures regardless of their size or position in a window. This method has several advantages over existing microaneurysms detectors: the size of the lesions can be unknown, it automatically distinguishes lesions from the vasculature in general, and it provides fair microaneurysm localization without post-processing the candidates with machine-learning techniques, facilitating the training phase. Figure 14.13 shows an example of the current algorithm output.

Exudates are another visible sign of DR and a marker for the presence of retina edema. As mentioned, various authors have developed segmentation algorithms for the automated detection of DR, however, to our knowledge, none has explicitly addressed the problem of analyzing a retina with a high

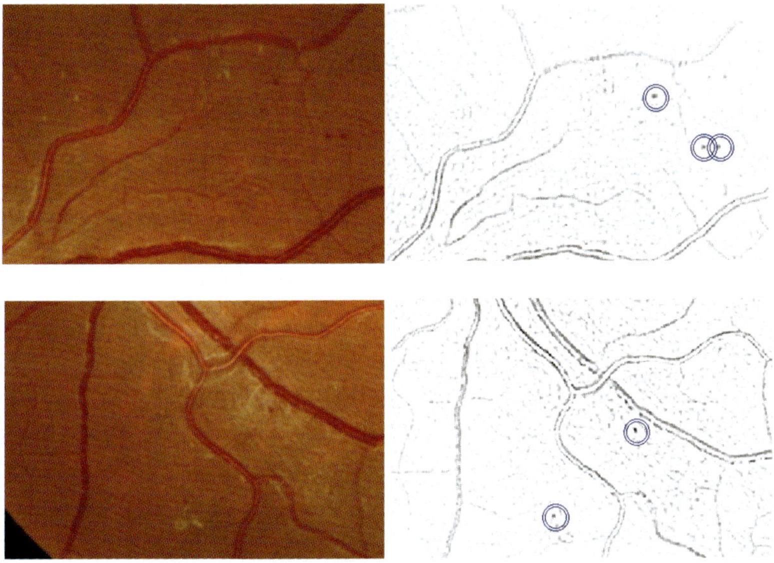

Fig. 14.13. Microaneurysm detection using Radon Cliff operator.

degree of reflecting artifacts due to the nerve fiber layer (NFL). This structure is particularly visible in young patients, especially on dark pigmented retinas such as seen in African American patients. We have developed a technique to detect bright lesions in patients with a high degree of reflective NFL. First, the candidate bright lesions are detected using image equalization and histogram analysis. Then, a classifier is trained using texture descriptors (multi-scale local binary patterns) and other statistical features in order to remove the false positives in the lesion detection. Finally, the area of the lesions is used to diagnose DR. More details are covered in Ref. [64]. Currently, the algorithm has been refined using a wavelet background subtraction approach similar to the one proposed for the microaneurysms segmentation. Figure 14.14 shows an example of this segmentation.

14.2.6. *Overall Fundus Description and Stratification*

Our overall fundus description consists of a measurement of 170 features related to the lesions detected (such as histograms of the sharpness of the lesion edges, their intensity, shape properties, etc.), the vascular density

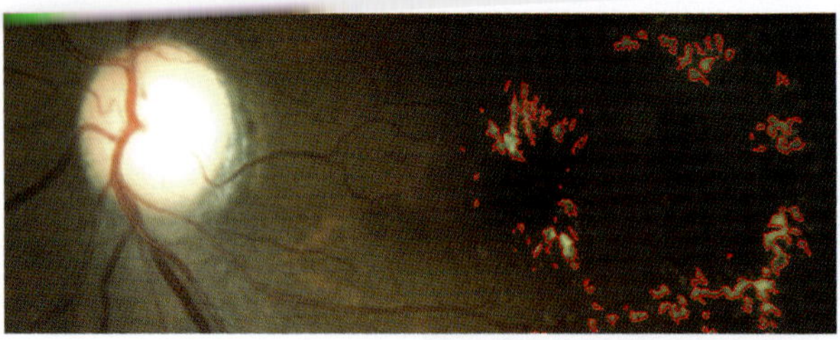

Fig. 14.14. Exudate detection using wavelet analysis and texture descriptors.

within the lesion population, population moments, and invariant moments relative to the central macula. Our goal is to determine an index that can be used to locate similar imagery by comparison. This determination can be made if we can produce an index that provides discrimination between the different disease states of interest in the archive (which were ground-truthed by the oversight physician in the process of building the archive). We use linear discriminant analysis (LDA) to project our image features in 170-dimensional space onto a smaller subspace which ideally optimally discriminates between our defined disease (and normal) disease states. For this discussion, we are applying a traditional multiple LDA method based on the well-known Fischer discriminant.[65]

To perform retrievals from the archive using the image index, we determine a similarity between a query image that has also been subjected to feature extraction through lesion segmentation and lesion population feature extraction. When the archive becomes very large, fast, and efficient methods for searching must be used. However, in our initial stage of development, our database size is still small and therefore, the indexing methods are not required.

We have described our CBIR system and its performance on test data sets in Refs. [66, 67] and summarize these key results here for completeness. The developed CBIR method uses the retrieval response to our query image, represented by an index vector $\mathbf{q}$, to estimate the posterior probability, $P(\omega_i|\mathbf{v})$, of each defined disease state ω_i. The retrieval process is similar to a k-nearest neighbor (k-NN) method.[65] Nearest neighbor classifiers function by locating the population of labeled data points nearest to an unknown point in index space for a specified number of neighbors, k.

A posterior probability can be derived from a simple k-NN voting scheme or, as in our case, a weighted summation of similarities can be used.

In a simple k-NN estimate, the posterior probability is expressed as $P(\omega_i|v) = k_i/k$, where k_i represents the number of neighborhood vectors corresponding to disease state ω_i of the k neighbors retrieved. For our method, we use a weighted summation of the similarity between points to give,

$$P(\omega_i|\mathbf{q}) \pm \sigma = \frac{\sum_i S(\mathbf{q}, \mathbf{v}_i)^w}{\sum_k S(\mathbf{q}, \mathbf{v}_k)^w} \pm \frac{\sqrt{\sum_i S(\mathbf{q}, \mathbf{v}_i)^w}}{\sum_k S(\mathbf{q}, \mathbf{v}_k)^w}, \tag{14.1}$$

where the exponent, $w \geq 0$, increases the influence of the points closest to the query point. The estimate approaches a nearly optimal posterior estimate, as the number of records in the system increases, meaning the diagnostic performance of the archive will theoretically improve as the archive population increases.

Note that we have also incorporated a confidence value, σ, using Poisson statistics.[66] Poisson statistics can be applied to phenomena of a discrete nature, such as the rate of disease occurrence in patients. By using the property that the standard deviation in the sampling of the disease category, ω_i, is the square root of the number of counts in the category, we can estimate our confidence in the posterior probability of a particular disease state as indicated.

We review our results from Refs. [66, 67] here. The authors of these sources used two independent sets of image data. The first set is an image archive of 1,355 macula-centered images obtained from a DR-screening program in the Netherlands,[68,69] which we abbreviate as NL. The second image set of 98 images was used for comparison purposes and originated from a Canadian Native American population.[70] We designate them as C.

Table 14.2 shows consolidated results for a number of different trials in which the two parameters of image quality and posterior probability confidence were varied for the independent data set. In the first row describing the performance of the patient archive containing 1,355 records, we have applied a statistical hold-one-out (HOO) procedure to determine the expected performance of the system based on the internal consistency of the data. The HOO performance values of sensitivity and accuracy are listed for the archive as 90%, and 95%, respectively. Since HOO performance often presents slightly higher results than is generally noted from truly

Table 14.2. Performance results generated for two data sets. The top row shows the expected performance of the NL archive using a hold-one-out method. For comparison, another independent data set was used in the bottom three rows, with variable confidence (σ) and image quality values. From Ref. [67] — S = sensitivity and A = accuracy.

Data	σ	Q	Records	% Below σ	S(%)	A(%)
NL	0	0	1355	0	89.5	94.8
C	0	0.5	81	0	66.7	76.5
C	3	0	98	25.5	78.0	87.7
C	3	0.5	81	22.2	82.0	88.9

independent data, we have assembled an entirely independent test population as described earlier, represented by the results in the bottom three rows of Table 14.2. For this test population, we have 98 records as shown. We initially specified the quality metric to only accept images of value >0.5, resulting in sensitivity and accuracy of 67% and 77%, respectively. Next, we set the quality threshold back to 0 and set a confidence level of 3σ, resulting in a slightly higher sensitivity, and accuracy of 78% and 88%. Finally, we set the image quality threshold to accept >0.5 and 3σ confidence to result in the best sensitivity and accuracy of 82% and 89%.

These results show tests of our system using a separate, completely independent set of data collected under unrelated conditions. Despite the disparity in the two data sets, we have shown a level of robustness resulting in sensitivity to disease discrimination of 82% and overall accuracy of 89%.

14.2.7. *Patient Demographics and Statistical Outcomes*

Since its initial deployment at the Church Health Center in Memphis, TN, in February 2009, and more recently in Internal Medicine Clinics at the University of North Carolina, the ocular telehealth network has provided diagnostic reports on 1,373 eyes from 669 patients. According to the patient demographics, the population ranged in age from 20 to 91, with a mean age of 55.4 years. The patient population was predominantly female (64.28%)

and the vast majority of the patients evaluated had Type II diabetes (93.42%). Type I diabetes comprised 3.58% of patients, and 2.86% of patients were unrecorded. Ethnicity profiles showed that 59.8% of patients were African American, 31.4% Caucasian, and approximately 3.8% were Hispanic or unrecorded.

14.2.8. *Disease State Assessment*

An assessment of the disease images is summarized in Table 14.3. A total of 1,036 eyes, comprising 75.46% of all patient eyes examined, had no evidence of DR. These patients did not require further evaluation by an eye-care specialist for diabetic eye disease and will be managed in the primary care clinic with follow-up retinal photography in 12 months. The incidence

Table 14.3. Epidemiology of DR in a population of diabetic patients examined for retinal lesions using the ocular telehealth network in Memphis, TN and Chapel Hill, NC.

	Disease state	OD	OS	Total	Percentage	
1	No DR	514	522	1036	75.46	86.38 (rows 1–2)
2	NPDR mild/minimal−CSME	75	75	150	10.92	
3	NPDR mild/minimal+CSME	17	22	39	2.84	6.55 (rows 3–11)
4	NPDR moderate−CSME	5	4	9	0.66	
5	NPDR moderate+CSME	15	14	29	2.11	
6	NPDR serve−CSME	3	0	3	0.22	
7	NPDR serve+CSME	2	1	3	0.22	
8	PDR−CSME	3	2	5	0.36	
9	PDR+CSME	0	1	1	0.07	
10	PDR+HRC−CSME	0	1	1	0.07	
11	PDR+HRC+CSME	0	0	0	0.00	
12	AMD Grade 1	2	2	4	0.29	
13	AMD Grade 2	2	4	6	0.44	
14	AMD Grade 3	2	1	3	0.22	
15	AMD Grade 4	0	0	0	0.00	
16	Other retinal diseases	37	47	84	6.12	
Total		677	696	1373	100.00	

of any DR in the population was 17.47% and is consistent with the known epidemiology of DR in the region. Of the patients with DR, 62.50% (10.92% of total eyes examined) had minimal or mild DR without fluid leakage and did not require a formal eye examination. These patients will be rescreened in the primary care clinic in 6–12 months (12 months for patients with only rare microaneurysms). Thus, 86.38% of all diabetic eyes screened to date will continue to be followed for DR in the primary care setting. A total of 6.55% of all eyes had DR severe enough to warrant referral and treatment. An additional 84 eyes (6.12%) from 68 patients (10.16%) had retinal findings, exclusive of DR, that warranted a formal ophthalmic evaluation. These findings included macular degeneration, vascular occlusions, optic disc findings suggestive of glaucoma, and other problems or diseases.

14.2.9. *Image QA*

The QA module computes the image quality value and issues a "good" or "bad" evaluation result by thresholding. The QA not only provides immediate feedback of the image quality to the photographers after image acquisition, but also provides a quantitative indication of statistical changes for monitoring the image quality. The overall image quality steadily improved since initial deployment (Fig. 14.10). More recently, significant variance in overall image quality was noted, as a new clinic came on board from UNC Chapel Hill. This fluctuation of the statistical image quality can help to identify the training needs for the photographers when new people are enrolled into the system.

14.2.10. *Physician Oversight Based on Quality and Confidence Levels*

We conclude this chapter with a review of the concept of physician oversight. Throughout this work, we have explained that we employ automatic processing and seek to establish confidence levels that can be used to invoke physician oversight in a truly automated system. Physician oversight is invoked in three main areas. First, the quality level must exceed a basic threshold before it is deemed acceptable for submission to the network. Second, once an image is received, another, higher quality level threshold must be satisfied to deem the image of sufficient quality for automatic screening. We proved the above in Ref. [50], in which we showed that ON estimation improved based on the quality of the image. We showed that diagnosis improved when the quality of the image improved.[67] Third, although not in "chronological" processing order, the confidence of the ON detection can be judged by the complementary method, which is more accurately a "degree of agreement." Images that fail this metric may still be evaluated but will most likely be passed to the reviewing physician. Last, the use of Poisson statistics allows us to attach a level of confidence to automatic diagnosis. Improved diagnosis accuracy is achieved at the cost of fewer automatic screenings, but again the role of the oversight physician can be to improve the system performance.

One can imagine other thresholds of interest; for example, since we assume the images are macula centered, images with estimates of the ON location or macula location that stray too far from a mean position may be flagged for manual review. Another possibility that remains for exploration is the goodness of a fit to the parabolic model of the vascular tree. Our network employs some issues of practice that are rarely (if ever) invoked. Since our system is under development, any software bugs in the process automation would invoke physician oversight immediately. (This immediate alert is of course not an issue in our current system, since a physician reviews all images.) Any nonlinear computational issues that are iterative by nature are also set with timing thresholds so that if convergence is not reached they will be set to automatic physician review. These issues are also not expected to be common, but they are essential to ensuring robust operation and maximum patient care.

Finally, any future progress in the use of computer-aided and automated telemedical applications for the diagnosis of DR and other retinal diseases must also consider the "Telehealth Practice Recommendations for Diabetic Retinopathy" proposed by the Ocular Telehealth Special Interest Group (SIG) of the American Telemedicine Association.[71] The SIG recognizes four categories of validation for telehealth for DR relative to the ETDRS 35-mm slide reference standards.

Category 1 validation can separate patients into those with none or very mild nonproliferative DR (ETDRS level 20 or below), and those with greater than ETDRS level 20 (stratifying DR by yes/no criteria for more than minimal disease). Category 2 validation should be able to accurately determine if sight-threatening DR including any level of macular edema, ETDRS level 53 (or worse), or proliferative DR is present. Category 2 validation distinguishes patients with sight-threatening DR who require prompt referral for possible laser surgery. Category 3 validation applies to a system that can distinguish ETDRS-defined levels of nonproliferative DR (mild, moderate, or severe), proliferative DR (early, high-risk), and macular edema with accuracy sufficient to determine appropriate follow-up and treatment strategies (equivalent to a retinal examination through dilated pupils). Category 4 validation applies to a system that matches or exceeds the ability of the ETDRS photos to determine levels of DR. Ultimately, we believe that the CBIR method has the potential to perform at a Category 3 validation level to stratify and effectively manage DR at an expert level using nonmydriatic images taken in the primary care setting.

Acknowledgments

These studies were supported in part by grants from Oak Ridge National Laboratory, the National Eye Institute (EY017065), the Health Resources and Services Administration, by an unrestricted UTHSC Departmental grant from Research to Prevent Blindness, New York, NY, Fight for Sight, New York, NY, and by The Plough Foundation, Memphis, TN.

References

1. International Diabetes Federation. Available at http://www.idf.org.
2. Klein, R., Klein, B.E., Moss, S.E., Davis, M.D., and DeMets, D.L. The Wisconsin epidemiologic study of diabetic retinopathy. II. Prevalence and risk of diabetic retinopathy when age at diagnosis is less than 30 years. *Arch Ophthalmol* **102**:520–526, 1984.
3. Klein, R., Klein, B.E., Moss, S.E., and Linton, K.L. The beaver dam eye study. Retinopathy in adults with newly discovered and previously diagnosed diabetes mellitus. *Ophthalmology* **99**:58–62, 1992.
4. McGlynn, E.A., Asch, S.M., Adams, J., Keesey, J., Hicks, J., DeCristofaro, A., and Kerr, E.A. The quality of health care delivered to adults in the United States. *N Engl J Med* **348**:2635–2645, 2003.
5. Newcomb, P.A. and Klein, R. Factors associated with compliance following diabetic eye screening. *J Diabet Complications* **4**(1):8–14, 1990.

6. DRS Report Number 8. The Diabetic Retinopathy Study Research Group. Photocoagulation treatment of proliferative diabetic retinopathy. Clinical application of Diabetic Retinopathy Study (DRS) findings. *Ophthalmology* **88**:583–600, 1981.
7. Early Treatment Diabetic Retinopathy Study Report Number 1. Photocoagulation for diabetic macular edema: *Arch. Ophthalmol* **103**:1796–1806, 1985.
8. Age-Related Eye Disease Study Report Number 3. Risk factors associated with age-related macular degeneration. A case-control study in the age-related eye disease study. *Ophthalmology* **107**:2224–2232, 2000.
9. Larsen, M., Godt, J., and Grunkin, M. Automated detection of diabetic retinopathy in a fundus photographic screening population. *Invest Ophthal Vis Sci* **44**:767–771, 2003.
10. Teng, T., Lefley, M., and Claremont, D. Progress towards automated diabetic ocular screening: a review of image analysis and intelligent systems for diabetic retinopathy. *Med Biol Eng Comput* **40**:2–13, 2001.
11. Foracchia, M., Grisan, E., and Ruggeri, A. Luminosity and contrast normalization in retinal images. *Med Image Anal* **9**:179–190, 2005.
12. Martin-Perez, M.E., Hughes, A.D., Stanton, A.V., *et al.* Segmentation of retinal blood vessels based on the second directional derivative and region growing. *IEEE Inte Conf Image Process* 173–176, 1999.
13. Hoover, A., Kouznetsova, V., and Goldbaum, M. Locating blood vessels in retinal images by piecewise threshold probing of a matched filter response. *IEEE Trans Med Imaging* **19**:203–210, 2000.
14. Hoover, A. and Goldbaum, M. Locating the optic nerve in a retinal image using the fuzzy convergence of the blood vessels. *IEEE Trans Med Imaging* **22**:951–958, 2003.
15. Foracchia, M., Grisan, E., and Ruggeri, A. Detection of the optic disk in retinal images by means of a geometrical model of vessel structure. *IEEE Trans Med Imaging* **23**:1189–1195, 2004.
16. Abramoff, M.D., Alward, W.L., Greenlee, E.C., Shuba, L., Kim, C.Y., Fingert, J.H., and Kwon, Y.H. Automated segmentation of the optic disc from stereo color photographs using physiologically plausible features. *Invest Ophthalmol Vis Sci* **48**:1665–1673, 2007.
17. Morris, D.T. and Donnison, C. Identifying the neuroretinal rim boundary using dynamic contours. *Image Vis Comput* **17**:169–174, 1999.
18. Li, H. and Chutatape, O. Automated feature extraction in color retinal images by a model based approach. *IEEE Trans Biomed Eng* **51**:246–254, 2004.
19. Tobin, K.W., Karnowski, T.P., Arrowood, L.F. *et al.* Content-based image retrieval for semiconductor process characterization. *EURASIP J Appl Signal Processing* **7**:704–713, 2002.
20. Karnowski, T.P., Tobin, K.W., Ferrell, R.K., and Lakhani, F. Using an image retrieval system for image data management. Design, Process Integration, and Diagnostics in IC Manufacturing, *Proceedings of the SPIE* **4692**:120. doi:10.1117/12.475648, 2002.
21. Chaum, E., Karnowski, T.P., Govindasamy, V.P., Abdelrahamen M., and Tobin K.W. Automated diagnosis of retinopathy by content-based image retrieval. *Retina* **28**:1463–1477, 2008.
22. Wei, Z., Wu, Y., Deng, R.H., Yu, S., Yao, H., Zhao, Z., Ngoh, L.H., Han, L.T., and Poh, E.W.T. A secure and synthesis tele-ophthalmology system. *Telemed E Health* **14**:833–845, 2008.

23. Li, Y., Karnowski, T.P., Tobin, K.W., Giancardo, L., and Chaum, E. A network infrastructure for automated diagnosis of diabetic retinopathy. *Proc American Telemedicine Association Annual Meeting*, 2009.
24. Cavallerano, A.A., Cavallerano, J.D., Katalinic, P., Tolson, A.M., Aiello, L.P., and Aiello, L.M. Joslin Vision Network Clinical Team, Use of Joslin Vision Network digital-video nonmydriatic retinal imaging to assess diabetic retinopathy in a clinical program. *Retina* **23**:215–223, 2003.
25. Fransen, S.R., Leonard-Martin, T.C., Feuer, W.J., and Hildebrand P.L. Inoveon Health Research Group. Clinical evaluation of patients with diabetic retinopathy: accuracy of the Inoveon diabetic retinopathy-3DT system. *Ophthalmology* **109**:595–601, 2002.
26. Wei, J., Valentino, D., Bell, D., and Baker, R. A web-based telemedicine system for diabetic retinopathy screening using digital fundus photography. *Telemed E Health* **12**:50–57, 2006.
27. Zahlmann, G., Schubert, M., Obermaier, M., and Mann, G. Concept of a knowledge based monitoring system for glaucoma and diabetic retinopathy using a telemedicine approach. *Proc. 18th Annual International Conference of the IEEE Engineering in Medicine and Biology Society*, Amsterdam, pp. 1230–1231, 1996.
28. Niemeijer, M., Abramoff, M.D., and van Ginneken, B. Information fusion for diabetic retinopathy CAD in digital color fundus photographs. *IEEE Trans Med Imaging* **28**: 775–785, 2009.
29. Wang, L., Bhalerao, A., and Wilson, R. Analysis of retinal vasculature using a multiresolution Hermite model. *IEEE Trans Med Imaging* **26**:137–152, 2007.
30. Hoover, A., Kouznetsova, V., and Goldbaum, M. Locating blood vessels in retinal images by piecewise threshold probing of a matched filter response. *IEEE Trans Med Imaging* **19**:203–210, 2000.
31. Chaudhuri, S., Chatterjee, S., Katz, N., Nelson, M., and Goldbaum, M. Detection of blood vessels in retinal images using two-dimensional matched filters. *IEEE Trans Med Imaging* **8**:263–269, 1989.
32. Jiang, X. and Mojon, D. Adaptive local thresholding by verification-based multithreshold probing with application to vessel detection in retinal images. *IEEE Trans Pattern Anal Mach Intell* **25**:131–137, 2003.
33. Soares, J.V.B. *et al.* Retinal vessel segmentation using the 2-D Gabor wavelet and supervised classification. *IEEE Trans Med Imaging* **25**:1214–1222, 2006.
34. Niemeijer, M., Staal, J., van Ginneken, B., Loog, M., and Abramoff, M.D. Comparative study of retinal vessel segmentation methods on a new publicly available database. *Proc SPIE* **5370**:648–656, 2004.
35. Niemeijer, M., Abramoff, M.D., and van Ginneken, B. Segmentation of the optic disc, macula and vascular arch in fundus photographs. *IEEE Trans Med Imaging* **26**:116–127, 2007.
36. Zana, F. and Klein, J. Segmentation of vessel-like patterns using mathematical morphology and curvature evaluation. *IEEE Trans on Image Processing* **10**:1010–1019, 2001.
37. Fang, B., Hsu, W., and Lee, M.L. Reconstruction of vascular structures in retinal images. *Proc ICIP'03* **II**:157–160, 2003.
38. Vincent, L. Morphological grayscale reconstruction in image analysis: applications and efficient algorithms. *IEEE Trans Image Processing* **2**:176–201, 1993.

39. Fleming, A.D., Philip, S., Goatman, K.A., Olson, J.A., and Sharp, P.F. Automated assessment of diabetic retinal image quality based on clarity and field definition. *Invest Ophthalmol Vis Sci* **47**:120–1125, 2006.
40. Lalonde, M., Gagnon, L., and Boucher M.C. Automatic visual quality assessment in optical fundus images. *Proc Vision Interface*, 2001.
41. Niemeijer, M., Abramoff, M.D., and van Ginneken, B. Image structure clustering for image quality verification of color retina images in diabetic retinopathy screening. *Med Image Anal* **10**:888–898, 2006.
42. Usher, D.B., Himaga, M., and Dumskyj, M.J. Automated assessment of digital fundus image quality using detected vessel area. *Proc Medical Image Understanding and Analysis, British Machine Vision Association (BMVA)*, pp. 81–84, 2003.
43. Giancardo, L., Abramoff, M.D., Chaum, E., Karnowski, T.P., Meriaudeau, F., Tobin, K.W. Elliptical local vessel density: a fast and robust quality metric for fundus images. *Conf Proc IEEE Eng Med Biol Soc* **1**:3534–3537, 2008.
44. Giancardo, L. Quality analysis of retina images for the automatic diagnosis of diabetic retinopathy. Master's Thesis, Msc Erasmus Mundis in VIBOT, 2008.
45. Youssif, A.-R., Ghalwash, A.Z., and Ghoneim, A.-R. Optic disc detection from normalized digital fundus images by means of a vessels' direction matched filter. *IEEE Trans Med Imaging* **27**:11–18, 2008.
46. Hoover, A. and Goldbaum, M. Locating the optic nerve in a retinal image using the fuzzy convergence of the blood vessels. *IEEE Trans Med Imaging* **22**: 951–958, 2003.
47. Tobin, K.W., Chaum, E., Govindasamy, V.P., and Karnowski, T.P. Detection of anatomic structures in human retinal imagery. *IEEE Trans Med Imaging* **26**:1729–1739, 2007.
48. Karnowski, T.P., Govindasamy, V.P., Tobin, K.W., and Chaum, E. Locating the optic nerve in retinal images: comparing model-based and Bayesian decision methods. *Proc. 28th Annual International Conf. of the IEEE EMBS*, 2006.
49. Karnowski, T.P., Aykac, D., Chaum, E., Giancardo, L., Li, Y., Tobin, K.W., and Abràmoff, M.D. Evaluating the accuracy of optic nerve detections in retina imaging by using complementary methods. *Proc World Congress of Medical Physics and Biomedical Engineering*, Munich, 2009.
50. Karnowski, T.P., Aykac, D., Chaum, E., Giancardo, L., Li, Y., and Tobin, K.W., Jr., Abramoff, M.D. Practical considerations for optic nerve location in telemedicine. *Proc 31st Annual International Conf of the IEEE EMBS*, 2009.
51. Marquardt, D. An algorithm for least-squares estimation of nonlinear parameters. *SIAM J Appl Math* **11**:431–441, 1963.
52. Huang, D., Kaiser, P., Lowder, C., and Traboulsi, E. *Retina Imaging*, Elsevier, 2006.
53. Patton, N., Aslam, T.M., MacGillivray, T., Deary, I.J., Dhillon, B., Eikelboom, R.H., Yogesan, K., and Constable, I.J. Retinal image analysis: concepts, applications and potential. *Prog Retin Eye Res* **25**:99–127, 2006.
54. Niemeijer, M., van Ginneken, B., Staal, J., Suttorp-Schulten, M.S., and Abramoff, M.D. Automatic detection of red lesions in digital color fundus photographs. *IEEE Trans Med Imaging* **24**:584–592, 2005.
55. Huang, K. and Yan, M. A local adaptive algorithm for microaneurysms detection in digital fundus images. *Proc Computer Vision for Biomedical Image Applications: First International Workshop*, Beijing, China, 2005.

56. Raman, B., Bursell, E.S., Wilson, M., Zamora, G., Benche, I., Nemeth, S.C., and Soliz, P. The effects of spatial resolution on an automated diabetic retinopathy screening system's performance in detecting microaneurysms for diabetic retinopathy. *Proc 17th IEEE Symposium on Computer-Based Medical Systems*, p. 128, 2004.
57. Hafez, M. and Azeem, S.A. Using adaptive edge technique for detecting microaneurysms in fluorescein angiograms of the ocular fundus. *Proc IEEE MELECON* 479–483, 2002.
58. Pallawala, P.M.D.S., Hsu, W., Lee, M.L., and Goh, S.S. Automated microaneurysm segmentation and detection using generalized eigenvectors. *Seventh IEEE Workshops on Application of Computer Vision*, pp. 322–327, 2005.
59. Sbeh, Z.B., Cohen L.D., Mimoun, G., and Coscas, G. A new approach of geodesic reconstruction for drusen segmentation in eye fundus images. *IEEE Trans Med Imaging* **20**:1321–1333, 2001.
60. Osareh, A., Mirmehdi, M., Thomas, B., and Markham, R. Automated identification of diabetic retinal exudates in digital colour images. *Br J Ophthalmol* **87**:1220–1223, 2003.
61. Walter, T., Klein, J.-C., Massin, P., and Erginay, A. A contribution of image processing to the diagnosis of diabetic retinopathy — detection of exudates in color fundus images of the human retina. *IEEE Trans Med Imaging* **21**:1236–1243, 2002.
62. Niemeijer, M., Abramoff, M.D., and van Ginneken B. Fast detection of the optic disc and fovea in color fundus photographs. *Med Image Anal* **13**:859–870, 2009.
63. Giancardo, L., Meriaudeau, F., Karnowski, T.P., Chaum, E., Tobin, K. W., and Li, Y. Microaneurysms detection with the Radon Cliff Operator in retinal fundus images. *Proceedings of SPIE Medical Imaging*, San Diego, 2010.
64. Giancardo, L., Chaum, E., Karnowski, T.P., Meriaudeau, F., Tobin, K.W., and Li, Y. Bright retinal lesions detection using color fundus images containing reflective features. *Proceedings of World Congress of Medical Physics and Biomedical Engineering*, Munich, 2009.
65. Duda, R.O., Hart, P.E., and Stork, D.G. *Pattern Classification*. Second Edition, John Wiley and Sons, Inc., New York, 2001.
66. Tobin, K.W., Abdelrahman, M., Chaum, E., Govindasamy, V.P., and Karnowski, T.P. A probabilistic framework for content-based diagnosis of retinal disease. *Proc 29th Annual International Conf of the IEEE EMBS*, Lyon, France, 2007.
67. Tobin, K.W., Abramoff, M.D., Chaum, E., Giancardo, L., Govindasamy, V.P., Karnowski, T.P., and Tennant, M.T.S. Using a patient image archive to diagnose retinopathy. *Proc 30th Annual International Conf of the IEEE EMBS*, Vancouver, Canada, 2008.
68. Abramoff, M.D. and Suttorp-Schulten, M. Web-based screening for diabetic retinopathy in a primary care population: the eyecheck project, *Telemed E Health* **11**:668–674, 2005.
69. Abràmoff M.D., Niemeijer, M., Suttorp-Schulten, M.S., Viergever, M.A., Russell, S.R., and van Ginneken, B. Evaluation of a system for automatic detection of diabetic retinopathy from color fundus photographs in a large population of patients with diabetes. *Diabetes Care* **31**:193–198, 2008.
70. Rudnisky, C.J., Tennant, M.T.S., Weis, E., Ting, A., Hinz, B.J., and Greve, M.D.J. Web-based grading of compressed stereoscopic digital photography versus standard slide

film photography for the diagnosis of diabetic retinopathy. *Ophthalmology* **114**:1748–1754, 2007.

71. American Telemedicine Association, Ocular Telehealth Special Interest Group, and the National Institute of Standards and Technology Working Group. Telehealth Practice Recommendations for Diabetic Retinopathy. *Telemed J E Health* **10**:469–482, 2004.

Index